AF356152

Linear Motors and Direct-Drive Technology

Linear Motors and Direct-Drive Technology

Editors

Qinfen Lu
Xinyu Fan
Cao Tan
Jiayu Lu

Basel • Beijing • Wuhan • Barcelona • Belgrade • Novi Sad • Cluj • Manchester

Editors

Qinfen Lu
Zhejiang University
Hangzhou
China

Xinyu Fan
Jiangsu University of Science
and Technology
Zhenjiang
China

Cao Tan
Shandong University of
Technology
Zibo
China

Jiayu Lu
Shandong University of
Technology
Zibo
China

Editorial Office
MDPI
St. Alban-Anlage 66
4052 Basel, Switzerland

This is a reprint of articles from the Special Issue published online in the open access journal *Actuators* (ISSN 2076-0825) (available at: https://www.mdpi.com/si/actuators/D69FL10A0O).

For citation purposes, cite each article independently as indicated on the article page online and as indicated below:

Lastname, A.A.; Lastname, B.B. Article Title. *Journal Name* **Year**, *Volume Number*, Page Range.

ISBN 978-3-7258-0783-3 (Hbk)
ISBN 978-3-7258-0784-0 (PDF)
doi.org/10.3390/books978-3-7258-0784-0

Contents

Article

Design and Analysis of a High Power Density Permanent Magnet Linear Generator for Direct-Drive Wave Power Generation

Xinyu Fan [1,*], Changkun Wang [1], Zhibing Zhu [2] and Hao Meng [1]

[1] School of Energy and Power, Jiangsu University of Science and Technology, Zhenjiang 212100, China
[2] Power Plant Division, Shanghai Marine Diesel Engine Research Institute, Shanghai 201108, China
* Correspondence: fanxy@just.edu.cn

Abstract: Wave energy is a new type of clean energy. Aiming at a low wave energy density and small wave height in China's coastal areas, a tubular permanent magnet linear generator (PMLG) with a short stroke, small volume, and high power density is designed for wave power generation. Firstly, the generator's electromagnetic parameters are analyzed by the analytical method, and the magnetic circuit topology and basic structure of the generator are analyzed by the equivalent magnetic circuit method (EMCM). Then, the finite element method (FEM) is used to analyze the influence law of the generator's basic structural parameters on the output electromotive force (EMF) and its sinusoidal characteristics. The multi-factor and multi-level analysis is carried out based on the orthogonal test method to study the size parameters of the above analysis, and the optimal structure parameter combination for the generator is obtained. Finally, the prototype is trial-produced and tested for steady-state and transient performance to confirm the accuracy of the simulation calculations, and the output performance under no-load and load conditions is examined. The results show that both the optimized prototype's power density and the output EMF's sinusoidal properties have been improved under the proposed scheme.

Keywords: wave power; permanent magnet linear generator; total harmonic distortion; optimization design

Citation: Fan, X.; Wang, C.; Zhu, Z.; Meng, H. Design and Analysis of a High Power Density Permanent Magnet Linear Generator for Direct-Drive Wave Power Generation. *Actuators* **2022**, *11*, 327. https://doi.org/10.3390/act11110327

Academic Editor: Zongli Lin

Received: 24 September 2022
Accepted: 7 November 2022
Published: 10 November 2022

Publisher's Note: MDPI stays neutral with regard to jurisdictional claims in published maps and institutional affiliations.

1. Introduction

With the worsening of world's energy crisis, wind, solar [1,2], wave [3], and other new energy sources have become research priorities in recent years. The wave energy is a kind of energy formed by the wind blowing across the sea surface [4]. The marine areas in different geographical locations have different wave characteristics [5]. The novel direct-drive wave power system with a simple structure and an improved efficiency adopts a linear generator [6,7] as the energy converter, which can directly match the up and down reciprocating motion of the wave [8,9]. Among the existing studies, the main linear generators applied to wave power generation are hybrid magnetic field modulated linear generators (HMFMLG) [10], switched reluctance linear generators (SRLG) [11,12], and permanent magnet synchronous linear generators (PMSLG) [13,14], etc.

Although HMFMLG has a superior output performance and a higher use efficiency of permanent magnets (PM), its complicated construction is a big disadvantage that is hard to overcome. The SRLG's structure is simple and inexpensive to produce, but it produces a lot of noise and has poor stability. The PMSLG [15–17] has the benefits of high power, a compact structure, and high efficiency. Liu et al. [18] proposed a two-body direct-drive wave energy converter, in which the translator and stator of the PMLG were mounted on two buoys to extract energy from the relative motion between two oscillating buoys. Zhang et al. [19] and Jang, S.M. et al. [20] studied a PMLG of the Halbach PM array to improve the air gap magnetic field of the generator, in which the effects of different slot and pole pitches on the detent force were analyzed, but the waveform of the EMF was not analyzed. Zhou et al. [21] proposed a calculation model based on the EMCM and used it to analyze the magnetic circuits corresponding to the stator core, PM, and air gap, respectively. Finally, the partial magnetic circuits were combined to obtain a complete equivalent magnetic circuit model for this motor. Arslan et al. [22] analyzed the correlation between the variable structure parameters of the motor and the efficiency and volume based on the effect surface optimization method and multi-objective genetic algorithm. Finally, the comprehensive performance of the motor was improved, and the optimal value of the motor was obtained.

This paper designed a high power density PMLG based on the Halbach array, which is suitable for operation in the near coastal environment of China. The PMLG is designed preliminarily by the analytical method and its electromagnetic parameters are determined. In order to increase power density, improve power production efficiency, and decrease the THD of the EMF, the variation law of induction EMF's peak and the THD under various parameters from the perspectives of yoke thickness, tooth width, and PM thickness was examined in depth by the FEM. The multi-parameter optimization is carried out based on the orthogonal design method to obtain the optimal parameter combination. Finally, the steady-state and dynamic performance of the protype are tested. The results show that the optimized PMLG has the advantages of high power density, high generation efficiency, and a small THD of the EMF.

2. Analytical Calculation of PMLG

Aiming at a low wave energy density and small wave height in China, this study develops a small Halbach array PMLG that uses the fundamental structure of a long secondary and short primary. As a wave generator, the primary is relatively fixed, and the secondary is directly connected to the wave collecting apparatus. The secondary is driven by the wave to perform reciprocating linear motion to produce electricity. Based on the wave characteristics, the generator performance requirements are shown in Table 1. According to the generator design index, the electromagnetic properties of the generator are calculated and analyzed using the analytical method.

Table 1. Performance requirements of PMLG.

Parameter	Value	Parameter	Value
Rated voltage U_N/V	25	Stoke H/mm	≤ 40
Rated speed V/m/s	0.5	Outer diameter D_{s_out}/mm	≤ 100
Rated frequency f_N/Hz	10	Phase m	1

2.1. Electromagnetic Calculation of PMLG

The main dimensions of the rotary motor are the armature diameter and the axial effective length. The linear motor evolves from the rotary motor and its corresponding main dimensions are the primary effective length and the inner diameter.

The primary effective length l_{ef} is denoted as:

$$l_{ef} = 2P\tau \tag{1}$$

Speed:

$$V = 2\tau f_N \tag{2}$$

According to the characteristics of small wave power in summer, taking $V = 0.5$ m/s, the polar distance $\tau = 25$ mm can be calculated. In this paper, the PMLG based on the Halbach PM array [23–26] adopts the structure of three slots and four poles, so the primary effective length of the generator can be obtained. Generally, the calculated power is determined by the rated index of the generator:

$$P' = mEI \tag{3}$$

Phase electromotive force:

$$E = 4.44\, f N K_{dp} \Phi_\delta K_\Phi \tag{4}$$

$$E = (1 + \Delta U) U_N \tag{5}$$

Rated power:

$$P_N = m U_N I \cos \varphi \tag{6}$$

Air gap magnetic flux:

$$\Phi_\delta = B_\delta \alpha_i \pi D_{s\_in} \tau \tag{7}$$

Electric load:

$$A = \frac{2mNI}{2P\tau} \tag{8}$$

Tidy up bring in:

$$2P\tau D_{s\_in} = \frac{P'}{13.94 \alpha_i K_{dp} K_\Phi A B_\delta V} \tag{9}$$

where, P' is the calculated power, E is the phase EMF, I is the phase current, N is the number of winding turns, K_{dp} is the fundamental winding coefficient, K_Φ is the air gap flux waveform coefficient, B_δ is the air gap magnetic density, α_i is the calculated pole arc coefficient, P is the pole logarithm, ΔU is the voltage regulation, D_{s_in} is the stator's inner diameter and φ is the power factor.

It is clear that the generator's primary dimensions are dependent on the calculated power, electromagnetic load, and air gap magnetic density.

2.2. Stator Design

According to the above formula, the stator's effective length can be calculated. To determine the turns per coil turns, the generator's EMF and fundamental winding coefficient are used as guides, and their relationship is as follows:

$$N = \frac{E}{4.44 f \Phi_\delta K_{dp} K_\Phi} \tag{10}$$

According to the diameter of the coil, the area occupied by the coil can be calculated, and the size of the teeth and grooves can be further determined.

The crack ratio K_S is defined as the ratio of the stator inner diameter D_{s_in} to outer diameter D_{s_out}, and the size of the primary outer diameter is determined by the crack ratio. Current research indicates that when $K_S = 0.45$, the output power and power density of the generator are the maximum. The preliminary design parameters of the PMLG can be derived using the aforementioned analytical procedure, and its structure is shown in Figure 1.

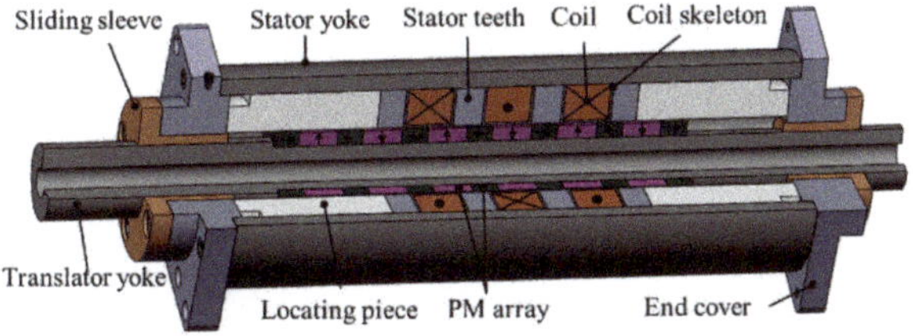

Figure 1. PMLG's structure.

3. Analysis of Equivalent Magnetic Circuit Model

The distribution of the magnetic field inside the motor is determined by converting the calculation of the motor magnetic field into the calculation of the magnetic circuit using the

EMCM. The magnetic circuit corresponding to the model is shown in Figure 2. The air gap allows the magnetic force lines produced by the radially magnetized PM to penetrate the stator teeth and yoke before moving on to the adjacent PM. Finally, the axial magnetized PM returns to the initial PM to form a closed loop. The magnetic circuit includes the magnetomotive force source F_c, the PM reluctance R_m, the air-gap reluctance R_g, the stator tooth reluctance R_c, the stator yoke reluctance R_o, and the translator yoke reluctance R_i.

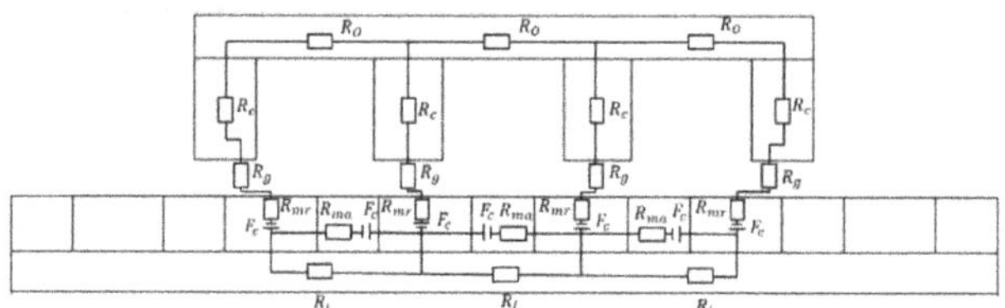

Figure 2. Equivalent magnetic circuit method.

Where, the magnetomotive force is generated by the PM:

$$F_C = H_C l = \frac{B_r l}{\mu_0 \mu_r} \tag{11}$$

H_C is the PM's coercivity, l is the length of the PM along the magnetization direction, B_r is the PM's remanence, μ_0 is the vacuum permeability, and μ_r is the relative permeability of the PM.

The Carter coefficient k_δ is introduced, and its computation follows. It takes into account how stator slotting affects the effective air gap length:

$$k_\delta = \frac{(w_s + w_t)(5g + w_t)}{(w_s + w_t)(5g + w_t) - w_t^2} \tag{12}$$

The air gap reluctance R_g is obtained from the reluctance equation as:

$$R_g = \int_{r_2}^{r} \frac{k_\delta dr}{2\pi r \mu_0 w_s} = \frac{k_\delta(\ln r - \ln r_2)}{2\pi \mu_0 w_s} \tag{13}$$

The magnetoresistance of PM is:

$$R_{mr} = \int_{r_1}^{r_2} \frac{dr}{2\pi r \mu_0 \mu_r \tau_r} = \frac{\ln r_2 - \ln r_1}{2\pi \mu_0 \mu_r \tau_r} \tag{14}$$

$$R_{ma} = \frac{\tau_a}{\mu_0 \mu_r \pi (r_2^2 - r_1^2)} \tag{15}$$

The internal resistance of the yoke and stator teeth is negligible in comparison to the air-gap resistance. According to the magnetic circuit theorem, the PM's magnetic flux in the air gap can be expressed as follows:

$$\Phi = \frac{F_C}{2R_g + R_{mr} + R_{ma}} = \frac{H_C(r_2 - r_1) + H_C \tau_a}{\frac{k_\delta(\ln r - \ln r_2)}{\pi \mu_0 w_s} + \frac{\ln r_2 - \ln r_1}{2\pi \mu_0 \mu_r \tau_r} + \frac{\tau_a}{\pi \mu_0 \mu_r (r_2^2 - r_1^2)}} \tag{16}$$

$$B_\delta = \frac{\Phi}{2\pi r \tau_r} = \frac{H_C(r_2 - r_1) + H_C \tau_a}{\frac{2r\tau_r k_\delta(\ln r_3 - \ln r_2)}{\mu_0 w_s} + \frac{r(\ln r_2 - \ln r_1)}{\mu_0 \mu_r} + \frac{2r\tau_r \tau_a}{\mu_0 \mu_r (r_2^2 - r_1^2)}} \tag{17}$$

where w_s is the tooth width, w_t is the notch width, g is the air gap length, r is the radius of air gap, r_1 is the inside radius of the PM, r_2 is the outside radius of the PM, r_3 is the inside radius of the stator teeth, τ_r is the length of the radial magnetized PM, and τ_a is the length of the axial magnetized PM.

The EMCM is used to investigate the PMLG's magnetic circuit topology and analyze the rationality of the magnetic circuit structure, which has certain theoretical guiding significance. Meanwhile, EMCM explains the relation between the air-gap magnetic density and reluctance, and analyzes the forming mechanism of high power density, providing a guiding reference for the subsequent finite element analysis and optimization.

4. Finite Element Analysis and Structure Optimization of PMLG

4.1. Model Establishment and Result Analysis

The three-dimensional finite element model of PMLG is established in electromagnetic simulation analysis software, which is used to analyze the electromagnetic characteristics of the generator. The translator yoke, stator yoke, and stator teeth are all made of steel-08 with high permeability. The PM array adopts high-performance NdFeB material (model: N45SH); the coil skeleton adopts high strength engineering plastic, and the coil is copper core enameled wire. The whole model uses tetrahedral automatic mesh, and the mesh is encrypted around the model air gap to improve the accuracy of the simulation calculation. Figures 3 and 4 show the static magnetic induction intensity cloud diagram and the static air gap magnetic density curve of the PMLG, respectively.

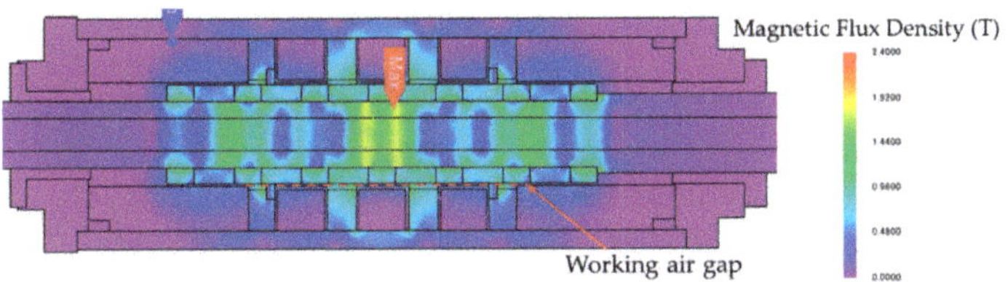

Figure 3. Cloud diagram of magnetic flux intensity.

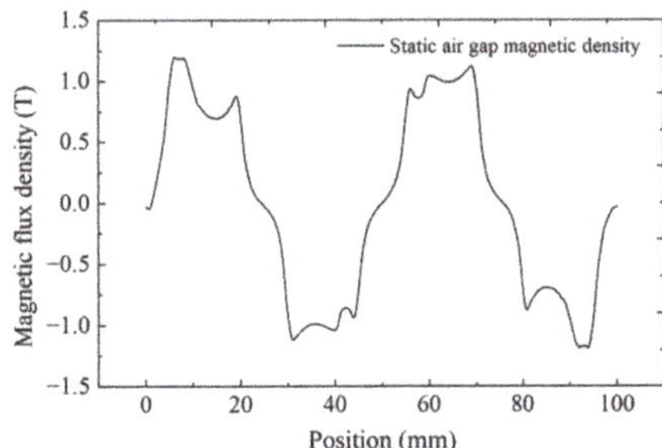

Figure 4. Static air gap magnetic density curve of PMLG.

The traveling wave magnetic field produced by the Halbach array moves relative to the coil at a constant speed when the translator moves at a speed of 0.5 m/s. According to Faraday's law of electromagnetic induction, the induced EMF is generated in the coil. Even while the entire waveform is sinusoidal, there are some localized distortions in the site where the peak and trough are more noticeable. The waveform's examination using the fast Fourier transform (FFT) reveals that the EMF's peak is 20.22 V and its THD is 4.01%.

4.2. Translator Structure Analysis

The theoretical analysis was conducted to make preliminary determinations of several important motor dimensions. In order to maximize the use of the PM and subsequently enhance the performance of the motor, it is typically required to study the impact of various structural dimensions on the EMF [27]. This paper analyzes the influence rule of PMLG's size change by taking the EMF's peak and THD as the targets, in order to improve the power density and facilitate the subsequent work of grid connection.

The translator yoke is a component of the generator secondary, which is repeatedly examined in various thicknesses. Figure 5a demonstrates that the EMF's peak steadily rises as yoke thickness increases, but the rising amplitude is getting smaller. The THD of the EMF gradually drops until it settles at a constant value of 3.62%. The generator's translator yoke is made of the magnetically conductive steel-08 material when it uses the Halbach array, which can increase the generator's power density.

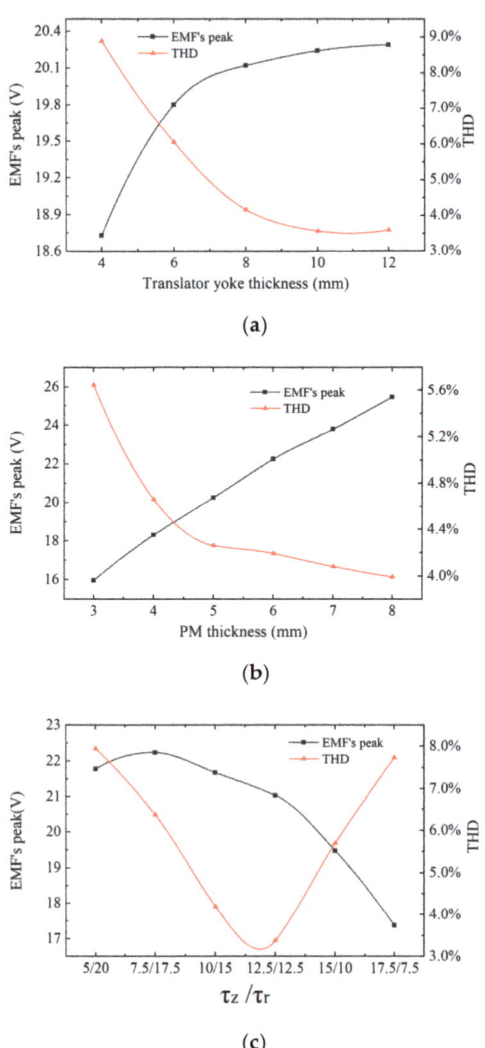

Figure 5. Effect of translator on EMF. (**a**) Translator yoke thickness. (**b**) PM thickness. (**c**) τ_z / τ_r

The influence of the PM's thickness on the performance of the generator was analyzed. The PM's thickness is taken as 3 mm, 4 mm, 5 mm, 6 mm, 7 mm, and 8 mm in turn. As can be seen in Figure 5b, the EMF increment is steady on the whole but exhibits a very minor downward trend. With the increase of PM thickness, the product of its volume and magnetic energy increases. However, due to the magnetic saturation occurring in ferromagnetic material, the increment of the EMF is steadily reducing. As PM thickness increases, the THD of the EMF steadily declines until it stabilizes eventually.

The generator performance is analyzed by setting τ_z / τ_r to 5:20, 7.5:17.5, 10:15, 12.5:12.5, 15:10, and 17.5:7.5, respectively. As shown in Figure 5c, the EMF's peak drops as the length of the radial magnetized PM decreases, since the magnetic induction line created by the radial magnetized PM in a Halbach array runs perpendicularly to the direction of motion. The magnetic field line produced by the axial magnetized PM is parallel to the direction of motion. That is, the radial magnetized PM plays a major role in the Halbach array. The THD of the EMF first falls and subsequently increases as the length of the axial magnetized PM increases. When $\tau_z = \tau_r$, the THD is at its lowest level. It can be seen that the THD will decrease with a decrease in size difference between radial and axial magnetization.

4.3. Stator Structure Analysis

The tooth is separated into edge teeth and middle teeth based on their distinct positions, the edge teeth are found at both ends of the linear generator, one side of which forms a groove with the stator yoke and the other side of which is broken; the middle teeth have grooves on both sides.

As shown in Figure 6a, the peak and THD of the EMF initially increase and then decrease as the width of the middle teeth increases. The THD is greater when the width between the middle teeth is near to 12.5 mm. When the size of the middle tooth is closer to the size of the PM, the EMF's peak is larger. A portion of the PM's magnetic field line will leak into the air if the middle tooth's size is too small, which reduces the main magnetic flux and, consequently, the EMF's amplitude. As can be seen in Figure 6b, with a 1.5 mm increase in the edge teeth, the EMF's peak increases by around 2 V. However, the rise in the EMF is slightly mitigated by the wider edge tooth spacing. The THD steadily decreases as edge tooth width increases.

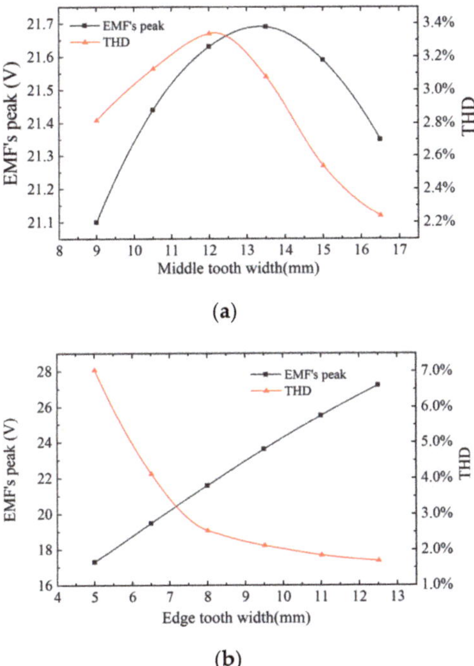

(a)

(b)

Figure 6. Effect of tooth width on EMF. (a) Middle tooth width; (b) edge tooth width.

4.4. Orthogonal Design Optimization Analysis

The above analysis process was conducted under a single factor, and the optimal parameters of the generator were not obtained. The orthogonal test was carried out to assess and research several factors of the stator and translator simultaneously. In orthogonal design, partial tests are used instead of full tests, with the results of the partial tests being analyzed to comprehend the full tests and ultimately to determine the ideal set of parameters. The thickness of the PM, the ratio of τ_z/τ_r, and the width of the edge teeth have a greater impact on the EMF according to the aforementioned single factor analysis. Next, the three-factor and three-level orthogonal design analysis was conducted, as shown in Table 2. This method has the advantages of having a more simple operation and a higher efficiency.

Table 2. Values of structural parameters.

Parameter	Code Name	Level Value 1	Level Value 2	Level Value 3
h_M	A	5	6	7
τ_z/τ_r	B	11.5/13.5	12.5/12.5	13.5/11.5
Edge tooth width	C	8	9.5	11

Table 3 shows the results of the orthogonal design, the data are processed and analyzed using the visual analysis approach in accordance with the previous orthogonal design results. The effect curve method is utilized in this research to more clearly visualize the outcomes of the aforementioned optimization analysis.

Table 3. Orthogonal design results.

Operation Number	Combination	Amplitude of EMF (V)	THD
1	A1B1C1	19.94	2.62%
2	A1B2C2	21.23	2.30%
3	A1B3C3	22.21	2.28%
4	A2B1C2	24.09	2.06%
5	A2B2C3	25.51	1.85%
6	A2B3C1	21.04	3.06%
7	A3B1C3	28.48	1.64%
8	A3B2C1	23.68	2.35%
9	A3B3C2	25.38	2.11%

According to the variation amplitude of each curve in Figure 7a, it can be shown that the PM thickness has the greatest influence on the EMF's peak, followed by the edge tooth width and the ratio of τ_z/τ_r. The EMF's peak displays an increasing trend when PM thickness and edge teeth width increase, and the EMF's peak shows a decreasing tendency as the length of the axial magnetized PM in the Halbach array increases. As can be seen in Figure 7b, the THD of the EMF steadily rises with the increase of the length of the axial magnetized PM, but falls with the increase of PM thickness and edge tooth width. In combination with Figure 7a,b, the optimal combination of PM thickness, τ_z/τ_r ratio, and edge tooth width is h_M = 7 mm, τ_z/τ_r = 11.5/13.5, and an edge tooth wide of 11 mm. The optimized PMLG size parameters are shown in the Table 4.

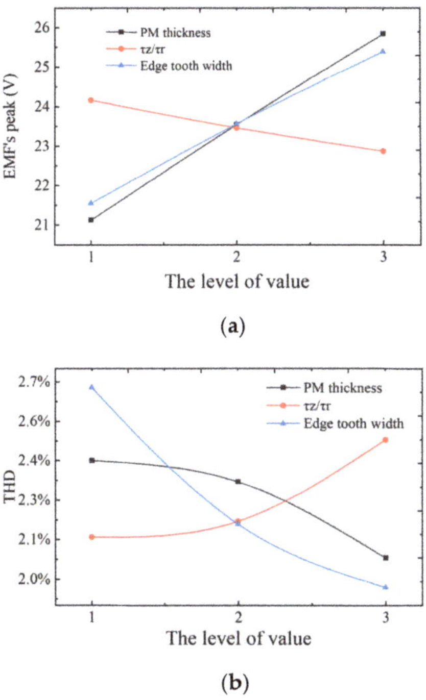

(a)

(b)

Figure 7. Effect curve of orthogonal design results. (**a**) Effect curve of EMF; (**b**) Effect curve of THD.

Table 4. PMLG structure and dimension parameters.

	Parameter	Symbol	Value
	Outer diameter	D_{s_out} (mm)	90
	Length	l_{ef} (mm)	100
Stator	Edge tooth width	ω_s (mm)	11
	Groove width	ω_t (mm)	19
	Winding turn	N	660
	Winding resistance	R_r (Ω)	2.82
	PM thickness	h_M (mm)	7
	Length	l_r (mm)	175
Translator	Outer diameter	D_{r_out} (mm)	38
	Radial magnetized PM	τ_r (mm)	13.5
	Axial magnetized PM	τ_a (mm)	11.5

5. Experimental Analysis

A single-phase winding unit is created by connecting the three coils in series, and the outside surface of the stator teeth is grooved to arrange the connection of the coils to one another. In the Halbach array, the axial magnetized PM is processed into a ring, and the radial magnetizing PM is spliced into a ring by using eight magnetic tiles to achieve radial radiation magnetization. Anaerobic glue is used to adhere the PM and the translator yoke together as a whole, as shown in the Figure 8.

5.1. Experimental Analysis of Steady State Performance

Detent force, as a performance index of a linear motor, includes end force and cogging force. The end force is produced by the linear motor's limited length, and the cogging force is caused by the different magnetic conductance of the teeth and the grooves. A large detent force will produce vibration and noise during motor operation, thus affecting the motor's performance. Stator edge teeth with an inner step structure can greatly reduce the detent force. The detent force is tested at various locations within 20 mm on the left and right sides of the initial position, with the middle axial magnetized PM in the middle stator slot serving as the initial position.

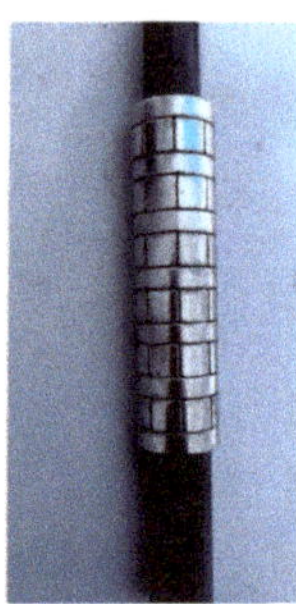

Figure 8. Winding unit and Halbach array of PMLG.

The detent force of the prototype is measured using the static displacement method. Using a ball screw slide to adjust the position of the translator, the translator's precise position is measured by the laser displacement sensor (optoNCDT2300). When the position of the translator is shifted by 1 mm, the measurement signal of the force sensor (T320A-S) is collected by the integrated controller (TMS320F2812), and the operation of the control system, as well as the processing of the measurement data, are carried out on the upper computer. Through point-by-point measurement, the detent force of the translator at different positions can be measured. As shown in Figure 9, the trend of the measurement results is consistent with that of the simulation results, which verifies the accuracy of the simulation.

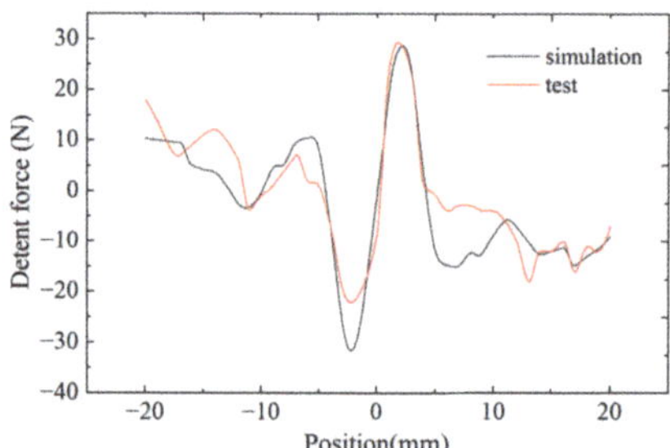

Figure 9. Detent force diagram of the prototype.

5.2. Dynamic Experimental Analysis under No Load Condition

Figure 10 is a photo of the experimental device. In the dynamic experiment, a simple harmonic motion is generated by using a rotating motor and an eccentric shaft connecting rod to simulate the motion law of ocean waves. The connecting rod is attached to the left side of the prototype translator, which is used to drive the reciprocating linear motion. The translator displacement function is:

$$x = A \sin(\omega t + \varphi) \tag{18}$$

where A stands for the translator's motion stroke, which is determined by the eccentric distance of the eccentric shaft, and ω stands for the translator's motion frequency, which is determined by the rotational speed of the rotary motor.

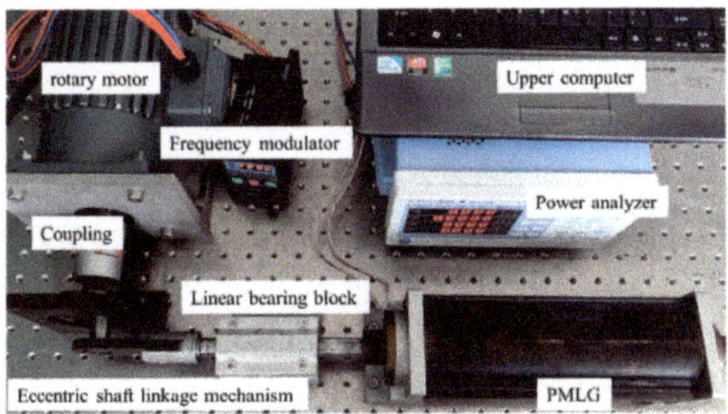

Figure 10. Prototype test system diagram.

When the motion frequency is 5 Hz, the prototype's EMF is evaluated using various motion strokes, and the strokes are 20 mm, 30 mm, and 40 mm, respectively. It can be seen from Figure 11, as the stroke increases, that the peak and the THD of the EMF tend to increase, while the waveform period remains unchanged. When the stroke increases, the motion speed of the translator increases, so the EMF's peak increases. The pole distance of the prototype PM array is 25 mm and the THD of the EMF, which is 8.07% when the stroke is 20 mm, is minimal. The THD of the EMF is 26.29% for a 40 mm stroke, which clearly shows distortion. Thus, it can be concluded that the sine characteristics of the EMF waveform are better when the translator moves within a pole distance range, and the waveform will be slightly distorted when the motion stroke surpasses the pole distances. By FFT of the EMF's waveform, the proportion of the fundamental wave and harmonic wave can be obtained. Based on 100% fundamental wave analysis, as shown in Figure 12, the second harmonic makes up nearly half of the entire harmonic, occupying the biggest percentage in the harmonic. The proportion of the second harmonic steadily rises as the motion stroke increases, although the change in the other harmonics is minimal. It is concluded that the THD of the EMF waveform is mainly related to the proportion of the second harmonic. By comparison, it can be seen that the simulated value of the EMF is basically consistent with the experimentally measured value, indicating the feasibility of the scheme.

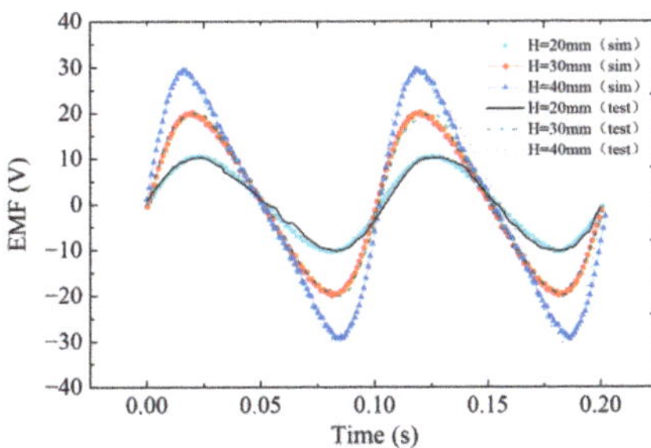

Figure 11. EMF curve of PMLG under different strokes.

Under the condition that the motion stroke is 20 mm, the EMF of the prototype is evaluated under various motion frequencies. The prototype is driven by the rotating motor through the eccentric shaft connecting rod mechanism, the prototype reciprocates motion once while the rotating motor circles around. The output speed of the rotating motor is adjusted by the frequency converter so as to adjust the movement frequency of the prototype. The exercise frequencies are set as 2.5 Hz, 3.3 Hz, and 5 Hz for the test. According to the EMF waveform in Figure 13, the EMF's peak steadily declines as the movement frequency reduces. The waveform period grows as the motion frequency declines, which is in fact 0.5 times related to the motion period. The THD of the EMF changes insignificantly at different frequencies. According to the waveform's FFT in Figure 14, second harmonic makes up the majority of the EMF waveform at varied frequency, with the proportions of the other harmonics essentially remaining unchanged.

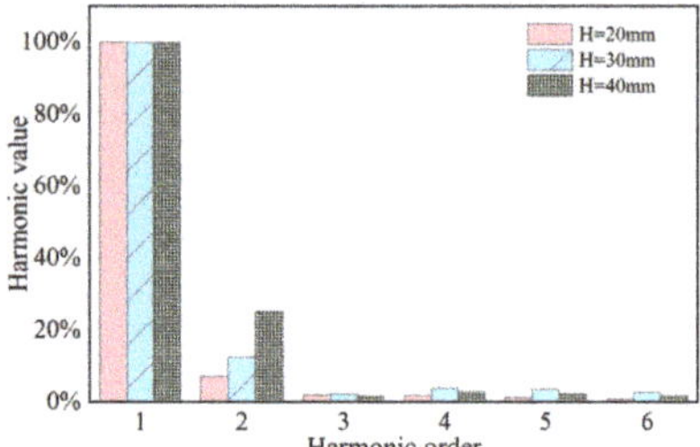

Figure 12. Harmonic distribution of EMF under different strokes.

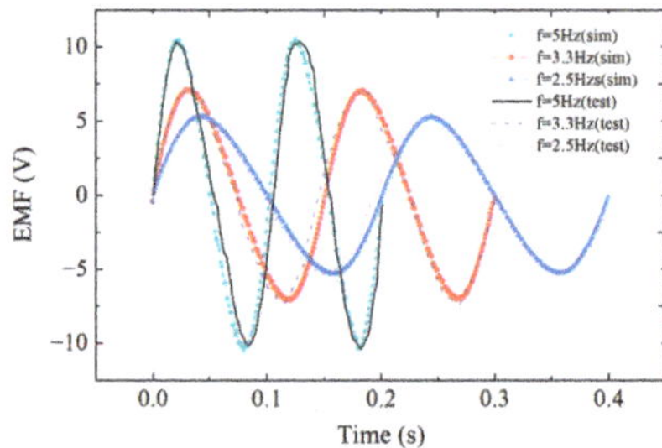

Figure 13. EMF curve of PMLG under different periods.

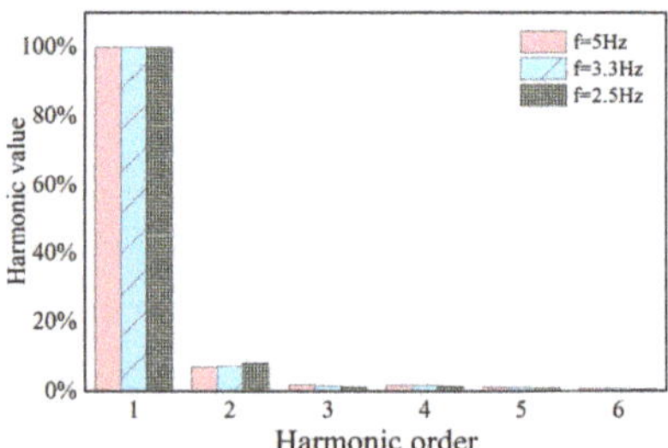

Figure 14. Harmonic distribution of EMF under different periods.

Therefore, it is clear that the prototype's EMF peak is influenced by the motion stroke and frequency. When the frequency is constant, EMF's peak increases with the increase of the motion stroke, when the stroke is constant, EMF's peak increases with the increase of the movement frequency. The EMF waveform period increases with the decrease of the motion frequency, which is actually 0.5 times related to the motion period, independent of the motion stroke. The THD of the EMF rises as the stroke rises, but it is independent of the frequency.

5.3. Dynamic Experimental Analysis under Load Condition

Load analysis is an essential link of motor performance testing. The output performance of PMLG is analyzed by simulation and experiment under different loads and movement frequencies with a motion stroke of 40 mm. PMLG's output characteristics are shown in Table 5.

As shown in Figure 15 and Table 5, when the movement frequency is constant, the power first increases then decreases as the load increases; when the external load equals to the PMLG's internal resistance, the power reaches its maximum level. The circuit's resistance increases as a result of the increased external load, leading to the decrease in the circuit's current and output power; the amount of voltage that the load separates increases as the external load increases relative to the internal resistance of the motor, leading to a gradual increase in the EMF output. When the load is constant and combined with the prior examination of the no-load characteristics, it is evident that, as the frequency reduces, the movement speed decreases, the EMF created lowers, and the output EMF and power steadily decline.

Table 5. The output characteristics of PMLG under different frequency and loads.

Load (Ω)	Frequency (Hz)	Simulation EMF RMS (v)	Test EMF RMS (v)	Simulation Output Power (W)	Test Output Power (W)
2.5	5	8.21	7.26	26.86	21.03
	3.3	5.66	4.92	12.76	9.73
	2.5	4.25	3.56	7.16	5.06
5	5	11.52	10.35	26.46	21.35
	3.3	7.84	6.89	12.18	9.53
	2.5	5.83	4.75	6.62	4.54
10	5	14.34	13.70	20.53	18.71
	3.3	9.67	8.68	9.26	7.54
	2.5	7.26	5.86	5.21	3.52
15	5	15.55	15.16	16.09	15.28
	3.3	10.45	9.63	7.36	6.31
	2.5	7.85	6.68	4.06	2.93
20	5	16.50	16.12	13.49	12.91
	3.3	10.82	10.17	5.81	5.15
	2.5	8.15	7.18	3.19	2.52

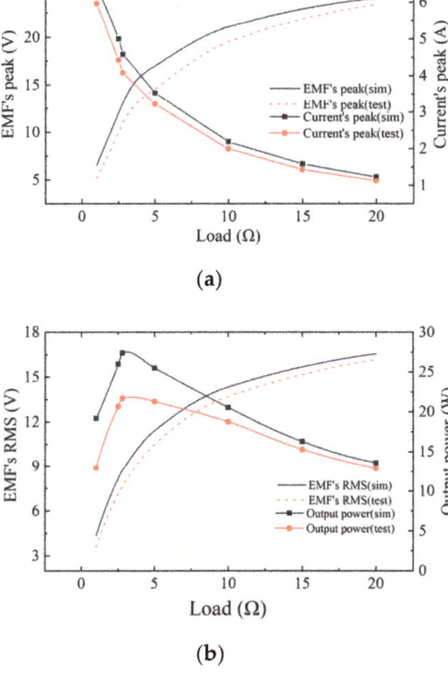

Figure 15. Output performance of simulation and test. (**a**) Output EMF and current; (**b**) output EMF and power.

The experimental value of PMLG's output is lower than the simulation value, because the iron loss is not calculated during the simulation process. When the external load is small, the split voltage ratio inside the generator is higher, and the loss accounts for a considerable portion of the PMLG's total electric energy. Therefore, the discrepancy between the measured value and the simulated value is large. With the increase of external load, the split voltage ratio inside the generator decreases, the proportion of loss decreases, and the deviation between simulation and measurement decreases. Figure 15a,b show the variations of the power, voltage, and current with load. When the load is small, the power deviation between measured results and simulation results is large, because the errors of the voltage and current are large. When the load is set to 20 Ω, the errors of the EMF, current, and power are 2.6%, 9.2%, and 5.6%, respectively, and the experimental results are essentially identical to the simulation results.

6. Conclusions

In this study, a high power density PMLG is proposed and its electromagnetic design and optimization are studied. The changes in EMF magnitude and waveform under various strokes and frequencies are investigated using a simulation and experiment. The main research conclusions are as follows:

1. The thickness of the PM, the width of the edge tooth, and the axial length ratio of the axial and radial magnetized PM have a greater influence on the peak and the THD of the EMF. With the increase of PM thickness and edge tooth width, the EMF's peak increases proportionally, and the THD decreases gradually. In the Halbach array, when the length of the radial magnetized PM is closer to the length of the intermediate tooth, the peak of the EMF is larger. The smaller the axial size difference between the axial and radial magnetized PM, the smaller the THD will be. The THD will be the smallest when the axial and radial magnetized PM are equal in size.
2. The orthogonal design method is used to study the influences of three factors, including the thickness of the PM, the width of the edge teeth, and the ratio of the axial length of the axial and radial magnetized PM. The ideal structural parameters are $h_M = 7$ mm, $\tau_z/\tau_r = 11.5/13.5$, and an edge teeth width of 11 mm. Finally, the optimal combination of PMLG structural parameters is obtained.
3. The EMF's peak is related to the movement stroke and frequency; in fact, it increases as the movement stroke and frequency increases. The EMF waveform period increases with the decrease of the motion frequency, which in fact is 0.5 times related to the motion period. The EMF's THD is correlated with the stroke, that is, the THD is low, and the sine characteristic of the waveform is better when the motion stroke is less than the PM pole distance. The waveform will show some distortions when the motion stroke is greater than the pole distance.

Author Contributions: Conceptualization, X.F. and C.W.; methodology, X.F., C.W. and Z.Z.; software, formal analysis, and investigation, X.F., C.W. and H.M.; data curation, X.F., Z.Z. and H.M.; writing—original draft preparation, X.F., C.W. and H.M.; writing—review and editing, X.F. and Z.Z. All authors have read and agreed to the published version of the manuscript.

Funding: This work was funded by the National Natural Science Foundation of China, grant number 52205271 and Natural Science Foundation of Jiangsu Province, China, grant number BK20190972.

Institutional Review Board Statement: Not applicable.

Informed Consent Statement: Not applicable.

Data Availability Statement: Not applicable.

Conflicts of Interest: The authors declare no conflict of interest.

References

1. Li, Y.; Luo, K.; Tao, R.; Wang, Z.; Chen, D.; Shao, Z. A new concept and strategy for photovoltaic and thermoelectric power generation based on anisotropic crystal facet unit. *Adv. Funct. Mater.* **2020**, *30*, 2002606. [CrossRef]
2. Li, Y.; Wang, Z.; Tao, R.; Fan, Y.; Xu, J.; Yu, L.; Ren, N.; Wu, J.; Chen, D.; Shao, Z. Preparation strategies of p-type cuprous oxide and its solar energy conversion performance. *Energy Fuels* **2021**, *35*, 17334–17352. [CrossRef]
3. Dong, F.; An, X.; Li, C. Performance evaluation of wind power industry chain based on three-stage DEA. *J. Renew. Sustain. Energy.* **2021**, *13*, 033313. [CrossRef]
4. Wan, Y.; Zheng, C.; Li, L.; Dai, Y.; Esteban, M.D.; López-Gutiérrez, J.S.; Qu, X.; Zhang, X. Wave energy assessment related to wave energy convertors in the coastal waters of China. *Energy.* **2020**, *202*, 117741. [CrossRef]
5. Qiu, S.; Liu, K.; Wang, D.; Ye, J.; Liang, F. A comprehensive review of ocean wave energy research and development in China. *Renew. Sustain. Energy Rev.* **2019**, *113*, 109271. [CrossRef]
6. Faiz, J.; Nematsaberi, A. Linear electrical generator topologies for direct-drive marine wave energy conversion-an overview. *IET Renew. Power Gener.* **2017**, *11*, 1163–1176. [CrossRef]
7. Tan, C.; Lu, Y.; Ge, W.; Li, B.; Lu, J. Depth fuzzy sliding-mode active disturbance rejection control method of permanent magnet linear motor for direct drive system. *J. Xi'an Jiaotong Univ.* **2023**, *1*, 1–9, *in press*.
8. Hamim, M.; Ibrahim, T.; Nor, N.M. Modeling and analyze a single-phase halbach magnetized tubular linear permanent magnet generator for wave energy conversion. In Proceedings of the 2014 International Conference on Power and Energy (PECon), Kuching, Malaysia, 1–3 December 2014; pp. 87–92.
9. Bode, C.; Schillingmann, H.; Henke, M. A free-piston PM linear generator in vernier topology using quasi-Halbach-excitation. In Proceedings of the 2014 International Conference on Electrical Machines (ICEM), Berlin, Germany, 2–5 September 2014; pp. 1950–1955.
10. Li, W.; Chau, K.T.; Jiang, J.Z. Application of linear magnetic gears for pseudo-direct-drive oceanic wave energy harvesting. *IEEE Trans. Magn.* **2011**, *47*, 2624–2627. [CrossRef]
11. Pan, J.F.; Li, Q.; Wu, X.; Cheung, N.; Qiu, L. Complementary power generation of double linear switched reluctance generators for wave power exploitation. *Int. J. Electr. Power Energy Syst.* **2019**, *106*, 33–44. [CrossRef]
12. Di Dio, V.; Franzitta, V.; Milone, D.; Pitruzzella, S.; Trapanese, M.; Viola, A. Design of bilateral switched reluctance linear generator to convert wave energy. *Adv. Mater. Res.* **2014**, *860*, 1694–1698.
13. Feng, N.; Yu, H.; Zhao, M.; Zhang, P.; Hou, D. Magnetic field-modulated linear permanent-magnet generator for direct-drive wave energy conversion. *IET Electr. Power Appl.* **2020**, *14*, 742–750. [CrossRef]
14. Zamri, N.A.M.; Ibrahim, T.; Nor, N.M. Direct drive linear generator designs with aluminium spacer and alternate slot winding for wave energy conversion system. *Int. J. Adv. Sci. Eng. Inf. Technol.* **2017**, *7*, 1282. [CrossRef]
15. Meng, B.; Xu, H.; Liu, B.; Dai, M.; Zhu, C.; Li, S. Novel magnetic circuit topology of linear force motor for high energy utilization of permanent magnet: Analytical modelling and experiment. *Actuators* **2021**, *10*, 32. [CrossRef]
16. Abdalla, I.I.; Ibrahim, T.; Nor, N.M.; Perumal, N. Optimal design of a single-slotted permanent-magnet linear generator for wave energy conversion. *Int. J. Appl. Electromagn. Mech.* **2018**, *56*, 21–34. [CrossRef]
17. Tan, C.; Ren, H.; Li, B.; Lu, J.; Li, D.; Tao, W. Design and analysis of a novel cascade control algorithm for braking-by-wire system based on electromagnetic direct-drive valves. *J. Frankl. Ins.* **2022**, *9*, 006. [CrossRef]
18. Liu, C.; Chen, R.; Zhang, Y.; Liu, W.; Wang, L.; Qin, J. Design and test of a novel two-body direct-drive wave energy converter. *Int. J. Appl. Electromagn. Mech.* **2021**, *65*, 527–544. [CrossRef]
19. Zhang, J.; Yu, H.; Chen, Q.; Hu, M.; Huang, L.; Liu, Q. Design and experimental analysis of AC linear generator with Halbach PM arrays for direct-drive wave energy conversion. *IEEE Trans. Appl. Supercond.* **2013**, *24*, 1–4. [CrossRef]
20. Jang, S.M.; Lee, S.H.; Cho, H.W.; Cho, S.K. Design and analysis of helical motion permanent magnet motor with cylindrical Halbach array. *IEEE Trans. Magn.* **2003**, *39*, 3007–3009. [CrossRef]
21. Zhou, T.; Huang, Y.; Dong, J.; Guo, B.; Zhang, L. Design and modeling of axial flux permanent magnet machine with yokeless and segment armature using magnetic equivalent circuit. In Proceedings of the 17th International Conference on Electrical Machines and Systems (ICEMS), Hangzhou, China, 22–25 October 2014; pp. 618–623.
22. Arslan, S.; Gurdal, O.; Akkaya Oy, S. Design and optimization of tubular linear permanent-magnet generator with performance improvement using response surface methodology and multi-objective genetic algorithm. *Sci. Iran.* **2020**, *27*, 3053–3065. [CrossRef]
23. Fan, X.; Yin, J.; Lu, Q. Design and analysis of a novel composited electromagnetic linear actuator. *Actuators* **2021**, *11*, 6. [CrossRef]
24. Tan, Y.; Lin, K.; Zu, J.W. Analytical modelling of Halbach linear generator incorporating pole shifting and piece-wise spring for ocean wave energy harvesting. *AIP Adv.* **2018**, *8*, 056615. [CrossRef]
25. Li, B.; Li, D.; Ge, W.; Tan, C.; Lu, J.; Song, A. Design and analysis of a novel cascade control algorithm for braking-by-wire system based on electromagnetic direct-drive valves. *China J. Highw. Transp.* **2021**, *34*, 121–132.
26. Li, Z.; Wu, Q.; Liu, B.; Gong, Z. Optimal design of magneto-force-thermal parameters for electromagnetic actuators with Halbach array. *Actuators* **2021**, *10*, 231. [CrossRef]
27. Faiz, J.; Amini-Valeshani, S.; Ghods, M. Design and performance of linear Vernier generators—The state of the art and case study. *Int. Trans. Electr. Energy Syst.* **2021**, *31*, e12723. [CrossRef]

Article

Design of Valve Seating Buffer for Electromagnetic Variable Valve System

Qingya Zhou, Liang Liu *, Cong Zheng, Zhaoping Xu and Xianhui Wang

School of Mechanical Engineering, Nanjing University of Science and Technology, Nanjing 210014, China
* Correspondence: l.liu@njust.edu.cn

Abstract: An electromagnetic variable valve (EMVV) system can significantly reduce pumping loss and discharge loss of the engine by enabling variable valve timing and variable valve lift. However, the valve seat easily produces a larger impact collision with the engine cylinder head because of fast valve seating velocity, greatly decreasing engine life. Therefore, in this paper, a valve seating buffer (VSB) is designed to solve the problem of large electromagnetic valve seating impact. Firstly, a scheme of an EMVV system with embedded buffer is proposed, the collision model is established to resolve the problem of the soft landing of the valve and the effectiveness of the model is verified by experiment. In addition, the structure, material and dimension parameters of the proposed buffer are designed, and some key parameters of the buffer are optimized by the Nelder–Mead (N–M) algorithm. Finally, a co-simulation model of the actuator and the buffer is built, and the valve seating performance is analyzed. The co-simulation results show that the valve seating velocity and rebound height of the EMVV system with the designed buffer are reduced by 94.8% and 97%, respectively, which verifies the advantages of the designed VSB.

Keywords: electromagnetic variable valve; seating buffer; soft landing; design and optimization

1. Introduction

Due to the rapid development of new energy vehicles, improving the performance of traditional internal combustion engine vehicles can greatly ease the competition within the automotive industry and the crisis of energy form change. Variable valve trains significantly reduce pumping loss and discharge loss of engine by enabling a continuously variable phase and lifting of the valve. At present, a variable valve train mainly includes either an electro-hydraulic drive, electric drive or an electromagnetic drive [1]. Compared with the first two, the electromagnetic variable valve (EMVV) system does not need a transfer medium, and it has a simple structure and high power density. There have been many related research cases in the world, and it has become a research hotspot in related fields [2,3]. Although an EMVV system can realize the continuous adjustment and movement of the valve opening and closing, the external force easily interferes with valve opening and closing because of its structural characteristics, and the control accuracy is also affected. In addition, due to the fast velocity of the valve seating, it is easy to have a large impact collision with the engine cylinder head, and it can produce a large noise; it also has a greater impact on the engine life. Therefore, the scheme of the valve seating buffer (VSB) is extremely important for an EMVV system.

Currently, the scheme of the VSB of an electromagnetic drive valve train mainly includes control strategy research and an external mechanism. Yang et al. [4] designed an energy compensation control for an optimized engine electromagnetic valve; valve "zero" seating velocity was achieved by neutralizing positive and negative work in the armature stroke. Paden et al. [5] proposed a new type of EMVV structure, and the rapid response of the valve between opening and closing state was realized through the implementation of a control strategy while ensuring low valve seating velocity and power consumption.

Citation: Zhou, Q.; Liu, L.; Zheng, C.; Xu, Z.; Wang, X. Design of Valve Seating Buffer for Electromagnetic Variable Valve System. *Actuators* **2023**, *12*, 19. https://doi.org/10.3390/act12010019

Academic Editor: Ioan Ursu

Received: 30 November 2022
Revised: 28 December 2022
Accepted: 30 December 2022
Published: 1 January 2023

Compared to traditional, complex control strategies, the method of using external buffers has lower difficulty and does not require study of sophisticated control strategies; it can achieve valve seat buffering in the most simple and reliable way. Tu et al. [6] used a one-way throttle valve to buffer the electro-hydraulic drive valve mechanism; the best buffer effect was achieved by changing throttling area and throttling stroke of the throttle valve, and the seating velocity was reduced, and the dynamic response of the valve was improved in this way. Pan et al. [7] added a one-way valve to the upper end of the valve plunger to buffer the valve seating; the valve was simultaneously subjected to the oil pressure and spring force of the valve system during seating, which effectively decreased the valve seating velocity. Xu et al. [8] introduced the disc spring into the EMVV system, so that it acted between the valve and the coil, which could store most of the kinetic energy of the coil and effectively reduced the impact stress by about 50%. The faster the seating velocity, the more obvious the effect. Nowadays, most of the solutions for valve seating buffer are realized by control methods, but, because the valve seating is fast, and the time is short, the control is really a difficult method.

Buffering, as an important part of shock absorption, is often used in vehicles, bridges, aerospace and other fields. Magnetorheological buffers have attracted much attention in recent years due to their low energy consumption, fast response and large output damping force. The working principle of magnetorheological fluid is to control the magnetic induction intensity in the flow channel by adjusting the current intensity of the coil and then changing the shear yield strength of the magnetorheological fluid, affecting the pressure drop at both ends of the buffer flow channel so as to achieve the purpose of outputting the buffer force. It has the advantages of stable performance, low input voltage and large yield stress. David Case et al. [9] designed a small magnetorheological buffer for upper limb orthosis and model building, and experimental verification of the buffer was carried out, and the feasibility of applying small magnetorheological buffer to tremor motion attenuation was discussed. T.M. Gurubasavaraju et al. [10] proposed a scheme applying a magnetorheological buffer to a semi-active suspension system; a quarter of the vehicle models were analyzed under different road conditions. The results showed that the semi-active suspension with magnetorheological buffer had better regularity and road stability. Alan Sternberg et al. [11] proposed a large magnetorheological buffer for reducing the vibration of high-rise buildings; the design, manufacture and testing of the buffer prototype were completed. The test results were in good agreement with the established finite element model. In the actual design of magnetorheological buffers, most researchers often obtain some key parameters through experience, such as the size of the flow channel gap, the piston, etc. Therefore, it is necessary to optimize the structural parameters of the magnetorheological buffer after the structural design. Mao et al. [12] proposed an optimistic method of the structuring of a magnetorheological buffer by using a nonlinear flow model; the test showed that the performance of the optimized prototype was greatly improved. Guan et al. [13] used a multi-objective genetic algorithm, selected the output buffer force and adjustable multiple of the buffer as the optimization objectives, optimized the seven key variables of the magnetorheological buffer and, finally, obtained a large buffer force and highly adjustable multiple. Wu et al. [14] optimized the magnetic circuit of a magnetorheological buffer through the APDL parametric language built-in ANSYS and achieved the expected requirements. Yang et al. [15] proposed a novel structure of the annular multi-channel magnetorheological valve, designed its magnetic circuit and improved the pressure drop performance of the annular multi-channel magnetorheological valve.

In this paper, a scheme for adding a VSB is proposed to solve the problem of the large seating impact of electromagnetic valve, and the VSB is designed and optimized. An EMVV can effectively improve the power and fuel economy of an engine, but the existing integration layout of the system cannot meet the problem of limited setting space during actual installation, and the traditional modeling method cannot fully show the complex coupling structural characteristics of the EMVV system. Therefore, this paper firstly proposes a new structure for an EMVV system and then a collision model is established for the problem of soft landing, and the model is verified by Zheng's [16] test bench. The structure, material and size parameters of the proposed buffer are designed in detail, and some key parameters of the buffer are optimized by the Nelder–Mead (N-M) algorithm. Finally, a co-simulation model is built based on the coupling relationship between the actuator and the buffer of the EMVV system; the effectiveness and advantages of the designed VSB are verified by analyzing the valve seating performance of the system.

2. System Scheme

2.1. Overall Scheme

The structure of EMVV system, shown in Figure 1, mainly includes EMVA, VSB and valve components. The EMVA is the component that drives the valve opening and closing, and the performance of the EMVA determines whether the system can achieve rapid opening and closing. The valve seating buffer (VSB) is the component that reduces the valve seating velocity when the valve is seated. It needs to be able to output a larger buffer force and a faster response speed, so that it can reduce the velocity in a very short time and achieve valve seat buffering. The working principal of the EMVV system is: when the valve is opening, the coil of the actuator is activated. The coil is driven by the axial electromagnetic force and moves in the magnet field generated by the permanent magnet; thus, the valve is controlled to open. Continuously adjustable valve lift and transition time can be achieved by controlling the current of the coil. When the valve is seated, the coil of the buffer is soon activated. The buffer produces a controllable buffer force opposite to the direction of motion of the valve, thus, realizing the deceleration and the soft landing of the valve. In Figure 1, the VSB designed in this paper is shown; it consists of a magnetorheological fluid, piston, cylinder and coil, etc.

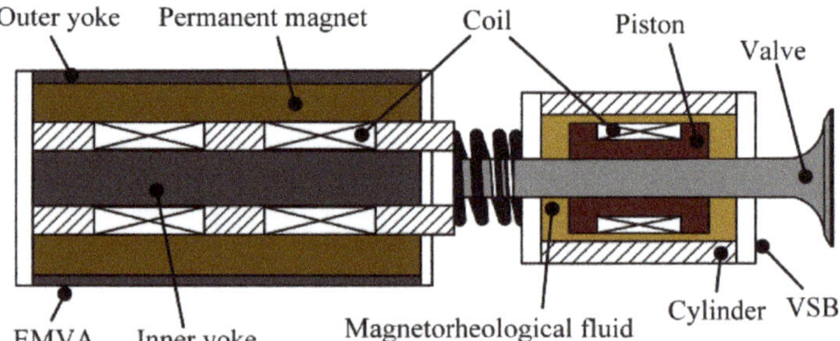

Figure 1. The structure of EMVV system.

2.2. System Design Scheme

Zheng et al. [16] proposed a parallel structure of EMVV, as shown in Figure 1. The scheme has a larger deficiency. On the one hand, the maximum buffer force that the buffer can output is 35 N, which cannot meet the needs of valve seating. On the other hand, the scheme needs greater setting space and aggravates the engine burden. Therefore, a new scheme is proposed in this paper, as shown in Figure 2. As the core components of the whole system, the actuator and the buffer adopt an integrated layout scheme. The buffer is placed in the hole pulled out by the inner yoke of the actuator. The piston rod is connected with the coil of the actuator and the valve components. The scheme greatly reduces the volume of the whole system, which is a tremendous boost for the engine in its limited setting space.

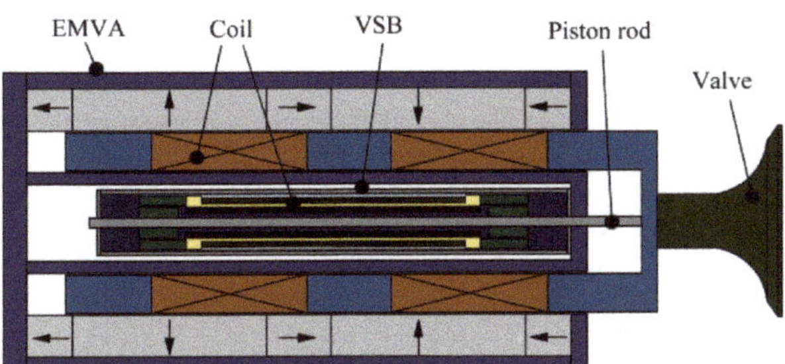

Figure 2. New structure of EMVV system.

2.3. System Modeling

2.3.1. Collision Modeling

The coil of the EMVA is driven by electromagnetic force. When the coil moves to the upper and lower ends of the stroke, it collides with the upper and lower yokes and produces a rebound. This phenomenon causes the valve to have a greater impact and noise on the engine cylinder head during valve seating and affects the engine life. In order to establish a complete simulation model with collision module of the system, it is necessary to identify the collision parameters between the coil skeleton and the yoke. According to collision theory [17], when two objects collide, the stiffness coefficient and damping coefficient are the main parameters affecting collision velocity and rebound height. The specific collision model is shown in Figure 3. The combination of spring and damper can be used to represent the collision model of the coil skeleton and yoke during collision. Among them, R represents that the slider moves in the track X with upper and lower boundaries. Kp and Kn represent the contact stiffness coefficients of the upper and lower boundaries, respectively. Dp and Dn represent the damping coefficients of the upper and lower boundaries, respectively. Gp and Gn represent the distances from the slider to the upper and lower boundaries. The coil stroke in the actuator is ($Gp + Gn$).

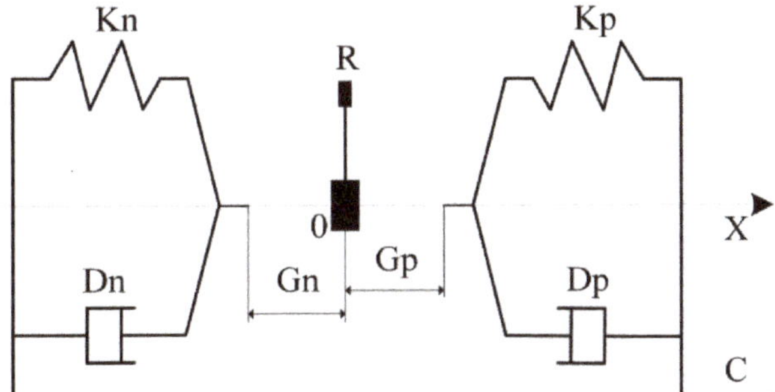

Figure 3. Collision model.

Based on the collision model, the stiffness coefficient and damping coefficient are used to establish the rebound module in the multi-physics finite element model of the system. In this paper, the recursive least-squares method is used to identify the collision parameters between the coil and the yoke. Compared to the traditional least-squares method, the inversion accuracy of the matrix in the recursive least-squares method does not affect the final identification results, and it can predict and update the data, significantly reducing computation. The main steps of collision parameter identification are as follows:

1. Analyze the dynamic principle during coil collision with yoke and establish the dynamic equation and finite element model of collision;

2. Build datasets of the stiffness coefficient and damping coefficient according to the calculation results of the dynamic equation;
3. Run 10 groups of different parameter combinations on the established, transient finite element model and collect velocity and rebound height and average them;
4. Compare the processed velocity and bounce height with test results to verify the correctness of the parameters;
5. Check whether the theoretical collision velocity and rebound height match the test; if not, check whether the finite element model and parameter equation are correct and identify again until the results meet the requirements.

According to the above steps, the stiffness coefficient is 5×10^7 N/m, and the damping coefficient is 1000 N·s/m. The finite element modeling of EMVA is carried out by using the identified collision parameters.

2.3.2. Model Simulation and Verification

Based on the above collision model, a sinusoidal current with an amplitude of 10 A and a frequency of 20 Hz is applied to the actuator. The motion state of the actuator coil is shown in Figure 4. As can be seen from the figure, the coil originally in the center of the air gap begins to move upwards under the action of electromagnetic force and collides with the upper yoke. The velocity at contact is 1.97 m/s, and the rebound height is 0.4 mm. Half a cycle later, the direction of the electromagnetic force becomes downward, and the coil moves downward (a stroke is 8 mm) and collides with the lower yoke; the collision velocity is 3.3 m/s, and the rebound height is 0.66 mm.

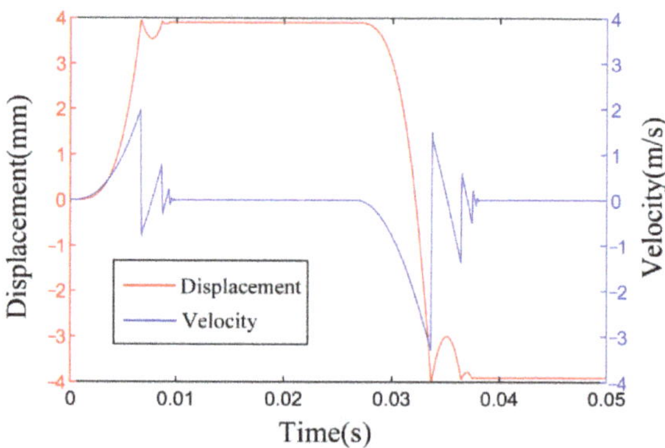

Figure 4. Displacement and velocity curves of coil.

The simulation model and the test bench of the valve seating buffer build by Zheng et al. [16] are used to verify the effectiveness of the collision model. The valve seating test of the EMVV system is carried out according to the established control strategy; the buffer is closed. The comparison results of simulation and experiment are shown in Figure 5; when the valve lift is 8 mm, the valve seating velocity is 0.71 m/s and 0.58 m/s, and the rebound height is 0.7 mm and 0.25 mm, respectively. The experiment verifies the effectiveness of the established collision model by simulating the impact on the cylinder head of the engine when the valve is quickly seated. Tests have found that reducing the impact can greatly reduce mechanical noise and extend the engine life. Therefore, the valve seating buffer is very important.

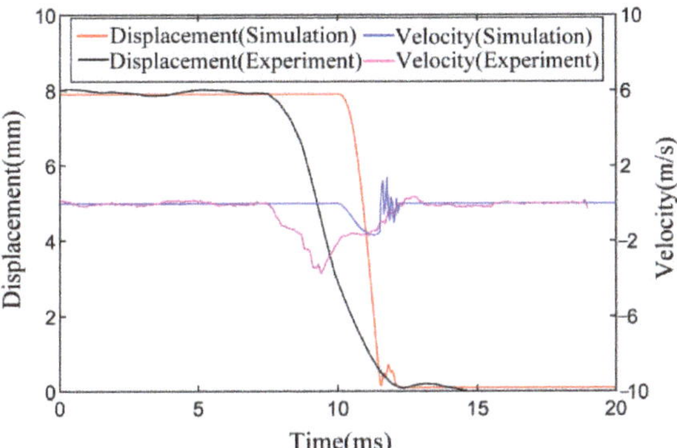

Figure 5. Comparison curves of the valve between experiment and simulation.

3. Design of Valve Seating Buffer

3.1. Structure Design

The magnetorheological buffer consists of a piston, piston rod, cylinder, cover, coil and magnetorheological fluid. When the buffer works, the piston rod drives the piston to move, the magnetorheological fluid with increased viscosity under the action of the magnetic field generated by the coil is squeezed, forms pressure difference and generates buffer force. In general, the magnetorheological buffer can be divided into a damping unit and a damping cylinder.

3.1.1. Channel Form

The flow channel of the damping unit is a channel that, through the magnetorheological fluid, affects the output buffer force. The main channel forms of the magnetorheological fluid are annular channel and disc channel. The annular channel has a simple structure and convenient installation, and it can decrease the volume of the buffer effectively. As for the disc channel, although the structure is relatively complex, its special structure can improve the utilization rate of the magnetic field and buffer force. Therefore, the annular channel and disc channel are combined as a mixed channel in this paper. Figure 6 is the structure of the mixed channel; it not only ensures small volume of the buffer, but also improves the performance of the buffer by promoting the utilization rate of the magnetic field.

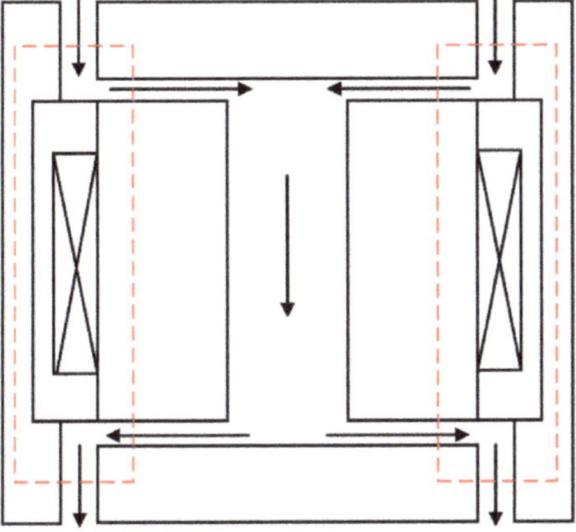

Figure 6. Mixed-channel structure.

3.1.2. Coil Number

As the input source of the buffer, the number of the electromagnetic coil has an important influence on the buffer performance. The coil has had single circle and multiple circles in recent research. A single circle is suitable for the magnetorheological buffer, which has a small volume; it can reduce control difficulty and improve space utilization. Although multiple circles have higher control difficulty, the diversity of the control strategies and the error-tolerant rate of the buffer are promoted. Because the VSB designed in this paper should have small volume and brief structure, the single circle is selected.

Coil turn is also a factor that has an effect on buffer force. In this paper, the coil turn is defined as the ratio of the area at the coil to the cross-sectional area of the single-turn coil. However, in the actual process of winding coil, a gap is generated between the coils, which leads to the actual turn number to be less than the calculated value. Thus, η is defined as a gap coefficient to eliminate the effect of the gap between coils on coil turn. The gap coefficient depends on coil diameter. The larger the wire diameter selected for the coil, the larger the gap and the greater η. After correction, the calculation formula of the coil turns is:

$$N = \eta \cdot \frac{CoilR \cdot CoilH}{S} \tag{1}$$

where N is coil turns, $CoilR$ is the groove depth at the coil, $CoilH$ is the groove width at the coil and the product of the two is the area at the coil. S is the cross-sectional area of wires. η is taken as 0.78 in this paper.

3.1.3. Combination of Damping Unit and Damping Cylinder

The combination modes of the damping unit and damping cylinder include embedded type and bypass type. Bypass type occupies a large volume, but it is more convenient to install and replace parts, and the piston neutral problem does not affect the buffer force. Embedded type is a piston with a coil moving in a sealed damping cylinder, changing the magnetic field of the magnetorheological fluid in the cylinder and squeezing the magnetorheological fluid to produce buffer force. An embedded buffer has a compact structure, small occupied volume and high magnetic field utilization, so the embedded scheme is selected as the combination of damping unit and damping cylinder in this paper.

3.2. Material Design

3.2.1. Magnetorheological Fluid

As the core material of magnetorheological buffer, the performance requirements of magnetorheological fluid mainly include low zero-field viscosity and high maximum shear yield strength. The viscosity of magnetorheological fluid increases and transforms into Bingham fluid in a very short time when it is affected by magnetic field, and it is a Newtonian fluid without a magnetic field. The typical Bingham model can be described as:

$$\begin{cases} \tau = \mu_0 \gamma + \tau_y, & \tau > \tau_y \\ \gamma = 0, & \tau \leq \tau_y \end{cases} \tag{2}$$

where $\tau(\mathrm{Pa})$ is the shear stress of magnetorheological fluid. $\mu_0(\mathrm{Pa \cdot s})$ is the zero-field viscosity of magnetorheological fluid, an important parameter to measure the performance of magnetorheological fluid. $\tau_y(\mathrm{Pa})$ is the shear yield strength of the magnetorheological fluid. $\gamma(\mathrm{s^{-1}})$ is the shear rate of the magnetorheological fluid, that is:

$$\gamma = \frac{du}{dy} \tag{3}$$

In the Bingham model, the magnetorheological fluid in the plug flow area is non-differentiable, which makes it difficult to solve in practical calculation. Therefore, through consulting the literature, this paper uses the approximate Bingham plastic model proposed by David Case [18]. The model becomes continuously differentiable by decomposing the plug flow area into continuous tiny regions, which can be described as:

$$\tau = \left(\frac{\tau_y tanh(\delta\gamma)}{\sqrt{\varepsilon^2 + \gamma^2}} + \mu_0 \right) \cdot \gamma \tag{4}$$

where ε is a constant used to eliminate discontinuities, usually approaching 0 in this model. δ is the proportional term of the slope, usually approaching infinity. The dynamic viscosity of the magnetorheological fluid obtained by deformation of Equation (4) is:

$$\mu = \frac{\tau_y tanh(\delta\gamma)}{\sqrt{\varepsilon^2 + \gamma^2}} + \mu_0 \tag{5}$$

MRF-122EG magnetorheological fluid produced by the American LORD Company is adopted in this paper. Its specific performance parameters are shown in Table 1.

Table 1. MRF-122EG performance parameters.

Parameter	Value	Unit
Density	2.28~2.48	g/cm^3
Zero-field viscosity	0.042	Pa·s
Mass fraction	72.00	%
Temperature	-40~130	°C
Color	Black gray	

In order to couple the electromagnetic field with the flow field, the relationship between the yield stress τ (kPa) and the magnetic induction B (T) of MRF-122EG is obtained by curve fitting:

$$\tau = 89.26B^4 - 237.1B^3 + 165.6B^2 + 17.22B - 0.31 \tag{6}$$

3.2.2. Valve Core Material

Valve core material is soft magnetic material that is prone to magnetization and demagnetization processes and has low coercive force. Industrial pure iron is one of the most widely used soft magnetic materials [19]; the performance of saturation magnetic induction and magnetic permeability in soft magnetic materials is excellent, but high conductivity also leads to high loss. Soft magnetic materials affect the response time of magnetorheological buffers through conductivity. According to the current research status of soft magnetic materials, new composite material has a large magnetic saturation induction intensity, so it does not reach the magnetic saturation in practical application and does not affect the maximum buffer force. Therefore, in order to improve the dynamic response velocity of the buffer, this paper chooses new composite material as the soft magnetic material of the valve core.

3.2.3. Cylinder Material

The cylinder body is one of the most important parts of the magnetic circuit of magnetorheological buffers; its magnetic property plays a key role in the magnetic performance. As for the selection of cylinder material, on the one hand, the material should have good magnetic properties. On the other hand, due to the diameter of the designed buffer being very small, the cylinder material should have better processing performance, which has the ability to maintain strength while being machined to a small diameter and small thickness. At present, the most widely used cylinder material is industrial pure iron, which has good magnetic properties but is not easy to process. Therefore, this paper chooses No. 10 steel as the cylinder material of the buffer, which has similar magnetic properties to industrial pure iron but has better machinability and lower price.

3.3. Dimension Design

After selecting the mixed-channel, embedded VSB structure scheme, the key dimensions need to be structurally designed; the structure of the design buffer is shown in Figure 7. The buffer is connected to the magnetic disk and the valve core is moved up and down by a piston rod. The magnetorheological fluid flows in the mixed channel. After the coil is energized, the magnetorheological effect occurs, so the viscosity of the magnetorheological fluid increases and outputs the controllable buffer force. According to the design requirements, the designed buffer can output buffer force of about 100 N with a maximum current of 2.5 A and a damping unit velocity of 0.5 m/s. The selected design variables and their ranges are shown in Table 2.

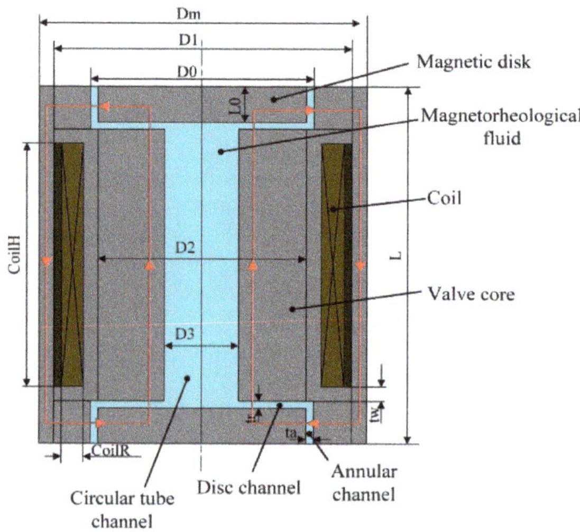

Figure 7. Structure parameters of the VSB.

Table 2. The range of designed variables for the VSB.

Variable	Physical Meaning	Range (mm)
D2	Outer diameter of spool	3~6
D3	Diameter of circular tube channel	1~3
ta	Gap of annular channel	0.1~0.3
tr	Gap of disc channel	0.5~2
CoilR	Depth of coil slot	0.5~2
CoilH	Width of coil slot	30~35

3.4. Optimization Design

At present, the most commonly used optimization algorithms in structural optimization include the multi-objective optimization algorithm [20], bound optimization BY quadratic approximation (BOBYQA) algorithm [21], Nelder–Mead (N-M) algorithm [22–24] and so on. In the actual optimization process, different optimization objectives cannot achieve the optimal value at the same time. Both BOBYQA and N-M optimize targets based on established finite element models. The optimization object of this paper is the buffer, and the optimization parameter is the dimension parameter of the buffer. The adjustable coefficient of buffer force is taken as the objective function of optimization, which is single-objective optimization. The ultimate goal is to make the adjustable coefficient of the buffer reach the maximum value. Because the modeling software used in this paper and the co-simulation platform COMSOL Multiphysics have a built-in N-M algorithm, the optimization can be directly processed after modeling, which can greatly improve efficiency and save time. Therefore, this paper uses the N-M algorithm to optimize the dimension of the buffer. The workflow of the N-M algorithm is shown in Figure 8.

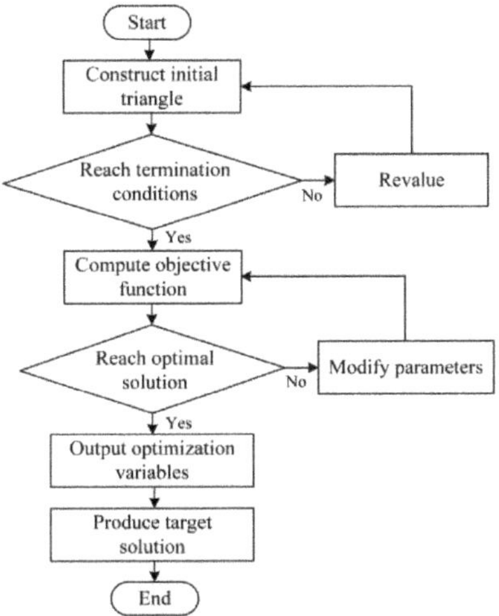

Figure 8. Workflow of N-M optimization algorithm.

1. Build an initial triangle based on the given data points. The best point $x_1^{(k)}$ is B (Best); the other two points $x_2^{(k)}$ and $x_3^{(k)}$ are G (Good) and W (Worse), respectively; k is current iteration times;

2. Generate point $x_r^{(k)}$ by reflecting the W (Worse):

$$x_r^{(k)} = \overline{x}^{(k)} + \alpha \left(\overline{x}^{(k)} - x_3^{(k)} \right) \tag{7}$$

   where α is reflection coefficient, usually taken as 1. $\overline{x}^{(k)}$ is the center of figure:

$$\overline{x}^{(k)} = \frac{1}{n} \sum_{i=1}^{n} x_i^{(k)} \tag{8}$$

   In two-dimensional space, n is 2. If $f\left(x_1^{(k)}\right) \le f\left(x_r^{(k)}\right) \le f\left(x_3^{(k)}\right)$, the direction of reflection is correct to the target value, then the iteration is terminated. If $f\left(x_1^{(k)}\right) > f\left(x_r^{(k)}\right)$, turn to the third step. Otherwise, take the fourth step;

3. Construct expansion point $x_e^{(k)}$ in the same direction of the reflection point:

$$x_e^{(k)} = \overline{x}^{(k)} + \beta \left(\overline{x_r}^{(k)} - \overline{x}^{(k)} \right) \tag{9}$$

   where β is the expansion coefficient, usually taken as 2. If $f\left(x_e^{(k)}\right) < f\left(x_r^{(k)}\right)$, replace $x_3^{(k)}$ with $x_e^{(k)}$ and terminate iteration. Otherwise, replace $x_3^{(k)}$ with $x_r^{(k)}$ and terminate iteration;

4. Choose shrinkage mode. If $f\left(x_2^{(k)}\right) \le f\left(x_r^{(k)}\right) \le f\left(x_3^{(k)}\right)$, point $x_c^{(k)}$ is generated:

$$x_c^{(k)} = \overline{x}^{(k)} + \gamma \left(x_r^{(k)} - \overline{x}^{(k)} \right) \tag{10}$$

where γ is shrinkage coefficient, usually taken as 0.5. If $f\left(x_c^{(k)}\right) \le f\left(x_r^{(k)}\right)$, replace $x_3^{(k)}$ with $x_c^{(k)}$ and terminate iteration. Otherwise, take the fifth step.

If $f\left(x_r^{(k)}\right) \ge f\left(x_3^{(k)}\right)$, iterate to obtain the point $x_{cc}^{(k)}$:

$$x_{cc}^{(k)} = \overline{x}^{(k)} - \gamma\left(x_r^{(k)} - \overline{x}^{(k)}\right) \tag{11}$$

If $f\left(x_{cc}^{(k)}\right) < f\left(x_3^{(k)}\right)$, replace $x_3^{(k)}$ with $x_{cc}^{(k)}$ and terminate iteration. Otherwise, take the fifth step;

5. Shrink point:

$$x_i^{(k+1)} = x_1^{(k)} + \delta\left(x_i^{(k)} - x_1^{(k)}\right) \tag{12}$$

where the range of shrinkage coefficient δ is 0~1, usually taken as 0.5, $\forall i \in \{2,..., n + 1\}$. Repeat steps 1 to 5 until the termination condition of the iteration is reached or the maximum number of iterations is reached. The final approximate solution of the unknown target value x is represented by the optimal solution $\hat{x}$.

According to the above-selected objective function, optimization variables and optimization algorithm, the multi-physics finite element model of the designed buffer is established, and the relevant settings are carried out in COMSOL software. It is found that the adjustable coefficient reaches a stable value of about 24.76 after about 280 iterations. The optimized variables and the rest key parameters of the buffer are shown in Table 3.

Table 3. Main structure parameters of the buffer after optimization.

Optimization Variable	Value (mm)	Rest Parameter	Value (mm)
D2	4	Dm	8.5
D3	1.25	D0	4.5
ta	0.25	D1	7.5
tr	1	tw	2
CoilR	1	L	50
CoilH	34	L0	5

The diameter of the optimized buffer is negligibly small, and the space of coil is small, so it is necessary to select the diameter of the copper winding to be as small as possible when the maximum current is 2.5 A. By querying the related information of varnished wire, the varnished wire of type QY-2/2200.100GB6109.6-88 is selected. According to the optimized structural parameters and Equation (1), the coil turns of the designed buffer are 136, and the optimized coil has two radial winding layers and 63 axial winding laps.

4. Simulation and Analysis

4.1. Simulation of the Designed VSB

Through the above design and dimension optimization of the VSB, the 3D structure of the designed buffer is as shown in Figure 9. Meanwhile, the geometric model and the magnetic field distribution of the designed VSB are as shown in Figure 10. The left side is the established 2D axisymmetric geometric model of the buffer, and the right side is the distribution of its magnetic field. It can be seen that the magnetic induction line of the buffer is evenly distributed, showing a symmetrical shape relative to the midpoint of the valve core, and the maximum value of the magnetic induction intensity appears at the place of valve core. Above all, the magnetic circuit and structural design of the buffer are reasonable.

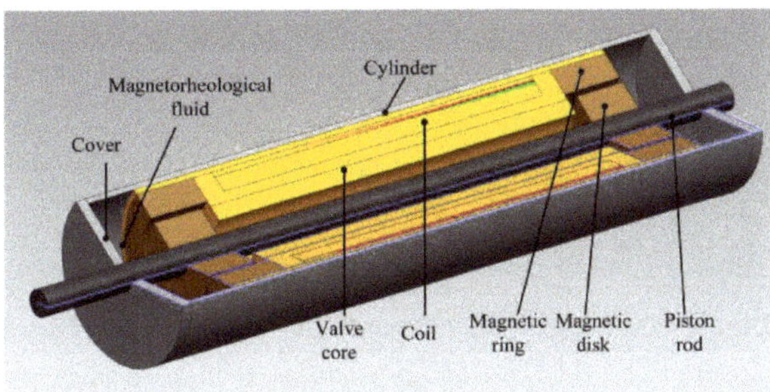

Figure 9. 3D structure of the designed VSB.

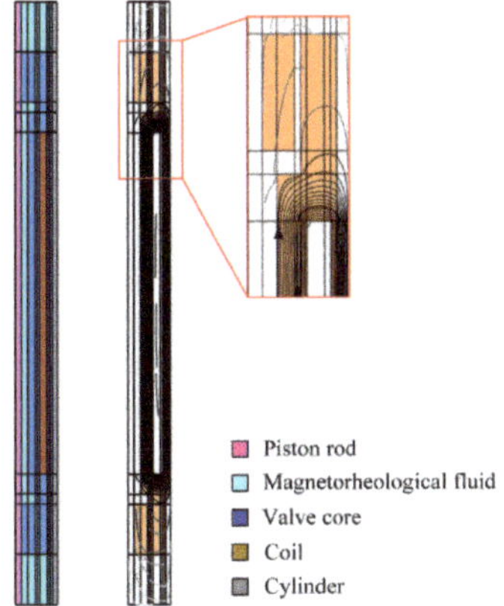

Figure 10. Geometric model and the magnetic field distribution of the designed VSB.

After establishing the multi-physics finite element model of the buffer, it is solved statically. When the current of coil is 2.5 A, the motion curve of the damping unit is a sine curve with an amplitude of 4 mm and a period of 0.05 s; the relationship between the output buffer force and the piston displacement is shown in Figure 11, and the relationship between the output buffer force and the velocity is shown in Figure 12. It can be seen that the output buffer force is positively correlated with the velocity of the piston and the current of the coil. The buffer force is 3.4 N when the coil does not have applied current, and the adjustable coefficient is 24.76. When the maximum input current is 2.5 A, the maximum buffer force reaches 84.2 N.

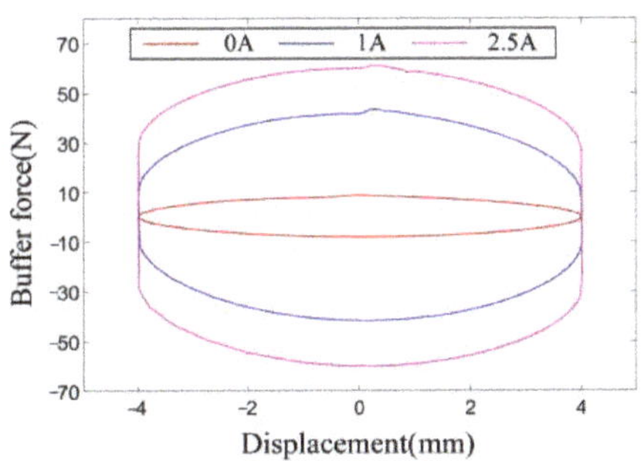

Figure 11. Curve between buffer force and displacement of the VSB.

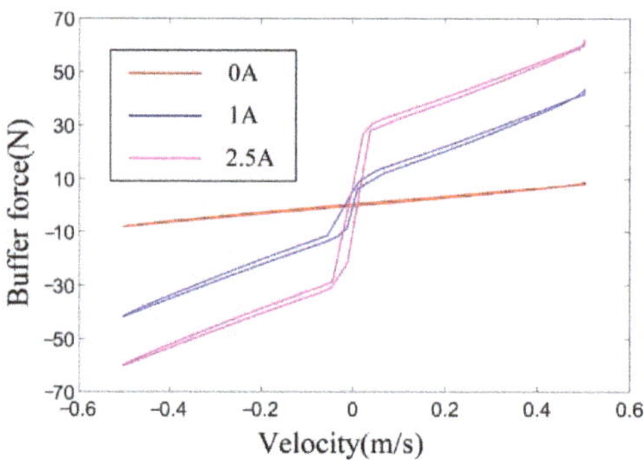

Figure 12. Curve between buffer force and velocity of the VSB.

In order to improve the response time of the designed buffer, transient simulation is needed. The stable current of the coil is set to 2.5 A, and the moving velocity of the damping unit is 0.1 m/s. The response time of the buffer force is shown in Figure 13. As can be seen from the figure, after optimizing the buffer dimension parameters, not only does the buffer force have a significant increase, but also the response time is improved from 9 ms to 1.3 ms. It is indicated that the work of structural optimization provides omni-bearing promotion for the overall performance of the buffer.

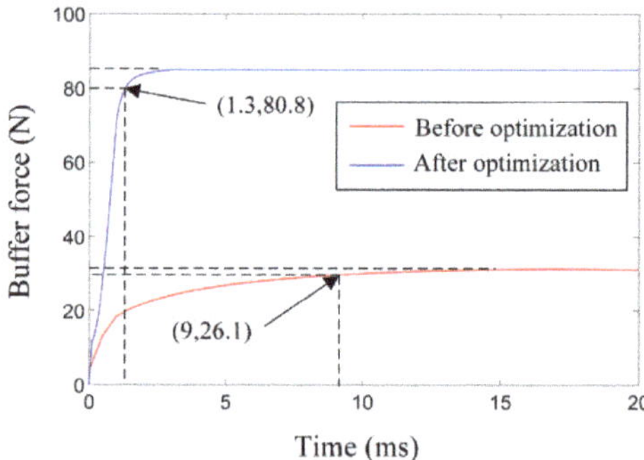

Figure 13. Buffer force curves of the designed VSB.

4.2. Construction of Co-Simulation Platform

In this paper, the coupling relationship of the designed EMVV system is studied, and a co-simulation platform is built to lay the foundation for studying the valve seating performance. At the macro level of the entire system, the EMVA and VSB have a complex coupling relationships, as shown in Figure 14. It can be seen that, in addition to the coupling relationship between the actuator and the buffer, they are also coupled to each other by the buffer force and the velocity of the coil. The buffer force output by the buffer and the electromagnetic force generated by the actuator are added together to drive the coil, and the velocity of the coil is also the input condition of the buffer for the piston motion.

The default data transfer channel between COMSOL and MATLAB/Simulink can realize the function of real-time data transmission. The latest version of COMSOL implements real-time transmission between the data and command by COMSOL Multiphysics with Simulink. Based on the data channel, the steps of building the co-simulation are:

1. Build the finite element models of the actuator and the buffer in COMSOL;
2. Set the input and the output of the actuator and the buffer in COMSOL. The input of the actuator is the voltage of the coil and the buffer force, and the output is the velocity and the displacement of the coil. The input of the buffer is the voltage of the

coil and the velocity of the piston, and the output is the buffer force. Then, save the finite element models as .slx files, recognizable by MATLAB/Simulink;

3. Import the .slx files of the actuator and the buffer to MATLAB/Simulink. Build the co-simulation model based on the coupling relationship between the actuator and the buffer;

4. Set the input and the output of the system and set calculation time and time step;

5. Solve to obtain the calculation results.

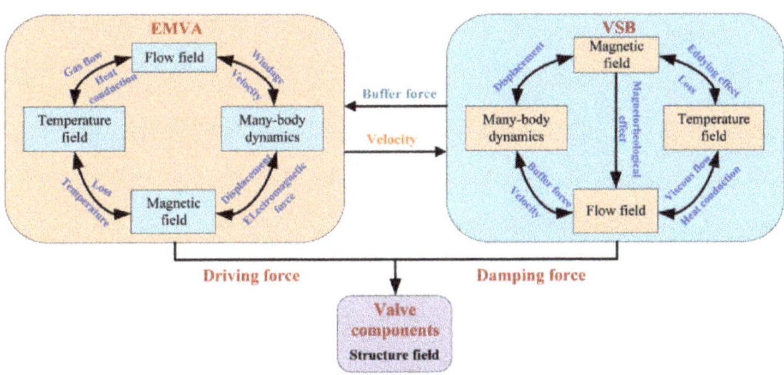

Figure 14. Coupling relationship of EMVV system.

The co-simulation of the EMVV system is carried out based on the above steps. Figure 15 is the co-simulation model of the entire system. After building the model, it is necessary to set the input voltage of the actuator and the buffer and adjust the computing time and step in MATLAB/Simulink. The solution process of the whole co-simulation is carried out in COMSOL. After each step of calculation, the data are transmitted to MATLAB/Simulink in real time, then the calculation is carried out according to the coupling relationship of the model. Finally, the calculated data are post-processed in MATLAB.

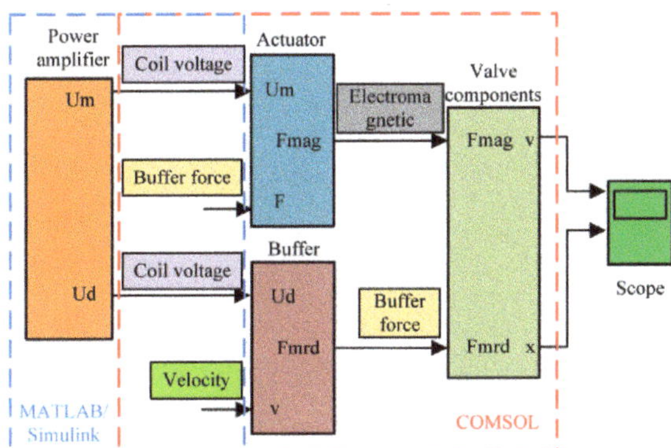

Figure 15. Co-simulation model of EMVV system.

4.3. Analysis of Valve Seating Performance

The electromagnetic valve seating is simulated under specific conditions based on the established co-simulation platform. The EMVA applies a sinusoidal current with an amplitude of 10 A and a frequency of 5 Hz, where the valve is at the top at the beginning of the simulation. The input excitation of the buffer is a pulse current with an amplitude of 0.5 A, a frequency of 10 Hz and a duty cycle of 5%, which ensures that the buffer can output stable buffer force during the period from the valve seating to the valve stability. In the meantime, through the dynamic analysis of the buffer, its response time is 1.3 ms, so opening the buffer 1.5 ms before valve seating is enough to make the buffer output a constant buffer force to reduce the valve seating velocity. The simulation results are shown in Figure 16. The lift of the valve is 8 mm. Before and after the opening of the buffer, the valve seating velocity decreases from 0.58 m/s to 0.03 m/s, and the valve rebound height decreases from 0.67 mm to 0.02 mm. Therefore, it is indicated that the design buffer can obviously reduce the seating velocity and the rebound height of the valve and achieve good valve seating performance.

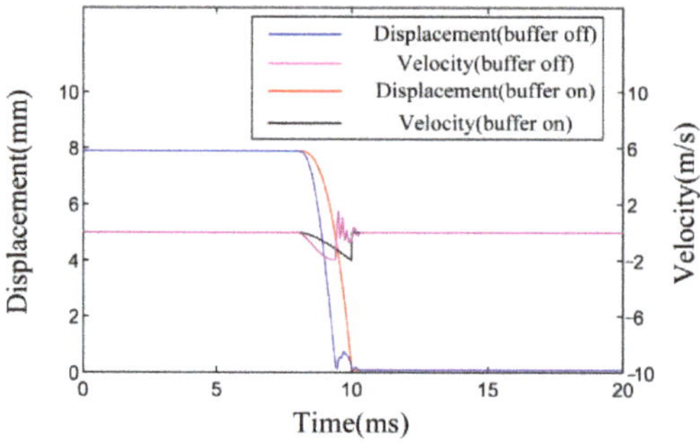

Figure 16. Co-simulation results of EMVV system.

5. Conclusions

In this paper, an external buffer is proposed, aiming to resolve the problem of large impact and noise of EMVV landing. The structure, material and dimension design process of the VSB rdescribed in detail, and the key size parameters of the buffer are optimized based on the N-M algorithm. Meanwhile, the coupling relationship between the actuator and the buffer in the EMVV system is studied; a scheme where the VSB is embedded in the EMVA is proposed. In addition, a COMSOL and MATLAB/Simulink co-simulation platform and co-simulation model are built based on the scheme. Finally, the results of the co-simulation are compared, and the valve seating performance is analyzed.

The integration scheme proposed in this paper not only improves the integration of the system, but also greatly reduces the volume of the whole system, which is a tremendous boost for the engine in its limited setting space. Furthermore, the valve seating velocity and the valve rebound height are reduced from 0.58 m/s and 0.67 mm to 0.03 m/s and 0.02 mm, and the reduction ratios are 94.8% and 97%, respectively. The co-simulation results show that opening the designed buffer before the valve seating can obviously relieve the valve seating impact, overwhelmingly improve the valve seating performance of the EMVV system and prolong the engine life. However, due to the small radial dimension of the designed buffer, machining accuracy and assembling mode are strongly required in processing. Therefore, it is necessary to further optimize the combination of the damping unit and the damping cylinder and the assembling mode of the damping unit of the buffer.

Author Contributions: Conceptualization, methodology, software, validation and formal analysis, Q.Z., L.L. and C.Z.; investigation, Q.Z., L.L., Z.X. and X.W.; data curation, Q.Z. and C.Z.; writing—original draft preparation, Q.Z. and L.L.; writing—review and editing, Q.Z. and L.L. All authors have read and agreed to the published version of the manuscript.

Funding: This work was funded by the National Natural Science Foundation of China, grant numbers 51975297 and 51875290.

Data Availability Statement: Not applicable.

References

1. Kuruppu, C.; Pesiridis, A.; Rajoo, S. *Investigation of Cylinder Deactivation and Variable Valve Actuation on Gasoline Engine Performance*; SAE World Congress and Exhibition: Detroit, MI, USA, 2014; pp. 7484–7492.
2. Reinholz, B.A.; Reinholz, L.; Seethaler, R.J. Optimal trajectory operation of a cogging torque assisted motor driven valve actuator for internal combustion engines. *Mechatronics* **2018**, *51*, 1–7. [CrossRef]
3. Fan, X.; Yin, J.; Lu, Q. Design and analysis of a novel composited electromagnetic linear actuator. *Actuators* **2022**, *11*, 6. [CrossRef]
4. Yang, Y.; Liu, J.; Ye, D.; Chen, Y.; Lu, P. Multiobjective optimal design and soft landing control of an electromagnetic valve actuator for a camless engine. *IEEE-ASME Trans. Mechatron.* **2013**, *18*, 963–972. [CrossRef]
5. Paden, B.A.; Snyder, S.T.; Paden, B.E.; Ricci, M.R. Modeling and control of an electromagnetic variable valve actuation system. *IEEE-ASME Trans. Mechatron.* **2015**, *20*, 2654–2665. [CrossRef]
6. Tu, B.; Tian, H.; Wei, H.; Pan, M. Research on buffering process of electro-hydraulic variable valve train. *China Mech. Eng.* **2016**, *27*, 2652. [CrossRef]
7. Pan, K.; Wang, Z.; Tian, F.; Chen, J. Simulation study on the buffering process of valve seating buffer mechanism. *Mod. Mach.* **2016**, *2*, 47–49. [CrossRef]
8. Xu, R.; Chang, S. Dynamic analysis and buffer structure for electromagnetic valve actuation. *Veh. Engine* **2015**, *06*, 8–12. [CrossRef]
9. Case, D.; Taheri, B.; Richer, E. Multiphysics modeling of magnetorheological dampers. *Int. J. Multiphys.* **2013**, *7*, 61–76. [CrossRef]
10. Gurubasavaraju, T.M.; Kumar, H.; Mahalingam, A. An approach for characterizing twin-tube shear-mode magnetorheological damper through coupled FE and CFD analysis. *J. Braz. Soc. Mech. Sci. Eng.* **2018**, *40*, 139. [CrossRef]
11. Alan, S.; René, Z.; Juan, C. Multiphysics behavior of a magneto-rheological damper and experimental validation. *Eng. Struct.* **2014**, *15*, 194–205.
12. Mao, M.; Choi, Y.T.; Wereley, N.M. Effective design strategy for a magnetorheological damper using a nonlinear flow model. In Proceedings of the Smart Structures and Materials 2005 Conference, San Diego, CA, USA, 7–10 March 2005; pp. 446–455.
13. Guan, X.; Guo, P.; Ou, J. Multi-objective optimization of magnetorheological fluid dampers. *Eng. Mech.* **2009**, *26*, 30–35.
14. Wu, J.; Wang, H. Magnetic circuit optimum design of a magneto-rheological damper based on ANSYS. *Mech. Electr. Eng. Mag.* **2008**, *25*, 74–76. [CrossRef]
15. Yang, X.; Chen, Y.; Liu, Y.; Zhang, R. Modeling and experiment of an annular multi-channel magnetorheological valve. *Actuators* **2022**, *11*, 19. [CrossRef]
16. Zheng, C.; Liu, L.; Guo, H.; Xu, Z. Optimization design of a fully variable valve system based on Nelder-Mead algorithm. *Proc. Inst. Mech. Eng. Part C-J. Mech. Eng. Sci.* **2022**, *236*, 5815–5825. [CrossRef]
17. Sheng, L.; Liu, J.; Yu, Z. Dynamic modeling of a flexible multi-body system with elastic impact. *J. Shanghai Jiaotong Univ.* **2006**, *40*, 1790–1793. [CrossRef]
18. Case, D.; Taheri, B.; Richer, E. Dynamical modeling and experimental study of a small scale magnetorheological damper. *IEEE-ASME Trans. Mechatron.* **2014**, *19*, 1015–1024. [CrossRef]
19. Li, Z.; Dong, Y.; Pauly, S.; Chang, C.; Wei, R.; Li, F.; Wang, X. Enhanced soft magnetic properties of Fe-based amorphous powder cores by longitude magnetic field annealing. *J. Alloys Compd.* **2017**, *706*, 1–6. [CrossRef]
20. Cha, Y.; Agrawal, A.K.; Phillips, B.M.; Spencer, B.F. Direct performance-based design with 200 kN MR dampers using multi-objective cost effective optimization for steel MRFs. *Eng. Struct.* **2014**, *71*, 60–72. [CrossRef]
21. Zheng, J.; Li, Y.; Wang, J. Design and multi-physics optimization of a novel magnetorheological damper with a variable resistance gap. *Proc. Inst. Mech. Eng. Part C-J. Mech. Eng. Sci.* **2017**, *231*, 3152–3168. [CrossRef]
22. Fakhouri, H.N.; Hudaib, A.; Sleit, A. Hybrid particle swarm optimization with sine cosine algorithm and Nelder–Mead simplex for solving engineering design problems. *Arab. J. Sci. Eng.* **2020**, *45*, 3091–3109. [CrossRef]
23. Zic, M.; Pereverzyev, S. Optimizing noisy CNLS problems by using Nelder-Mead algorithm: A new method to compute simplex step efficiency. *J. Electroanal. Chem.* **2019**, *851*, 113439. [CrossRef]
24. Abdelrahman, E.M.; Abdelazeem, M.; Gobashy, M. A minimization approach to depth and shape determination of mineralized zones from potential field data using the Nelder-Mead simplex algorithm. *Ore Geol. Rev.* **2019**, *114*, 103123. [CrossRef]

 *actuators*

Article

Design of an Improved Active Disturbance Rejection Control Method for a Direct-Drive Gearshift System Equipped with Electromagnetic Linear Actuators in a Motor-Transmission Coupled Drive System

Shusen Lin [1,*]**, Min Tang** [1]**, Bo Li** [2] **and Wenhui Shi** [1]

[1] School of Aeronautical Engineering, Taizhou University, Taizhou 318000, China
[2] School of Transportation and Vehicle Engineering, Shandong University of Technology, Zibo 255000, China
* Correspondence: linshusen1007@126.com

Abstract: In this study, a type of direct-drive gearshift system integrated into a motor-transmission coupled drive system is introduced. It used two electromagnetic linear actuators (ELAs) to perform gearshift events. The adoption of ELAs simplifies the architecture of the gearshift system and has the potential to further optimize gearshift performance. However, a number of nonlinearities in the gearshift system should be investigated in order to enhance the performance of the direct-drive gearshift system. An active disturbance rejection control (ADRC) method was selected as the principal shifting control method due to the simple methodology and strong reliability. The nonlinear characteristics of the electromagnetic force produced by the ELA were subsequently reduced using the inverse system method (ISM) technique. The ADRC approach also incorporated an acceleration feedforward module to enhance the precision of displacement control. The extended state observer (ESO) module used a nonlinear function in place of the original function to improve the ability to reject disturbances. Comparative simulations and experiments were carried out between the ADRC method and improved ADRC (IADRC) method. The outcomes demonstrate the effectiveness of the designed control method. The shift force fluctuates less, and the shift jerk decreases noticeably during the synchronization procedure. In conclusion, combined with the optimized IADRC method, the direct-drive gearshift system equipped with ELAs shows remarkable gearshift performance, and it has the potential to be widely used in motor−transmission coupled drive systems for EVs.

Keywords: electromagnetic linear actuator; direct−drive; active disturbance rejection control; gearshift system; nonlinear; motor−transmission coupled drive system

Citation: Lin, S.; Tang, M.; Li, B.; Shi, W. Design of an Improved Active Disturbance Rejection Control Method for a Direct-Drive Gearshift System Equipped with Electromagnetic Linear Actuators in a Motor-Transmission Coupled Drive System. *Actuators* **2023**, *12*, 40. https://doi.org/10.3390/act12010040

Academic Editor: Eihab M. Abdel-Rahman

Received: 13 December 2022
Revised: 9 January 2023
Accepted: 10 January 2023
Published: 12 January 2023

1. Introduction

The electromagnetic linear actuator (ELA) can output linear movement without any intermediate motion converter mechanism such as gearings. Because of this, it can directly drive the controlled objects, which is usually known as direct−drive technology. Less components, quicker dynamic performance, improved driving efficiency, and reliability are some benefits of using direct−drive technology. Hence, the direct−drive technology is widely applied recently in robotics, machine tools, wind electricity and wheel drive of electric vehicles [1–4]. ELA is one of the most important components of the direct−drive system. Recent studies focus on several areas, including parametrization structure design [5,6], precise motion control [7,8] and elimination of nonlinear disturbance [9,10].

Additionally, the ELAs' application fields now include the automotive industry area. In the reference [11], a fully variable valve system for a high−performance engine was built using ELAs. Li adopted linear motors as the actuating device of an electromagnetic active suspension system [12]. In order to obtain high precision servo control, Li created a sort of direct−drive−type AMT equipped with ELAs and investigated the servo dynamic stiffness of the ELAs [8,13]. The authors also employed two ELAs to actuate the gearshift events of automated manual transmission (AMT) in reference [14]. However, there is no discussion of the precise displacement control in the shifting process that takes nonlinearities and disturbances into account.

The applications of ELAs in AMT gearshift systems are explored in references [13] and [14], especially for electric vehicles (EVs). The reason for this is that multi−speed transmission−equipped EVs are considered as an appropriate and competitive solution for extended−range EVs and can also can enhance the EVs' capacity for accelerating and climbing. In addition, the size and cost of the driving motor for the can be reduced with the use of multi−speed transmission.

In comparison to other types of transmission such as automatic transmission, continuous variable transmission and dual clutch transmission, AMT has a superior dynamic performance, a simpler structure, and lower cost. Therefore, researchers prefer to integrate driving motors with AMT to extend the driving range for EVs. Power interruption problems during the gearshift process are one of the most fatal drawbacks of AMT. However, the capacity of the electric motor to quickly change its rotational speed and torque is useful to reduce the impact of power interruption on the gearshift performance. In order to minimize the rotational speed difference that must be synchronized during the gearshift process, the driving motor rotational speed is swiftly adjusted to the desired value. As a result, the power interruption time, which also serves as the gearshift time, noticeably decreases. The rotational speed difference can typically be minimized to 50−300 r/min depending on the operating conditions.

The applications of AMT in EVs and HEVs have been widely investigated. Experimental and theoretical studies on coordinated control of AMT and driving motor, structural innovation and control parameters optimization of passenger cars have been conducted more frequently recently. In reference [15,16], additional power delivery paths were attempted in an effort to structurally innovate a solution to the torque interruption problem. In reference [15], a brand-new dual input multi−speed AMT was designed for electric cars in order to implement power−on shifting. A low−speed driving motor was employed as the assisting motor installed on the final shaft to supply power during the gearshift process. A high−speed driving motor was used as the primary power source that directly linked to AMT. Song et al. develop a kind of seamless two-speed AMT. It consisted of a single planetary gear system, a disk friction clutch and a drum brake. According to simulation and test data, the transmission considerably increased the electric motor efficiency, vehicle dynamics, and energy consumption [16]. It also completely eliminated power interruption. A novel two−speed inverse AMT with an overrunning clutch was proposed in their further research for light electric vehicles [17]. Torque interruption was prevented by employing a controllable overrunning clutch mechanism while the synchronizer and shift−fork were removed. In addition, a novel inverse actuator using a worm gear and camshaft were developed for the clutch control [18]. The synchronizer was cancelled by using active motor control to produce a zero-speed difference. Furthermore, Walker et al. replaced the traditional cone clutch synchronizer with a harpoon−shift synchronizer to optimize the engagement process so that the driving comfort was improved [19].

In addition, researchers also use optimization methods to obtain better shift quality of motor−AMT coupled drive systems. This usually includes shift schedule optimization [20], matching optimization [21] and comprehensive optimization [22]. The motor−AMT coupled drive system for electric vehicles made noticeable advancements in both performance and structure. However, thorough, and deep, investigations are still required and to optimize the performance in order to match the higher standards for transient behavior and overall characteristics in EVs.

The majority of automotive systems exhibit nonlinearity. It is hard to obtain the analytic solution of the nonlinearity. Nonetheless, it compromises the effectiveness of all control systems. Therefore, it is important to identify the cause of nonlinearity in order to develop an effective control strategy. The nonlinearity in a motor−AMT coupled drive system typically manifests itself in a number of ways, including actuator characteristics, gearing mechanism, shaft vibration, segmented control schemes and disturbances. Specifically, as the driving motor attaches to the transmission input shaft directly, the nonlinear characteristics increase considerably. Wang took the nonlinear contact backlash of the gear and synchronizer into account to create the dynamic model of the shifting process [23]. In reference [24], the sliding mode control strategy was employed to lessen the impact of the nonlinearity in the position control due to the multistage shift process. Moreover, nonlinear dynamic characteristics of gear systems are usually neglected when researchers are developing a model. Demonstrated by references [25,26], the impact of the sleeve and gear ring, clearance, the bending-torsion coupling response and mesh stiffness of the gear system vary and present nonlinearity, and these factors have an impact on the dynamic performance during the gear engagement process. Gear vibration caused by pos five kinds of various engaging conditions of the sleeve and gear ring was investigated through simulations in reference [26]. Meanwhile, Song et al. developed and compared the linear and nonlinear multi−freedom torsional vibration models of a clutchless AMT in electric vehicle applications [27]. The results show that the nonlinear torsional vibration of gear system will increase shift time, shift jerk and friction work. Furthermore, reference [28] demonstrated the existence of nonlinear stiffness in a half−shaft by experiments since the half−shaft transmits the torque amplified by transmission. A bifurcation theory was adopted to analyze the drive-shaft model, and it was evident that the nonlinear stiffness may lead to fold bifurcation, which will cause resonance response instability.

Shift force is an important factor to shift quality. However, the nonlinearity of the shift force, which is usually produced by the shift actuator, is always ignored in gearshift systems. Due to the working principle of the ELA, it is almost inevitable that the output characteristics appear nonlinear. In reference [9], the emphasis is on the efficient compensation of the nonlinear electromagnetic field effect, allowing the linear motor to be used at higher accelerations or larger loads without compromising control performance. Li et al. studied the effect of nonlinearities including nonlinear friction force, ripple force, magnetic saturation of linear motors on electromagnetic active suspension performance, and the results indicated the electromagnetic nonlinearities of the linear motor reduce the effective force output and active suspension performance [12]. The literature [29] emphasizes the nonlinearity of the industrial linear motor caused by external disturbances, and a neural network learning adaptive robust controller is synthesized to achieve good tracking performance and excellent disturbance rejection ability. Obviously, the nonlinearities exist in ELAs exactly, and they have a distinct influence on the control performance. However, the nonlinearities of the direct-drive gearshift system have not been explored yet, and it is essential to develop effective control methods to weaken the effect of nonlinearities.

The aim of this study was to analyze the nonlinear problems of the motor−AMT coupled drive system, and design effective control methods to reduce or eliminate the influence of the nonlinearities to the shift quality and gearshift performance. To achieve this, a detailed dynamic model that considers nonlinearities such as the nonlinearity of gearshift system including ELAs, nonlinearity of control methods and nonlinear disturbances was constructed, and the mechanisms of the nonlinearities were analyzed. Then, an improved active disturbance rejection control method which considers the precise motion control and robustness was designed. Comparative simulation and experiment were carried out to verify effectiveness.

2. The Direct-Drive Gearshift System

DC motors are usually used to actuate the gearshift events in electric AMT shift system. Nevertheless, motion converter and intermediate mechanisms are necessary to transmit and amplify the shift force, which makes the gearshift system complicated. By using direct−drive technology which adopts electromagnetic linear actuators (ELAs) can optimize the structure of the gearshift system. Less components, less moving mass and more reliable driving system are all advantages of the application of ELAs. What is more, the dynamic response of the gearshift system will be improved.

In most cases, the motor-AMT coupled driving system usually uses two−speed AMT so that one DC motor is in charge of two gear ratios. In this work, two ELAs are employed, and each ELA can handle one or two gear ratios. With this gearshift system, two, three or four−speed AMT schemes are available for motor−AMT coupled driving system. In addition, the gear selection process, which is essential in a DC motor gearshift system, is cancelled, and hence, the total shift time decreases. Figure 1 shows the novel direct-drive gearshift system. It includes two ELAs, and the output shafts connect with the shift fork directly. Two−way movement of the ELAs will drive the synchronizer through the shift fork to engage a target gear or disengage with the current gear. The intermediate mechanisms such as motion converter and reduction gears which is necessary for a DC motor driven AMT are no longer required. Obviously, the direct−drive gearshift system equipped with ELAs is simpler, and it is beneficial for rapid and precise displacement control. Moreover, Figure 2 presents the inner structure of the ELA prototype.

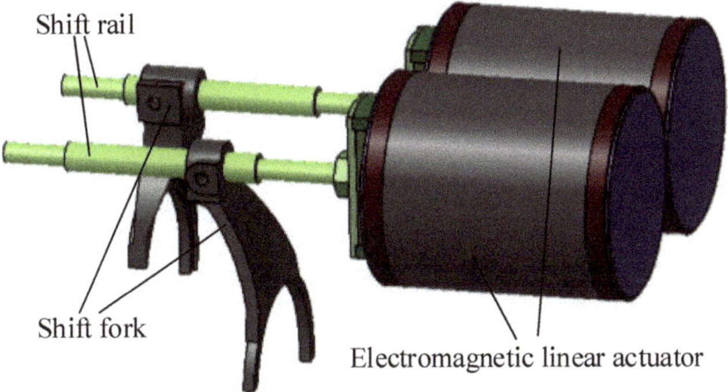

Figure 1. The direct drive gearshift system.

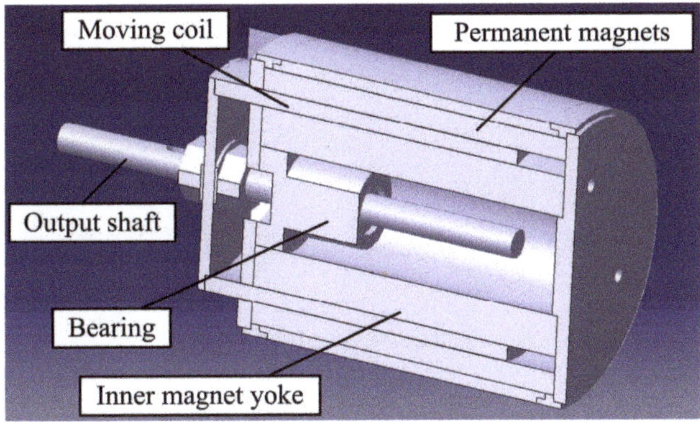

Figure 2. The structure of the ELA.

The test bench of motor−AMT coupled drive system is presented in Figure 3. It consists of ELAs, driving motor, 4−speed AMT, flywheel, torque and speed sensors, adjustable inertia plate and displacement sensors. Each side of the AMT has ELA attached, and 0.01 mm accurate displacement sensors are mounted on the ELAs.

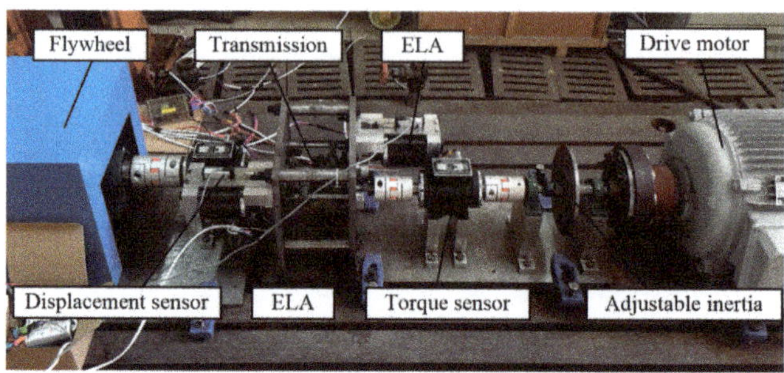

Figure 3. The test bench of motor-AMT coupled drive system.

3. Nonlinear Analysis and Modeling of Motor-AMT Coupled Drive System

Nonlinearities exist in numerous complex and integrated systems. However, the nonlinear feature of the motor−AMT coupled drive system has not been studied extensively and comprehensively yet. The output characteristics are frequently nonlinear regardless of the kind of motor−induction motor or permanent magnet synchronous motor. ELAs should be taken into account while modeling the coupled system because they are a source of nonlinearities because of their operating theory and technology limitations. It is also important to study the nonlinearities that will manifest in an AMT's gearings, shafts, and shift forks when enormous or rapidly fluctuating power is applied. Additionally, because the AMT shifting process is discrete, researchers have divided it into a number of steps in order to develop various control strategies [30]. As a result, the nonlinearities might be produced during the switch of different control strategies and transferring of control parameters.

3.1. Nonlinear Analysis of the ELA

Three subsystems−the electric, magnetic, and mechanical subsystems−can be used to characterize the ELA. Figure 4 displays the mathematical representation of ELA. Transfer lines are another way to show how the three subsystems are coupled. Obviously, the relationship between the input parameter voltage U and output parameter electromagnetic force F_m is nonlinear.

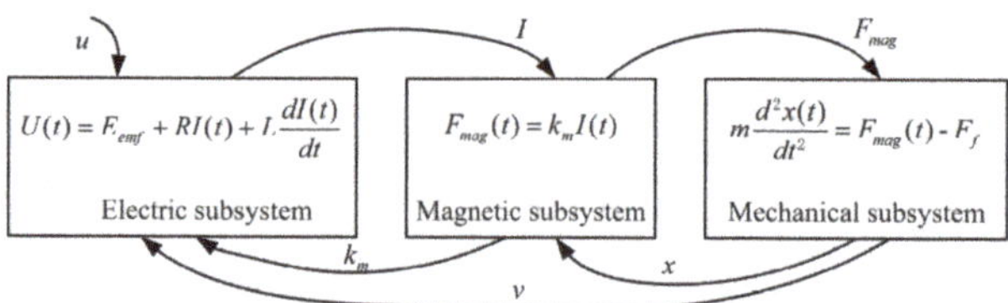

Figure 4. Coupled mathematical model of the ELA.

The permanent magnets inside the ELA are arranged using Halbach magnetized topology. Halbach magnetized topology is recognized that it can diminish the saturation of the magnet yoke to enhance the magnetic flux density so that the output electromagnetic force will be enlarged, and it is used to design the ELA. Figure 5a shows the topological structure. The magnetization direction of permanent magnets is represented by the arrowheads. In the space between the permanent magnets, the energized coil moves back and forth. Nevertheless, due to movements, such as those in positions 2 and 3, as shown in Figure 5b, the activated coil would not stay at a constant magnetic field. Hence, the electromagnetic force's characteristics are variable.

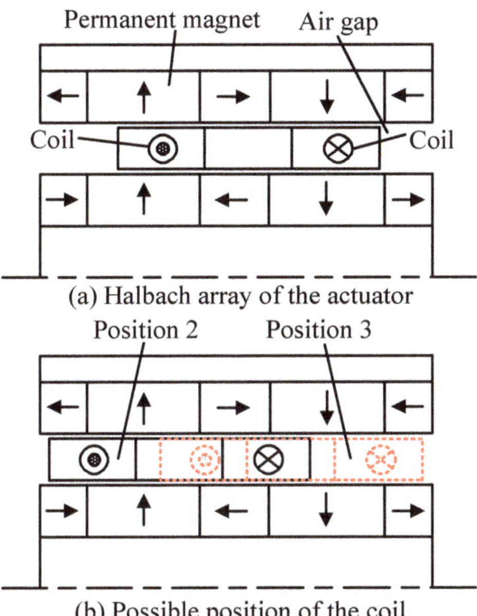

Figure 5. The Halbach structure of the ELA: (**a**)Halbach array of the actuator; and (**b**) Possible position of the coil.

Experiments are performed to examine the output characteristics of ELA. The experiments were implemented as follows,

1. Maintain the ELA shaft stationary in the middle and use a force sensor to detect the electromagnetic force as the coil current increases from 0 A to 30 A, increasing by 0.5 A each time;
2. Move the ELA shaft forward by 0.5 mm, then uses a force sensor to measure the electromagnetic force while the coil current increases from 0 A to 30 A, with each increment of 0.5 A;
3. Using the procedure described above, move the ELA shaft forward by 0.5 mm increments until the displacement reaches 10 mm. Then, measure and record all of the electromagnetic force data;
4. Using the procedure described above, move the ELA shaft back to the center position and then backward by 0.5 mm at a time until the displacement reaches 10 mm. Then, record all the electromagnetic force data.

Figure 6 shows the discovered correlations among current, position and electromagnetic force. The static force−displacement−current characteristics are indicated by electromagnetic force and electromagnetic force constant. The ELA has a 20 mm stroke. Obviously, the ELA force characteristics are nonlinear. The electromagnetic force varies with the change of displacement value while the current remains unchanged. The maximum force is produced when the ELA shaft remains in the middle and the displacement value is 10 mm. Additionally, when the current increases, the degree of nonlinearity increases. When the input current I is 10 A, the variation range of the k_m is between 40.1 and 43.4 N/A; when I is 20 A, the k_m is between 39.8 and 42.3 N/A; when I is 30 A, the k_m is between 38.9 and 41.7 N/A. In brief, the nonlinear characteristics of the ELA are influenced by not only the relative displacement of the ELA shaft, but also the variable current. Hence, the nonlinearity of the ELA would affect the precise displacement control during the gearshift process, and proper control method should be designed to eliminate the influence.

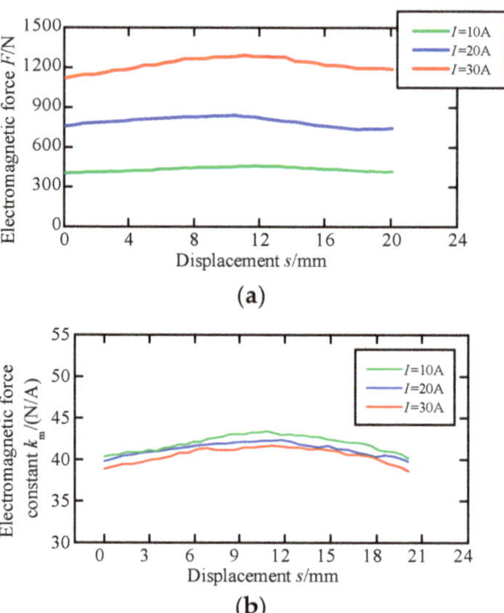

Figure 6. Static force-displacement−current characteristics: (**a**) static characteristics among current, displacement and electromagnetic force; and (**b**) static characteristics among current, displacement and electromagnetic force constant.

3.2. Gearshift Synchronization Process Analysis

Early in 2006, the eight phase equations for the gear engagement procedure of a Borg-Warner synchronizer were explored in reference [30]. The relative position of the spline chamfers and the required turning force were revealed by a thorough investigation of the second-bump and stick−slip phenomena. However, it concentrated on a manual transmission since it has distinct, noticeable characteristics when utilized with an EV's clutchless motor−AMT coupled drive system. It is necessary to explore the detailed gearshift synchronization process of the motor-AMT coupled drive system. In addition, the gearshift stroke is typically less than 10 mm, making it challenging to develop the proper control strategy for each of the eight distinct stages.

The shifting process is divided into four parts in the author's earlier work [14], as indicated in Figure 7, and a corresponding control mechanism was developed in accordance with the unique characteristics of each step. For example, the proportion−differentiation approach was developed to accomplish simple, quick, and no-overshot motion control in the first stage, which is to eliminate the gap between the sleeve and the synchronizer ring. However, switching control methods during various stages will result in a hazy transmission of control variables.

For instance, at the end of the first phase, the shift force is not zero, and it would be transmitted to the second phase. Such a transmit of control variables occurs during each of the four phases. Hence, the piecewise control method should be improved to obtain better shift quality. Wang et al. divide the gearshift process into two stages according to the movement state of the sleeve [24]. An integrated position and force switching control scheme including sliding mode control method for nonsynchronous stage and force control for synchronization is developed. Chen et al. use hybrid system theory to analyze the influence of various shifting force, relative rotational position, and speed of synchronization components to shift quality [31]. In fact, the gearshift synchronization process can be viewed as a nonlinear system since the process is discrete while the control variables and the control target are greatly different, and linearization theory should be employed to optimize the control performance.

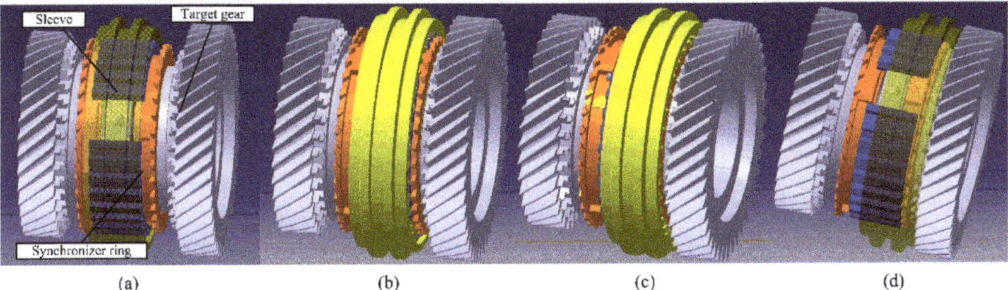

Figure 7. Four phases of the gearshift synchronization process: (**a**) the first phase, eliminate the gap; (**b**) the second phase, synchronization process; (**c**) the third phase, turn and cross the teeth of the synchronizer ring; and (**d**) the fourth phase, cross the teeth of the target gear and mesh with the target gear.

3.3. Shift Fork Deformation Analysis

The shift fork, which directly transmits the shift force to the synchronizer ring, may deform because the ELA designed for gearshift in this work can output more than 1000 N of electromagnetic force. To assess the influence, it is crucial to measure the shift fork's deformation values under various shift forces. The finite element model of the shift fork is built with ABAQUS, and two kinds of shift force 500 N and 1000 N are applied on the connection place with ELA shaft.

Figure 8 displays the mesh model and analysis outcomes. With values of 0.26 mm and 0.52 mm, respectively, and shift forces of 500 N and 1000 N, the shift fork's MAX labeled location experiences the greatest deformation. The distortion is minor when compared to the 9.5 mm space between gearshifts. Compared with the gearshift distance value 9.5 mm, the deformation is nonnegligible. Meanwhile, a larger shift force will result in greater deformation, thus it is best to carry out gearshift events when the rotational speed difference is small, which means that small shift force is transmitted to the shift fork so that the deformation will be negligible. However, it is contradictory to use a greater shift force to achieve a quicker gearshift. As a result, while gear shifting, the shift fork will continue to deform, intensifying the nonlinear properties of the shift force delivered to the synchronizer ring.

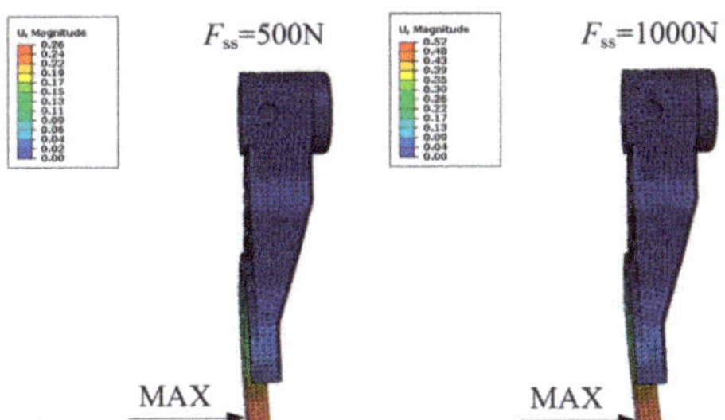

Figure 8. Analysis results of the shift fork deformation.

3.4. Modeling of the Powertrain System

Shift qualities usually include shift time, shift jerk, friction work and shaft torsional vibration. The shift jerk J is described as follow

$$J = \frac{da(t)}{dt} = \frac{r}{J_w} \frac{d(T_s i_d - T_L)}{dt} \tag{1}$$

where a is the accelerate speed of vehicle, r is the radius of wheel, J_w is the equivalent rotational inertia of the vehicle on wheel, T_s is the synchronizing torque, i_d is the gear ratio of main reducing gear, T_L is the load torque.

Apparently, the shift jerk J is mainly influenced by the T_s. Due to the nonlinear characteristics of the shift force, deformation of shift fork and variable control parameters, the value of J will vary fiercely. Moreover, there are two nonlinear components gear pairs and half shaft which the nonlinearities are usually neglected while analyzing the gearshift performance. The nonlinearity of gear pairs happens during the gear mesh process. According to the analysis of literature [27], the nonlinear dynamic characteristics of gear pairs will lower the shift quality, and the torsional vibration of gear pairs will also change the gear ratio. In fact, it is almost impossible to eliminate such influence by using effective method. Nevertheless, with appropriate robust control method, such influence can be weakened. In addition, due to the amplification of the transmission, the transmitted torque on the output shaft of the transmission and half shaft enlarges evidently. As a result, the shafts will show nonlinear vibration feature and it has been verified by He [28]. Such nonlinear vibration feature is more obvious in an EVs since the drive motor has faster torque regulation ability. Hence, the nonlinear feature of shafts should be considered in the simulation model of motor−AMT coupled drive system.

The improved powertrain model of a multi-speed motor−AMT coupled powertrain system is shown in Figure 9. Without the clutch, the powertrain model is simpler without clutch. For the motor, it can be described as

$$J_m \ddot{\theta}_m = T_m - T_{in} - c_m \dot{\theta}_m \tag{2}$$

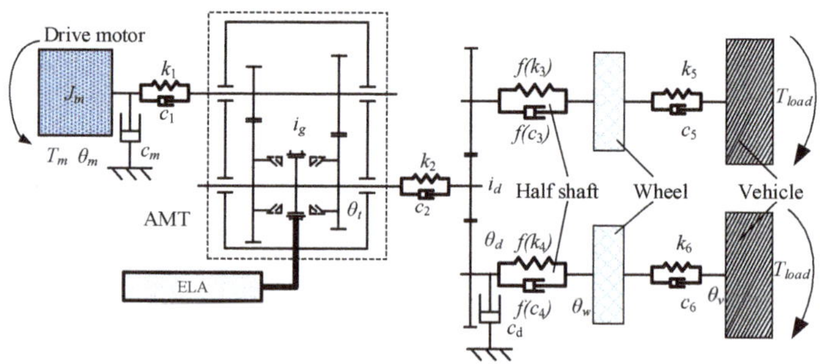

Figure 9. The vehicle model with motor-AMT coupled drive system.

For the AMT,

$$J_t \ddot{\theta}_t = T_{in} i_g - c_t \dot{\theta}_t - T_{out} \tag{3}$$

where J_m is the rotational inertia of the motor, θ_m is rotational angel, c_m is damping coefficient, T_{in} is the input torque of AMT, J_t is rotational inertia of AMT, θ_t is rotational angle of the AMT output shaft, c_t is damping coefficient, T_{out} is output torque of AMT, and $\theta_m = \theta_t i_g$.

Meanwhile, the T_{out} can be described as

$$T_{out} = k_2(\theta_t - \theta_d i_d) + c_2\left(\dot{\theta}_t - \dot{\theta}_d i_d\right) \tag{4}$$

For the final driver,

$$J_d \ddot{\theta}_d = i_d\left[k_2(\theta_t - \theta_d i_d) + c_2\left(\dot{\theta}_t - \dot{\theta}_d i_d\right)\right] - c_d\dot{\theta}_d - k_4(\theta_d - \theta_w) - c_4\left(\dot{\theta}_d - \dot{\theta}_w\right) \tag{5}$$

θ_d is the rotational angle of the final driver, and with $\theta_d = \theta_m/(i_t i_d)$, the equation can be transformed into

$$J_d\ddot{\theta}_m = i_d^2 i_t \left[k_2(\theta_t - \theta_d i_d) + c_2\left(\dot{\theta}_t - \dot{\theta}_d i_d\right) \right] i_d^2 i_t - c_d\dot{\theta}_m - i_t i_d \left[k_4(\theta_d - \theta_w) - c_4\left(\dot{\theta}_d - \dot{\theta}_w\right) \right] \tag{6}$$

θ_w is the rotational angle of the wheel. For the wheel, the influence of the tire is neglected to simplify the model. Hence, the wheel can be described as

$$(J_w + J_v)\ddot{\theta}_w = k_4(\theta_d - \theta_w) + c_4\left(\dot{\theta}_d - \dot{\theta}_w\right) - T_{\text{load}} = k_4\left(\frac{\theta_m}{i_t i_d} - \theta_w\right) + c_4\left(\frac{\dot{\theta}_m}{i_t i_d} - \dot{\theta}_w\right) - T_{\text{load}} \tag{7}$$

And the T_{load} is

$$T_{\text{load}} = [M_{\text{vehicle}}g\sin(\alpha) + fM_{\text{vehicle}}g + \frac{1}{2}C_d A_F \rho r^2 \dot{\theta}_v{}^2] \tag{8}$$

Suppose three state variable x_1, x_2, x_3 are

$$x_1 = \frac{\theta_m}{i_t i_d} - \theta_w, x_2 = \dot{\theta}_m, x_3 = \dot{\theta}_w \tag{9}$$

And the state space equation is obtained,

$$\begin{bmatrix} \dot{x}_1 \\ \dot{x}_2 \\ \dot{x}_3 \end{bmatrix} = \begin{bmatrix} 0 & \frac{1}{i_t i_d} & -1 \\ \frac{k_d}{J_{eq} i_t i_d} & \frac{c_4 + \frac{c_d}{i_t^2} + \frac{2c_d}{i_t^2 i_d^2}}{J_{eq}} & \frac{c_4}{i_t i_d} \\ \frac{k_4}{J_w} & \frac{c_4}{J_w i_t i_d} & -\frac{c_4}{J_w} \end{bmatrix} \begin{bmatrix} x_1 \\ x_2 \\ x_3 \end{bmatrix} + \begin{bmatrix} 0 & 0 & 0 \\ 0 & 1 & 0 \\ 0 & 0 & 1 \end{bmatrix} \begin{bmatrix} 0 \\ T_m \\ T_{\text{load}} \end{bmatrix} \tag{10}$$

where

$$J_{eq} = \frac{J_d}{i_t^2 i_d^2} + J_m + \frac{J_t}{i_t^2} \tag{11}$$

The damping coefficient and stiffness coefficient of the shafts are usually considered as constant in the model. However, the study results of references [28] prove that the nonlinearity of stiffness and damping exist in the shafts, and since the half shaft transmits the maximum torque among the shafts, it should consider the nonlinearity of the half shaft while study the dynamic performance of motor−AMT integrated powertrain system. Hence, the parameters k_4 and c_4 should be replaced by functions f_k and f_c [32]. Nonetheless, although it is a compromise way to take the nonlinearity into account, it is hard to implement analyzed results into practical application.

Obviously, several nonlinear influence factors which are neglected commonly exist in the motor−AMT coupled drive system, and according to the above analysis, such factors might cause damage to the gearshift performance. Hence, appropriate control strategy is essential to weaken or even eliminate the effect of nonlinearities.

4. Control Strategy Design

In the author's previous work, a kind of piecewise control method which contains four different control methods was proposed since the gearshift process can be divided into several phases and each phase has different feature [33]. Nonetheless, the switch of control methods and the transmit of variables among different control methods makes the process disordered and imprecise, and the integration of precise motion control and robust control makes the system complicated.

Afterwards, the piecewise control is replaced with the active disturbance rejection control (ADRC) approach. The basic topology is shown in Figure 10. It includes a tracking differentiator (TD), a nonlinear state error feedback law (NLSEF) and an extended state observer (ESO). It could be thought of as an improved PID approach with disturbance rejection. For the ELA gearshift system, which is influenced by linearities, it is difficult to establish quick and accurate motion control while yet guaranteeing robustness.

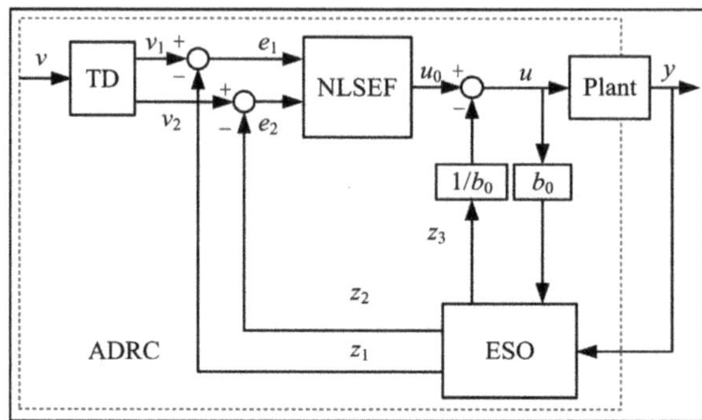

Figure 10. Basic topology of the ADRC.

Therefore, the ADRC approach needs to be enhanced to meet the control requirements. There are several improvements described as follows. First, the inverse system method (ISM) was used to minimize the nonlinear characteristics of the ELA output force. Second, an acceleration feedforward (AFF) module was introduced to the standard ADRC method to control the displacement of the gearshift process, since it can increase displacement control accuracy [34]. It can partially compensate the disturbance. Furthermore, it is challenging to calculate the compensation value because the shift fork's deformation varies with the shift force. As a result, the deformation of shift fork is considered as disturbance in this work. Similarly, the nonlinearity of gear pairs will lead to the variation of rotational speed of shafts, and it also can be regarded as external disturbance so that the ESO module can reject such disturbance. In order to enhance the disturbance rejection ability of the controller, the ESO module is modified to improve the rejection ability of the controller.

4.1. Basic Topology of ADRC

The typical form of TD is usually described as

$$
\begin{cases}
g = f(v_1 - v, v_2, r, h) \\
\dot{v}_1 = v_1 + h v_2 \\
\dot{v}_2 = v_2 + h g
\end{cases}
\tag{12}
$$

v_1 is a transitional trajectory, v_2 is the differential of v, and h is the sampling period, g is an intermediate variable. Parameter r influences the dynamic response of v_1, and the larger value of r will shorten the time which is taken by the v_1 of a specific v. The function $f(v_1-v, v_2, r, h)$ is a time−optimal function, which is defined as [34,35]

$$
f(v_1 - v, v_2, r, h) = \begin{cases}
d = r{\cdot}h, d_0 = h{\cdot}d, y = v_1 - v + h{\cdot}v_2 \\
a_0 = \sqrt{d^2 + 8r{\cdot}|y|} \\
a = \begin{cases}
v_2 + (a_0 - d){\cdot}\frac{sgn(y)}{2}, & |y| > d_0 \\
v_2 + \frac{y}{h}, & |y| \le d_0
\end{cases} \\
f = \begin{cases}
-rsgn(a), & |y| > d_0 \\
-\frac{ra}{d}, & |y| \le d_0
\end{cases}
\end{cases}
\tag{13}
$$

Function $fal(e, \alpha, \delta)$ is defined as

$$
fal(e, \alpha, \delta) = \begin{cases}
e{\cdot}\delta^{1-\alpha}, & |e| \le \delta \\
|e|^{\alpha}{\cdot}\text{sgn}(e), & |e| > \delta
\end{cases}
\tag{14}
$$

where α should satisfy $\alpha < 1$, $\delta = k_p{\cdot}h$, k_p is a positive integer.

The function of NLSEF is to generate the intermediate input variable u_0. It is described as

$$\begin{cases} e_1 = y_1 - z_1 \\ e_2 = y_2 - z_2 \\ u_0 = \beta_{11} \cdot fal(e_1, \alpha_1, \delta) + \beta_{12} \cdot fal(e_2, \alpha_2, \delta) \end{cases} \tag{15}$$

where e_1 and e_2 are estimate errors, z_1 and z_2 are estimated values, β_{11} and β_{12} are control parameters, y_1 and y_2 are the output values of the TD module, α_1 and α_2 satisfy the condition $0 < \alpha_1 < 1 < \alpha_2$.

ESO module is kernel module of the ADRC since it can estimate the internal and external disturbances according to the analysis of output variables. The discrete time form of the ESO module is

$$\begin{cases} e = z_1 - y \\ z_1 = z_1 + h(z_2 - \beta_{01} \cdot e) \\ z_2 = z_2 + h(z_3 - \beta_{02} \cdot fal(e, \alpha, \delta) + b_0 \cdot u) \\ z_3 = z_3 - h \cdot \beta_{03} \cdot fal(e, \alpha, \delta) \end{cases} \tag{16}$$

z_1, z_2 and z_3 are estimated values of the output variable y, e is estimate error, h is the sampling time, and any disturbance information is included. β_{01}, β_{02} and β_{03} are observer gains and they are usually selected as

$$\beta_{01} \approx 1/h, \quad \beta_{02} \approx 1/1.6h^{1.5}, \quad \beta_{03} \approx 1/(8.6h^{2.2}) \tag{17}$$

4.2. Design of Inverse System Method

Designing a suitable and reliable control system for a nonlinear system is more challenging than for a linear one. In order to construct a controller for a nonlinear system that will achieve the control aims, the inverse system method is utilized to build a pseudo-linear system for the nonlinear system. A pseudo−linear contains an inverse system and an original system, and the first step is to build the inverse system.

The coupled mathematics shown in Figure 4 allows the mathematical representation of the gearshift system to be rewritten as

$$\begin{cases} \dot{I} = -\frac{R}{L}I - \frac{k_m}{m}v + \frac{u}{L} \\ \dot{v} = \frac{k_m}{m}I - \frac{c}{m}v \\ \dot{S} = v \end{cases} \tag{18}$$

where I is coil current, R is coil resistance, L is coil inductance, k_m is force constant, v is sleeve's velocity, m is moving mass, c is viscous friction damping coefficient, S is displacement of the sleeve. The state variables are chosen as

$$x = \begin{bmatrix} x_1 & x_2 & x_3 \end{bmatrix}^T = \begin{bmatrix} I & v & S \end{bmatrix}^T \tag{19}$$

Consequently, the state equation is written as

$$\begin{bmatrix} \dot{x}_1 \\ \dot{x}_2 \\ \dot{x}_3 \end{bmatrix} = \begin{bmatrix} -\frac{R}{L} & k_m & 0 \\ \frac{k_m}{L} & -\frac{c}{m} & 0 \\ 0 & 1 & 0 \end{bmatrix} \begin{bmatrix} x_1 \\ x_2 \\ x_3 \end{bmatrix} + \begin{bmatrix} \frac{1}{L} \\ 0 \\ 0 \end{bmatrix} u, \quad y = \begin{bmatrix} 0 & 0 & 1 \end{bmatrix} \begin{bmatrix} x_1 \\ x_2 \\ x_3 \end{bmatrix} \tag{20}$$

where y is the system output.

According to the inverse system theory, the under equations

$$\begin{cases} y = x_3 \\ \dot{y} = \dot{x}_3 = x_2 \\ \ddot{y} = \dot{x}_2 = \frac{k_m}{m}I - \frac{c}{m}\dot{y} \\ \dddot{y} = \ddot{x}_2 = -\frac{c}{m}\ddot{y} + \frac{k_m}{m}\left(-\frac{R}{L}x_1 - \frac{k_m}{L}\dot{y} + \frac{u}{L}\right) \end{cases} \tag{21}$$

are obtained. The input variable u specifically only appears in the expression y, hence the inverse system is a third order system with a dimension equal to the state vectors. The system is hence reversible. One solution to the inverse system is

$$u = \frac{ml}{k_m}\dddot{y} + \frac{cl}{k_m}\ddot{y} + k_m\dot{y} + Rx_1 \tag{22}$$

The chosen state variables for the pseudo linear system are

$$w = \begin{bmatrix} w_1 & w_2 & w_3 \end{bmatrix}^T = \begin{bmatrix} y & \dot{y} & \ddot{y} \end{bmatrix}^T \tag{23}$$

and the pseudo−linear system's state space equations are derived as

$$\begin{cases} \begin{bmatrix} \dot{w}_1 \\ \dot{w}_2 \\ \dot{w}_3 \end{bmatrix} = \begin{bmatrix} 0 & 1 & 0 \\ 0 & 0 & 1 \\ 0 & 0 & 0 \end{bmatrix} \begin{bmatrix} w_1 \\ w_2 \\ w_3 \end{bmatrix} + \begin{bmatrix} 0 \\ 0 \\ 1 \end{bmatrix} \varphi = Aw + B\varphi \\ y = \begin{bmatrix} 1 & 0 & 0 \end{bmatrix} \begin{bmatrix} w_1 \\ w_2 \\ w_3 \end{bmatrix} = Cw \end{cases} \tag{24}$$

State feedback controllers are made to characterize the dynamic properties of control input and output for pseudo−linear systems. The target equation is created as

$$y^k(t) + a_{k-1}y^k(t) + \ldots + a_1y'(t) + a_0y(t) = r(t) \tag{25}$$

where $a_0, a_1, \ldots, a_{k-1}$ are real number, and $r(t)$ is reference input.

The following control law can be established using state feedback control theory because the pseudo linear system's state variables are three

$$\begin{cases} \varphi = r - y_f \\ y_f = a_0y + a_1\dot{y} + a_2\ddot{y} \end{cases} \tag{26}$$

where r is the desired value and y_f is the feedback value. The transfer function can be deduced as

$$\frac{Y(s)}{R(s)} = \frac{1}{s^3 + a_2s^2 + a_1s + a_0} \tag{27}$$

Consequently, the transfer function's characteristic equation is

$$\left(s^2 + 2\varsigma\omega_n + \omega_n^2\right)(s + \varsigma\omega_n) = 0 \tag{28}$$

where ς is the damping coefficient, ω_n is the natural frequency. The transition time can be calculated by equation $t_s \approx 4/\varsigma\omega_n$, and the standard damping coefficient is chosen to be 0.707. The natural frequency is 282.88 since the transition period is calculated to be 20 ms. Put ς and ω_n into the characteristic Equation (28), the following equation are obtained

$$s^3 + 600s^2 + 160021.1s + 16004218.9 = 0 \tag{29}$$

By applying Ackermann formula, $a_2 = 600$, $a_1 = 160021.1$, $a_0 = 16004218.9$ is obtained.

4.3. Improvement of ESO Module

The observation and estimation performance of the ESO is mainly decided by the nonlinear function *fal*(e, α, δ). Function *fal*(e, α, δ) is widely used due to its simple structure. However, the continuity and flatness of the *fal* function has potential to be improved especially near the original point [36]. A higher error feedback gain will improve the ESO module's ability to observe and convergence rate, but it will also amplify signal noise and other disturbances that threaten the stability of the control system [37]. As a result, a kind of novel nonlinear function *faln*(x, σ)vis employed to solve the problem. The *faln*(x, σ) is designed as

$$faln(x,\sigma) = \begin{cases} \frac{x}{\sigma^2}e^{-\frac{x^2}{2\sigma^2}}, |x| \ll 1 \\ \frac{1}{\sigma^2}e^{-\frac{1}{2\sigma^2}}, x > 1 \\ -\frac{1}{\sigma^2}e^{-\frac{1}{2\sigma^2}}, x < -1 \end{cases} \tag{30}$$

where σ is the regulation parameter. According to the analysis in literature [34], the *faln* function, which only has one parameter as opposed to the *fal* function's two, is superior to the *fal* function in terms of observation and noise rejection. Obviously, the control system design is simplified and the performance of ESO is enhanced. The designed controller is presented in Figure 11.

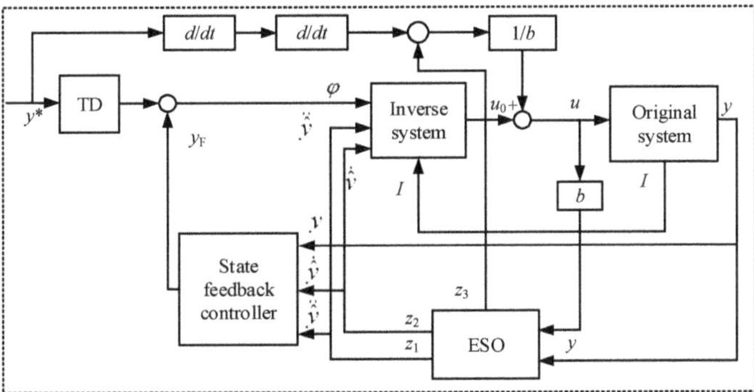

Figure 11. Improved topology of ADRC.

4.4. Comparative Simulation and Analysis

Comparative simulations were carried out to verify the performance of the designed method. First, displacement control performance of the ADRC method is simulated and meanwhile PID method is employed as a comparison group to realize the same displacement control. Figure 12 shows the comparative results. There are three different target displacement values chosen: 2 mm, 4 mm and 9.5 mm. When the displacement value is 4 mm, the control parameters of the two controllers are regulated and established, and they remain constant when the displacement value changes. It can be discovered that both controllers can reach the target displacement value quickly, however the ADRC controller is noticeable quicker than the PID controller. Furthermore, when the target value is 9.5 mm, the displacement of the PID technique is greater than the target value. Obviously, the ADRC controller has superior parameter-dependent stability than the PID controller. On the other hand, every displacement curve reaches the desired value in less than 25 ms, demonstrating the actuator's quick dynamic performance.

Figure 13 displays the disturbance rejection ability of three different types of controllers, including IADRC, ADRC and PID controllers. An external load torque is injected at 11 ms to observe the performance of the three controllers. PID controller presents the most distinct variation among the three controllers, and stead−state error of the displacement control is also produced. Only a little change in the displacement curve is observed when using the ADRC and IADRC controllers, which appear to have higher disturbance rejection capabilities. Additionally, IADRC controller has quicker adjusting ability than ADRC controller. Obviously, the AFF module enhances the IADRC controller's quick dynamic performance, and the modified ESO strengthens its capacity to reject disturbances.

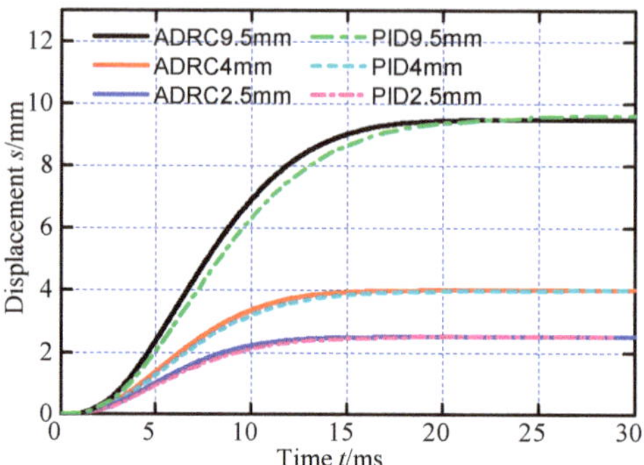

Figure 12. Comparison of displacement control.

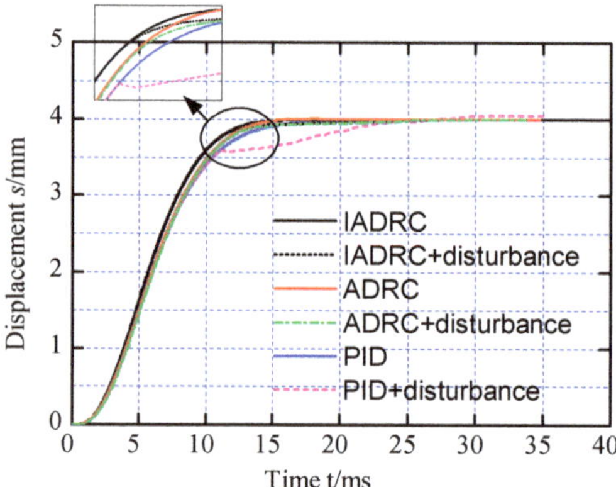

Figure 13. Comparison of disturbance rejection performance.

Figure 14 shows the displacement tracking performance. The target displacement curve is designed and divided into four stages as the real displacement variation of the gearshift process. Obviously, the IADRC tracks the target curve best both on the time and precision indexes. In brief, the IADRC controller presents faster dynamic performance, more accurate displacement control and better stability to the parameter variations and external disturbances.

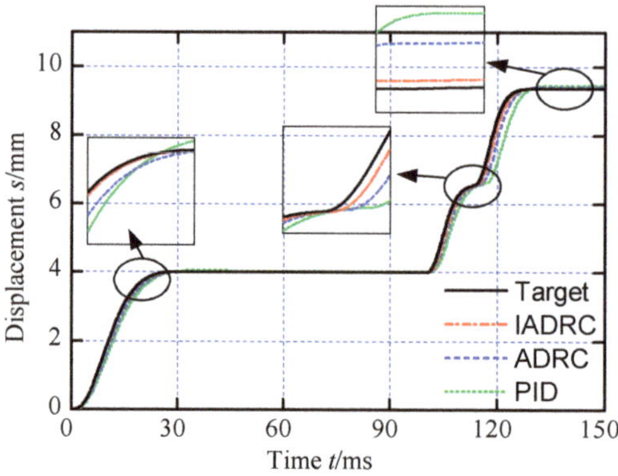

Figure 14. Comparison of displacement tracking performance.

5. Experiments and Analysis

Figure 3 displays the test bench in detail. The motor-transmission coupled drive system and the direct drive gearshift system that are proposed in this work are primarily for EVs. Since the drive motor is often under active speed regulation control, the speed difference is less than with regular AMT. As a result, two types of speed differences—100 and 300 r/min—are chosen to verify the gearshift performance of the direct drive gearshift system. As it is known, shift time, jerk, and friction work make up the majority of the gearshift performance indices. Due to the rapid development of the material and manufacturing technique, the service life of the synchronizer ring is distinctly extended. Meanwhile, the friction work apparently decreases since the speed difference is smaller in EVs. Friction work is therefore not the primary focus of this work. In earlier research, the performance of the ADRC approach was found to be superior to the PID method overall in terms of both the accuracy of displacement control and the capacity to reject disturbances. Therefore, in this work, the IADRC method and the ADRC approach are compared.

Figure 15 shows the gearshift displacement, shift force and jerk of the gearshift process when the speed difference is 100 r/min. It can be seen from the Figure 15a shows that the gearshift time for the two control methods is less than 100 ms, and the values are 93.4 ms and 100.1 ms respectively. One of the important reasons is that the IADRC approach performs dynamically more quickly, which reduces a few shift times during stage (a)(c)(d) that include displacement increments.

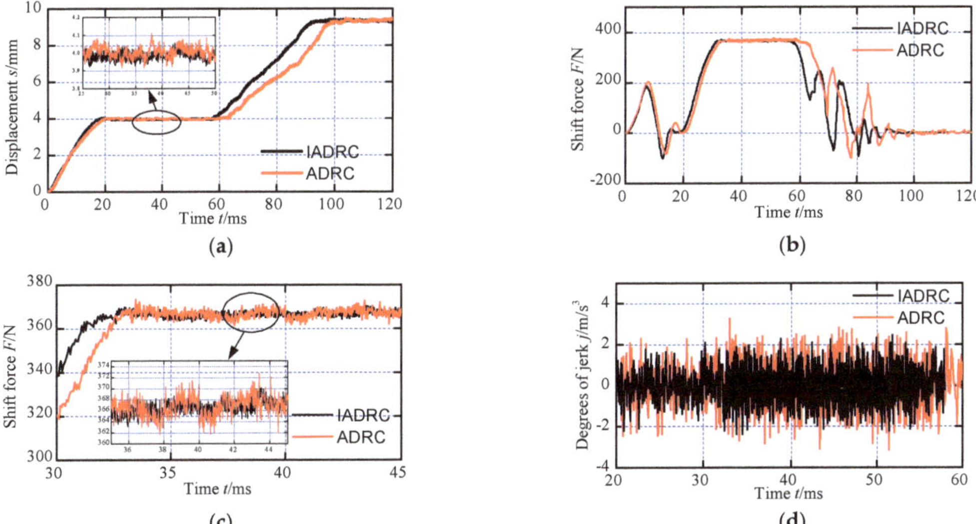

Figure 15. Gearshift experiment results (speed difference 100 r/min): (**a**) Displacement; (**b**) Shift force; (**c**) Amplified shift force during the synchronization process; (**d**) Degree of jerk.

Additionally, the partial enlarged drawing in Figure 15a illustrates the fluctuation of the displacement during the synchronization process. The fluctuation is mainly produced by the nonlinear of shift force, deformation of shift fork, accuracy error of displacement sensor and external disturbances including vibration, and it will influence the friction torque loaded on the synchronizer mechanisms so as to slow the synchronization process. All these disturbance factors lead to the fluctuation of the displacement. Apparently, the displacement fluctuation of the IADRC method is smaller than that of the ADRC method. Although the disturbance rejection ability of the ADRC method is excellent [34–36], the improved IADRC method can suppress the disturbances better.

In addition, the four stages described in Figure 7 are not very clear on the displacement curve of IADRC method in Figure 15a. The reason is that the teeth of the sleeve pass by the teeth of the target gear without distinct contact so that the displacement increases smoothly. On the contrary, there is a tiny standstill of the displacement on the curve of ADRC. It usually costs several seconds for the sleeve to revolve at just the right angle so as to cross the teeth of the target gear.

The shift force suggests that the two controllers are performing similarly. The amplified drawing of the synchronization process is presented in Figure 15c. Since the speed difference is comparatively small, the shift time will be small too. To achieve the proper shift jerk, the maximum shift force is therefore restricted to a certain value. The maximum shift force usually happens in the synchronization process. It is clear that both of the two curves in Figure 15c exhibit variance. However, the fluctuation amplitude of the of the shift force is smaller when the IADRC technique is adopted which let the ELA output force have better linear properties. For the ADRC approach, the fluctuation range is 362 to 373 N, whereas for the IADRC method, it is 363 to 370 N. Also, it is beneficial to decrease shift time. The shift force curve clearly shows the four steps. The second stage ends when the shift force starts to rapidly fall after the first stage, which takes over 20 ms. In the third stage, it usually needs a bump force to turn the sleeve so as to cross the teeth of the target gear. After that, the teeth of the sleeve will come into contact with those of the target gear, necessitating another bump force to cause the sleeve to turn at a slight angle once more in order to mesh with the gear. Therefore, there are two bump force existing on the shift force curve.

The degree of jerk is shown in Figure 15d. Due to the limitation of maximum shift force, the degrees of jerk are all acceptable for both methods. For the IADRC method and the ADRC method, the maximum degrees of jerk are 2.31 m/s^3 and 3.13 m/s^3, respectively. The shift jerk is directly caused by the drastic contact of two components, however, the deformation of shift fork, fluctuation of displacement, control parameter variations and external vibrations will also result in unpredictable shift jerk. The improve ESO module of IADRC method can estimate such factors from the output variables, and proper adjustment will produce to limit the influence of those disturbances to the shift fork. Moreover, the linearization of the ELA output force is also helpful to reduce the shift jerk. Apparently, it appears that the IADRC can restrain the jerk to a better degree than ADRC approach, and it demonstrate the better disturbance rejection ability and the effectiveness of the IADRC method.

Figure 16 depicts the gearshift displacement, shift force and jerk of the gearshift process when the speed difference is 300 r/min. The limitation of shift force increases to 600 N to guarantee small shift time and degree of jerk. For the IADRC method and the ADRC approach, the shift times are 121 ms and 137 ms, respectively. Both curves exhibit displacement fluctuation, however it is less evident when the IADRC approach is used. The displacement value during the synchronization process is nearly 4.1 mm while it is nearly 4 mm when the speed difference is 100 r/min. It is caused by the increase of the shift force during the synchronization process. Larger force will produce slight deformation of the shift fork, and the deformation will transfer to the displacement sensor. In addition, the amplitude of fluctuation of the shift force becomes bigger since the nonlinear output characteristics will be more obvious with the increase of shift force. Similarly, the IADRC method has better performance than ADRC method in terms of restraining the nonlinear output characteristics of the ELA. Although the degrees of jerk appear slight increase with the values presented in Figure 15, they are still acceptable since all the values are smaller than 4 m/s^3. The specific values of shift time, synchronization time and degrees of jerk are all given in Table 1.

Table 1. Gearshift indexes of 30 times gearshift events.

Speed Difference (r/min)	Inertia (kgm²)	Control Method	Gearshift Time (ms)	Synchronization Time (ms)	Maximum Jerk(m/s³)
	0.03	IADRC	90~94	39~42	2.31
	0.03	ADRC	95~101	44~48	3.13
100	0.05	IADRC	121~130	65~71	2.70
	0.05	ADRC	127~143	66~75	3.48
	0.03	IADRC	120~127	68~73	2.57
300	0.03	ADRC	128~145	74~82	3.91
	0.05	IADRC	165~180	111~119	2.79
	0.05	ADRC	169~188	112~126	3.96

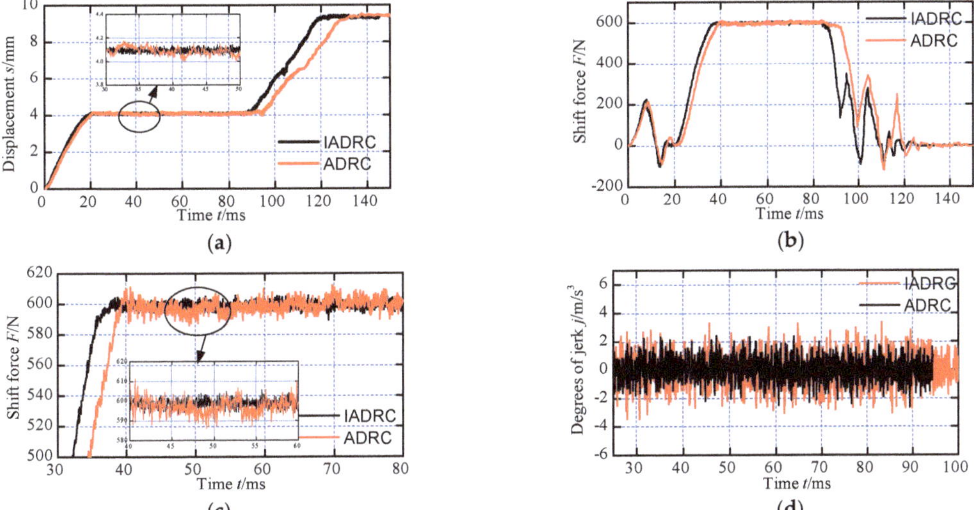

Figure 16. Gearshift experiment results (speed difference 300 r/min): (**a**) displacement; (**b**) shift force; (**c**) amplified shift force during the synchronization process; and (**d**) degree of jerk.

Another inertia value 0.05 kgm² is chosen to perform gearshift events in order to further confirm the effectiveness and advantages of the designed method, and the outcomes of 30 times of gearshift events are presented in Table 1. When the speed differential or inertia value grows, the gearshift time also grows. Similar change rules are presented by the maximum jerk and synchronization time. Meanwhile, the maximum values of friction work per unit during these gearshift events is 0.13 J/mm², which is smaller than the permission value 1.2 J/mm² evidently. It can be concluded that the IADRC method has better comprehensive control performance than ADRC method. The ISM technique can linearize the output force characteristics of the ELA so that both the shift jerk and fluctuation of displacement is smaller, and the optimized ESO module improve the disturbance rejection ability compared with that of the ADRC method. In short, the IADRC method can achieve better gearshift performance than ADRC method from Table 1, and the effectiveness of the IADRC method is also proved by experiments.

6. Conclusions

This paper introduced a type of direct−drive gearshift system that used two electromagnetic linear actuators (ELAs). The direct-drive gearshift system, which gains from the ELAs' powerful drive capabilities and the system's direct-drive design, has the potential to enhance shifting performance, including gearshift time and jerk. The gearshift system's nonlinear properties were taken into account and examined in order to build the best control strategies to lessen the impact of the nonlinearities.

A kind of improved active disturbance rejection control (IADRC) method was designed. It is derived from the ADRC method, and it has the advantages of a simple structure and strong robustness. The ELA's nonlinear output characteristics are lessened when the inverse system method (ISM) is used, and the IADRC method's ability to reject disturbances is improved by optimizing the ESO module. In addition, the use of an acceleration feedforward module improved the dynamic response of the controller.

Comparative simulations and experiments were conducted, and the findings indicated the effectiveness and improvement of the designed IADRC method. In conclusion, the IADRC method has better gearshift performance than the ADRC method. Moreover, with the application of the IADRC method, the direct-drive gearshift system employing two ELAs has excellent gearshift performance. Further studies will focus on the coordinated control of the motor−transmission coupled drive system to achieve a fast, comfortable and seamless EV drive system.

Author Contributions: Conceptualization, S.L. and B.L.; methodology, S.L. and M.T.; software, simulation, formal analysis and validation, S.L., M.T. and W.S.; writing—original draft preparation, S.L.; writing—review and editing, S.L. and B.L. All authors have read and agreed to the published version of the manuscript.

Funding: This work was funded by the National Natural Science Foundation of China, grant number 51905364 and 51975341.

Institutional Review Board Statement: Not applicable.

Informed Consent Statement: Not applicable.

Data Availability Statement: The data presented in this study are available on request from the corresponding author.

Conflicts of Interest: The authors declare no conflict of interest.

References

1. Ye, Y. Application & Development of the Linear Motor in the Modern Machine Tool Industry. *Electr. Mach. Technol.* **2010**, *3*, 1–5.
2. Gokhan, C.; Ozan, K. Axial Flux Generator with Novel Flat Wire for Direct-Drive Wind Turbines. *IET Renew. Power Gen.* **2021**, *1*, 139–152.
3. Liu, M.; Chen, P.; Ma, J. Structural Optimization Design and Research of Direct-Drive Quadruped Robots. *China Mech. Eng.* **2021**, *9*, 2246–2253.
4. Zhang, D.; Han, Q.; Zhang, X. Network-Based Modeling and Proportional-Integal Control for Direct-Drive-Wheel Systems in Wireless Network Environments. *IEEE Trans. Cybern.* **2020**, *6*, 2462–2474. [CrossRef]
5. Fan, X.; Yin, J.; Lu, Q. Design and Analysis of a Novel Composited Electromagnetic Linear Actuator. *Actuators* **2022**, *11*, 6. [CrossRef]
6. Tan, C.; Li, B.; Ge, W. Thermal Quantitative Analysis and Design Method of Bistable Permanent Magnet Actuators Based on Multiphysics Methodology. *IEEE Trans. Ind. Electron.* **2020**, *9*, 7727–7735. [CrossRef]
7. Sun, G.; Wu, L.; Kuang, Z.; Ma, Z.; Liu, J. Practical Tracking Control of Linear Motor via Fractional-Order Sliding Mode. *Automatica* **2018**, *94*, 221–235. [CrossRef]
8. Lu, J.; Li, B.; Ge, W.; Tan, C.; Sun, B. Analysis and experimental Study on Servo Dynamic Stiffness of Electromagnetic Linear Actuator. *Mech. Syst. Signal. Process.* **2022**, *169*, 108587. [CrossRef]
9. Chen, Z.; Yao, B.; Wang, Q. Accurate Motion Control of Linear Motors with Adaptive Robust Compensation of Nonlinear Electromagnetic Field Effect. *IEEE/ASME Trans. Mechatron.* **2013**, *3*, 1122–1129. [CrossRef]
10. Lu, Y.; Tan, C.; Ge, W.; Li, B.; Lu, J. Improved Sliding Mode-Active Disturbance Rejection Control of Electromagnetic Linear Actuator for Direct-Drive System. *Actuators* **2021**, *10*, 138. [CrossRef]
11. Zheng, C.; Liu, L.; Guo, H.; Xu, Z. Optimization Design of a Fully Variable Valve System Based on Nelder-Mead Algorithm. *Proc. Inst. Mech. Eng. C-J. Mech E.* **2022**, *11*, 5815–5825. [CrossRef]
12. Li, Y.; Zheng, L.; Liang, Y.; Yu, Y. Adaptive Compensation Control of an Electromagnetic Active Suspension System Based on Nonlinear Characteristics of the Linear Motor. *J. Vib. Control* **2020**, *21–22*, 1–13. [CrossRef]
13. Li, B.; Ge, W.; Li, Q.; Li, Y.; Tan, C. Gearshift Sensorless Control for Direct-Drive-Type AMT Based on Improved GA-BP Neural Network Algorithm. *Math. Probl. Eng.* **2020**, *2020*, 6456410. [CrossRef]

14. Lin, S.; Li, B. Shift Force Optimization and Trajectory Tracking Control for a Novel Gearshift System Equipped with Electromagnetic Linear Actuators. *IEEE/ASME Trans. Mechatron.* **2019**, *4*, 1640–1650. [CrossRef]
15. Liang, J.; Yang, H.; Wu, J.; Zhang, N.; Walker, P. Power-on Shifting in Dual Input Clutchless Power-Shifting Transmission for Electric Vehicles. *Mech. Mach. Theory* **2018**, *121*, 487–501. [CrossRef]
16. Nguyen, S.; Song, J.; Fang, S.; Song, H.; Tai, Y.; Li, F. Simulation and Experimental Demonstration of a Seamless Two-Speed Automatic Mechanical Transmission for Electric Vehicles. *J. Tsinghua Univ. Sci. Technol.* **2017**, *10*, 1106–1113.
17. Yue, H.; Zhu, C.; Gao, B. Fork-Less Two-Speed I-AMT with Overrunning Clutch for Light Electric Vehicle. *Mech. Mach. Theory* **2018**, *130*, 157–169. [CrossRef]
18. Tian, F.; Wang, L.; Sui, L.; Zeng, Y.; Zhou, X.; Tian, G. Active Synchronizing Control of Transmission Shifting without a Synchronizer for Electric Vehicles. *J. Tsinghua Univ. Sci. Technol.* **2020**, *2*, 1010–1108.
19. Mo, W.; Walker, P.; Zhang, N. Dynamic Analysis and Control for an Electric Vehicle with Harpoon-Shift Synchronizer. *Mech. Mach. Theory* **2019**, *133*, 750–766. [CrossRef]
20. Guo, L.; Gao, B.; Chen, H. Online Shift Schedule Optimization of 2-Speed Electric Vehicle Using Moving Horizon Strategy. *IEEE/ASME Trans. Mechatron.* **2016**, *6*, 2858–2869. [CrossRef]
21. Sheng, J.; Zhang, B.; Zhu, B.; Wang, M.; Jin, Q. Parameter Optimization and Experimental Comparison of Two-Speed Pure Electric Vehicle Transmission Systems. *China Mech. Eng.* **2019**, *7*, 763–770.
22. Chai, B.; Zhang, J.; Wu, S. Compound Optimal Control for Shift Processes of a Two-Speed Automatic Mechanical Transmission in Electric Vehicles. *Proc. IMechE Part D J. Automob. Eng.* **2018**, *8*, 2213–2231. [CrossRef]
23. Qi, X.; Yang, Y.; Wang, X.; Zhu, Z. Analysis and Optimization of the Gear-Shifting Process for Automated Manual Transmissions in Electric Vehicles. *Proc. IMechE Part D J. Automob. Eng.* **2017**, *13*, 1751–1765. [CrossRef]
24. Wang, X.; Li, L.; He, K.; Liu, Y.; Liu, C. Position and Force Switching Control for Gear Engagement of Automated Manual Transmission Gear-Shift Process. *J. Dyn. Syst. Meas. Contr.* **2018**, *140*, 081010. [CrossRef]
25. Wang, J.; Zhang, J.; Yao, Z.; Yang, X.; Sun, R.; Zhao, Y. Nonlinear Characteristics of a Multi-Degree-of-Freedom Spur Gear System with Bending-Torsional Coupling Vibration. *Mech. Syst. Signal Process.* **2019**, *121*, 810–827. [CrossRef]
26. Sui, L.; Tian, F.; Li, B.; Zeng, Y.; Tian, G.; Chen, H. Nonlinear Dynamics Analyses of Gear Shifting with Gear Vibrations. *J. Tsinghua Univ. Sci. Technol.* **2020**, *2*, 109–116.
27. Song, Q.; Sun, D.; Zhang, W. Shift Nonlinear Modeling and Control of Automated Mechanical Transmission in Pure Electric Vehicle. *J. Jilin Univ. Enginee. Technol.* **2021**, *3*, 810–819.
28. He, R.; Han, Q. Stability Analysis of Torsional Vibration of Vehicle Powertrain Based on the Nonlinear Drive-shaft Model. *J. Mech. Eng.* **2019**, *18*, 125–131.
29. Wang, Z.; Hu, C.; Zhu, Y.; He, S.; Yang, K.; Zhang, M. Neural Network Learning Adaptive Robust Control of an Industrial Linear Motor-Driven Stage with Disturbance Rejection Ability. *IEEE Trans. Ind. Electron.* **2017**, *5*, 2172–2183. [CrossRef]
30. Lovas, L.; Play, D.; Marialigeti, J.; Rigal, J. Mechanical Behaviour Simulation for Synchromesh Mechanism Improvements. *Proc. Inst. Mech. Eng. D-J. Aut.* **2006**, *7*, 919–945. [CrossRef]
31. Chen, H.; Tian, G. Modeling and Simulation of Gear Shifting in Clutchless Coupled Motor-Transmission system. *J. Tsinghua Univ. Sci. Technol.* **2016**, *2*, 144–151.
32. Talebitooti, R.; Morovati, M. Study on TVD Parameters Sensitivity of a Crankshaft Using Multiple Scale and State Space Method Considering Quadratic and Cubic Non-Linearities. *Lat. Am. J. Solids Stuct.* **2014**, *11*, 2672–2695. [CrossRef]
33. Lin, S.; Li, B.; Jiao, W. Gearshift Performance Improvement for an Electromagnetic Gearshift System Based on Optimized Active Disturbance Rejection Control Method. *Adv. Mech. Eng.* **2019**, *2*, 1–15. [CrossRef]
34. Shi, X.; Chang, S. Extended State Observer-Based Time-Optimal Control for Fast and Precise Point-to-Point Motions Driven by a Novel Electromagnetic Linear Actuator. *Mechatronics* **2013**, *23*, 445–451. [CrossRef]
35. Tan, C.; Lu, Y.; Ge, W.; Li, B.; Lu, J. Depth fuzzy sliding-mode active disturbance rejection control method of permanent magnet linear motor for direct drive system. *J. Xi'an Jiaotong Univ.* **2023**, *1*, 1–9.
36. Yang, W.; Lu, J.; Jiang, X.; Wang, Y. Design of Quadrotor Attitude Active Disturbance Rejection Controller Based on Improve ESO. *Sys. Eng. Electro.* **2022**, *12*, 3792–3799.
37. Tan, C.; Ren, H.; Li, B.; Lu, J.; Li, D.; Tao, W. Design and analysis of a novel cascade control algorithm for braking-by-wire system based on electromagnetic direct-drive valves. *J. Frankl. Ins.* **2022**, *9*, 6. [CrossRef]

 *actuators*

Article

A Numerical Study on the Transient Injection Characteristics of Gas Fuel Injection Devices for Direct-Injection Engines

Tianbo Wang, Hongchen Wang, Lanchun Zhang *, Yan Zheng, Li Li, Jing Chen and Wu Gong

School of Automotive and Traffic Engineering, Jiangsu University of Technology, Changzhou 213001, China
* Correspondence: zlc@jsut.edu.cn

Abstract: Natural gas has emerged as one of the preferred alternative fuels for vehicles owing to its advantages of abundant reserves, cleaner combustion and lower cost. At present, the gas supply methods for natural-gas engines are mainly port fuel injection (PFI) and direct injection (DI). The transient injection characteristics of a gas fuel injection device, as the terminal executive component of the PFI or DI mode, will directly affect the key performance of a gas fuel engine. Therefore, gas fuel injection devices have been selected as the research object of this paper, with a focus on the transient injection process. To explore the impacts of valve vibration amplitude, period, frequency and velocity on transient injection characteristics, one transient computational fluid dynamics (CFD) model for gas fuel injection devices was established. The findings thereof demonstrated that there is a linear relationship between the instantaneous mass flow rate and instantaneous lift during the vibration process. However, this relationship is somewhat impacted when the valve speed is high enough. A shorter valve vibration period tends to preclude a shorter period of flow-hysteresis fluctuation. The near-field pressure fluctuation at the throat of an injection device, caused by valve vibration, initiates flow fluctuation.

Keywords: gas fuel; gas fuel injection device; valve vibration; transient injection characteristics; flow hysteresis

Citation: Wang, T.; Wang, H.; Zhang, L.; Zheng, Y.; Li, L.; Chen, J.; Gong, W. A Numerical Study on the Transient Injection Characteristics of Gas Fuel Injection Devices for Direct-Injection Engines. *Actuators* **2023**, *12*, 102. https://doi.org/10.3390/act12030102

Academic Editors: Paolo Mercorelli and Ioan Ursu

Received: 7 December 2022
Revised: 16 February 2023
Accepted: 22 February 2023
Published: 25 February 2023

1. Introduction

The need for alternative fuels for conventional internal combustion engines (ICEs) has grown as a result of depletion of crude oil reserves, global warming and more rigorous emission regulations. Natural gas (NG), generally containing more than 90% methane, is considered one of the most promising alternative fuels owing to its advantages of abundant reserves, cleaner combustion and lower cost [1]. According to the International Energy Agency, the importance of NG can be compared with that of gasoline in terms of power generation, transportation, etc. [2]. Driven by current sustainability development policies, over one-quarter of global power generation is provided by NG [3]. In terms of road transport, NG has replaced other alternative fuels as the preferred choice for vehicles [4].

In recent years, natural-gas vehicles (NGVs) have been vigorously developed and employed around the world [5]. Until 2019, there were more than 28.5 million NGVs worldwide (including all land-based motor vehicles, from two-wheeled to off-road). Asian countries led the world in this respect with 20.5 million NGVs, followed by Latin American countries with 5.4 million NGVs [6]. In particular, NG is the main alternative fuel for long-haul and heavy trucks, such as long-distance transport vehicles and fleets that require centralized refueling [4].

The fuel supply system plays an important role in the performance of NG-fueled engines. Based on the supply mode, fuel supply systems can be classified as out-cylinder premixed method, port fuel injection (PFI) and direct injection (DI). The premixed method, which is the oldest NG supply method, works through the incorporation of a Venturi tube [7]. NG is premixed with air in the Venturi tube, and then the mixture is drawn into the cylinder because of the in-cylinder subatmospheric pressure during intake. This method leads to great power reduction, although it has the best mixing uniformity. The PFI mode can be divided into single-point injection (SPI) and multipoint injection (MPI). For the former mode, NG is injected near the intake manifold near the gas injector, while for the latter, NG is injected into the intake port of each cylinder, closer to the cylinder than in SPI. At present, the PFI mode is the most common fuel supply system for compressed natural gas (CNG) spark-ignition engines. However, it will lead to torque cutting due to reductions in volume efficiency. In contrast, for the DI mode, this problem can be avoided via injecting gaseous fuel after the intake-valve closing time (IVC).

The transient injection characteristics of a gas fuel injection device, as the terminal executive component of the PFI or DI mode, will affect the key performance of an NG engine directly. One outward-opening NG injector, driven with a piezo actuator, was developed by Siemens under support from the New Integrated Combustion System for Future Passenger Car Engines project (during the period of 2004 to 2012) [8,9]. Considering that the piezo actuator's stroke was rather small, a hydraulic stroke amplifier unit was designed to achieve the desired needle lift (about 200 μm) and mass flow rate. Based on this injector, Baratta et al. [10] used planar laser-induced fluorescence (PLIF) and the computational fluid dynamics (CFD) method to investigate the differences of transient NG jet patterns between different engine cycles. The NG jets were identified as having "cloud-like" and "umbrella-like" shapes.

In recent years, research of NG injectors (or NG injection devices) mainly focused on electromagnetic characteristics, structural design and structural optimization based on steady internal-flow characteristics [11–14]. A novel cascade control algorithm was used by Tan et al. to achieve multiobjective optimization of control parameters for electromagnetic linear actuators [11]. One new sensorless electronic closed-loop antibounce solution was proposed by Glasmachers et al. to effectively reduce bouncing and provide robust, soft landing for fuel injectors [12]. Single-shot X-ray radiography was used by Swantek et al. to investigate the steady-state behavior of an outward-opening gas injector [13]. Several CFD simulations of steady-state gas flow through various poppet-valve geometries were performed by Kim et al. to suggest design improvements for obtaining more efficient poppet valves with reduced stagnation pressure loss [14].

The transient injection characteristics of the gas fuel injector have, however, received little attention. The transient gas flow development of one outward-opening injector was studied by Deshmukh et al. [15,16] with the large eddy simulation (LES) method. The research results showed that an NG jet's features (such as axial penetration length, maximum jet width and volume) and subsequent mixture-formation process would be seriously affected by the transient motion process of an injector. In addition, there would be about a 30% disparity between the data of an LES simulation and the corresponding experiment if the motion process of an NG injector was ignored. One high-pressure NG injector was developed by Rogers et al. [17], based on BOSCH's gasoline injector, and the transient injection process of this NG injector was studied with the particle image velocimetry method. It was found that the injector would have a large degree of seating bounce when its valve was seated at the valve seat or the upper stop position because of the undesirable force characteristics of a traditional solenoid. A solenoid is generally characterized with greater force at both ends of the stroke, and it is difficult to achieve accurate displacement control with it. The maximum bouncing lift was almost 80% larger than the expected stable lift. In addition, the duration of the seating bounce was about 16% of the whole injection-pulse width of the NG injector [17]. It is obvious that these on/off transition processes of NG injectors will play a vital role in the process of gas fuel injection and subsequent in-cylinder mixture formation.

However, this research of the effects of transient valve motion on gas jet patterns is not comprehensive enough because the effects of key factors (such as injector-valve vibration magnitude, period and vibration time) have not been analyzed. As had been discussed in the literature cited above [11–14], researchers were accustomed to trying to reduce the seating impact of the nozzle valve in order to improve the control accuracy of the fuel supply of the gas fuel engine. Problems such as how much impact the valve-seating-bounce process has on the transient injection mass flow rate and which factor among vibration magnitude, period and vibration time has the greatest impact on jet pattern have not been discussed. In this paper, the affecting laws of these factors will be given focus. In order to explore the effects of injector-valve transient vibration on transient injection characteristics, one transient CFD model of an NG fuel injection device of the DI mode is proposed. Factors such as valve vibration amplitude, period, frequency and velocity are discussed. Finally, the CFD model is validated in both the aspects of transient gas fuel jet pattern and cumulative mass flow rate. The research results of this paper will provide guidance for future research on seating control of gas fuel injection devices.

2. CFD Model and Validation

2.1. Cases and CFD Model

At present, most gas fuel (mainly hydrogen, NG) injectors are driven with solenoids [11–17], resulting in valve seating problems, even if a piezo actuator is used [8–10]. One new type of NG injector [18] was designed by authors in previous studies to solve the seating problem of traditional gas fuel injectors. A moving-coil electromagnetic linear actuator (MCELA) was used as the driving aspect. It has a longer stroke, higher power density and better controllability than does a solenoid. Therefore, in theory, this new NG injector could achieve less rebound during seating process.

However, a softer seating advantage is usually achieved through closed-loop control, and the control signal is usually given via a displacement sensor. The scheme (working with a displacement sensor) may not be suitable for the DI supply mode because the installation volume limitation for injectors with the DI mode is very strict. Therefore, open-loop control was still used for the NG injector with the DI mode, but without the displacement sensor. In addition, a new sensorless soft-landing control strategy for improving the dynamic performance and fatigue life of NG injectors was proposed [19], as shown in Figure 1a. It was found that there was still a certain seating vibration even with the soft-landing control strategy.

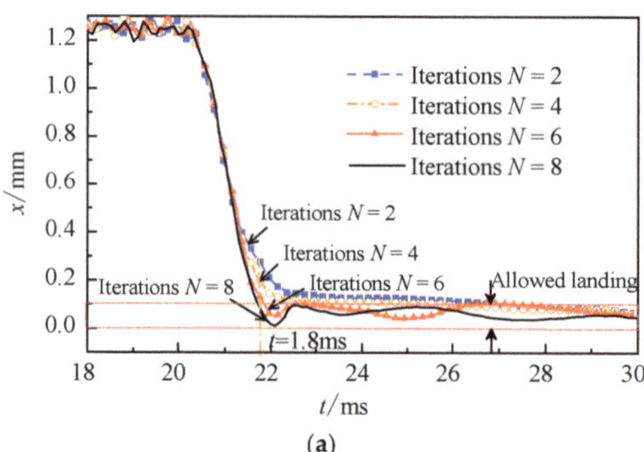

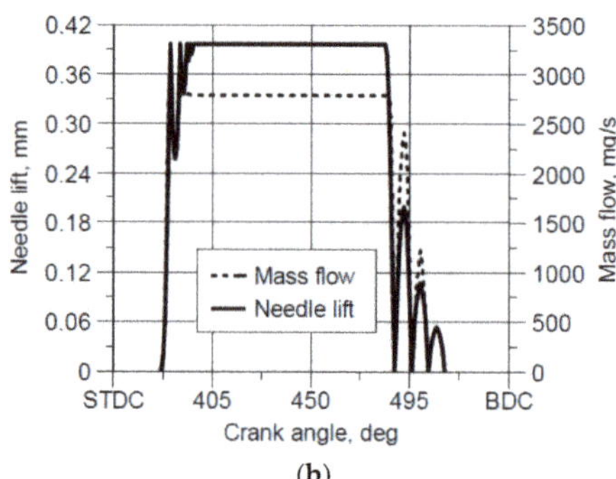

Figure 1. (a) Seating vibration of an injector driven with an MCELA (under the sensorless soft-landing control strategy). (b) Seating vibration of an injector driven with a solenoid [20].

On the other hand, the typical needle (or valve) lift curve of a conventional solenoid-driven injector, given by Bosch GmbH, is shown in Figure 1b [20]. In addition, the gas mass flow rate of this injector was calculated using the equation from Saint-Venant.

In this study, the effects of injector-valve lift-vibration amplitude, period and cycle time on the transient outflow rate and jet pattern are presented using three-dimensional CFD simulation software Fluent, based on the structure of the new NG injector proposed in our previous study [21]. The main dimensions of the NG injector and the applicable large-bore NG engine operating conditions are listed in Tables 1 and 2, respectively. Vibration amplitude is defined as the maximum vibration lift, the vibration period is the time in which the valve experiences one vibration cycle after seating or reaching the maximum lift and vibration-cycle time is the number of vibrations experienced before the valve stabilizes. Considering the measured lift data of an injector driven with an MCELA and an injector driven with a solenoid, case 2 was taken as the base case when the vibration lift was half of the maximum valve lift (1.4 mm), the vibration period was 0.4 ms and the vibration-cycle time was 1. In order to discuss the effect of vibration amplitude, amplitudes were taken as 25%, 50%, 75% and 100% of the maximum lift, respectively. In addition, in order to discuss the effect of the vibration period, periods were taken as 0.2 ms, 0.4 ms, 0.6 ms and 0.8 ms, respectively. Finally, based on case 2, the cycle times were set as 1, 2, 3 and 4, respectively, to discuss the effect of vibration-cycle time, as shown in Table 3 and Figure 2. The on/off process of the NG injector needed transition time, which was set at 1 ms for every case.

Table 1. Specifications of NG injector.

Injector Parameter	Value
Outlet Diameter (mm)	7
Valve Lift (mm)	1.5
Injection Pressure (MPa)	1.0
Injection Duration (CA)	54.5°CA (at 255 kW and 1900 rpm) [21]

Table 2. Specifications of engine.

Parameter	Value
Bore (mm) × Stroke (mm)	131 × 155
Displacement Volume (L)	12.53
Compression Ratio	11.5
Rated Power (kW)/Speed (rpm)	255/1900
IVO/IVC(CA)	30°BTDC/46°ABDC
EVO/EVC(CA)	78°BBDC/30°ATDC

Table 3. Parameters of each case.

	Vibration Amplitude	Vibration Period	Vibration Cycle
Case 1	25%	0.4 ms	1
Case 2	50%	0.4 ms	1
Case 3	75%	0.4 ms	1
Case 4	100%	0.4 ms	1
Case 5	50%	0.2 ms	1
Case 6	50%	0.6 ms	1
Case 7	50%	0.8 ms	1
Case 8	50%	0.4 ms	2
Case 9	50%	0.4 ms	3
Case 10	50%	0.4 ms	4

The transient CFD model of the NG injector is shown in Figure 3. The inlet and outlet boundaries of the model were both set as pressure boundary conditions. The injection pressure was set as 1.0 MPa to reduce the manufacturing-accuracy requirement of the injector and to make full use of the gas fuel in the tank [21]. A value of 1.0 MPa was determined according to the stagnation state formula and the in-cylinder pressure calculated from the engine model built in the authors' preliminary study, and this pressure could ensure that the injection flow rate was independent from the in-cylinder pressure (back pressure) [21]. The outlet pressure was set as the standard atmospheric pressure.

The natural gas used in this article was assumed to be 100% ideal methane for simplicity. Considering the characteristics of a supersonic jet and the great pressure/velocity gradient near the injector's throat position, the near-field meshes were refined. The minimum mesh size was 0.05 mm. In addition, considering the CFD calculation cost, the mesh size was gradually increased at the position farthest from the throat. In addition, the maximum mesh size was about 1 mm. The whole domain was assumed to be initially quiescent. The RNG k–ε turbulence model and the nonequilibrium wall function were used in this study. The turbulent Schmidt number took the fixed default value of 0.7. The coefficient $C_{1\varepsilon}$, used in the ε equation, took the value of 1.42; coefficient $C_{2\varepsilon}$ took the value of 1.68; and coefficient C_μ took the value of 0.0845. This program was based on the pressure-correction method and used the PISO algorithm. The first-order upwind differencing scheme was used for the momentum, energy and turbulence equations.

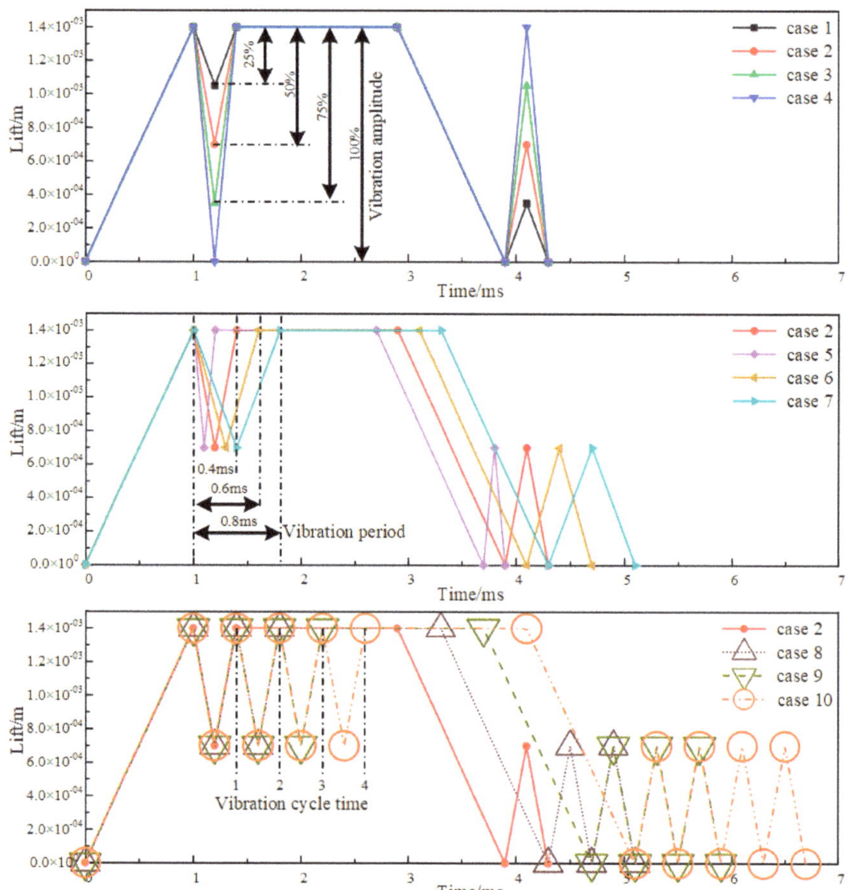

Figure 2. Lift curve of each case.

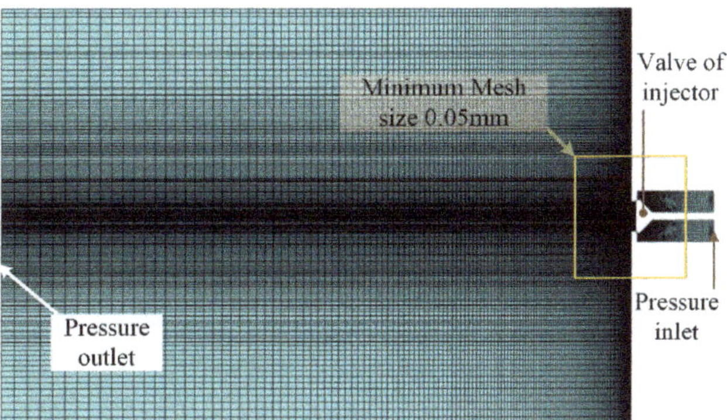

Figure 3. CFD calculation domain and mesh.

2.2. Transient CFD Model Verification

2.2.1. Verification with Cumulative Flow Rate

In order to test the steady volumetric flow rate of an NG injection device, one steady flow measurement bench, as shown in Figure 4, had been established previously by the authors of this paper [18]. In addition, the simulation accuracy of the steady CFD model of an NG injection device was verified based on this bench. However, its measuring pressure range was just 0.02–0.05 MPa, which is much lower than the injection pressure of the injector for the DI mode (gas fuel supply pressure is usually higher than 1 MPa). In order to meet the need for higher pressure, the spring of the pressure relief valve in the bench was replaced with a stronger spring, and the measured pressure range was expanded to 0.1–0.28 MPa. The measurement accuracy of the vortex flowmeter was ±1.0% of the measurement range. What is more, it was impossible to directly measure the cumulative flow rate of a single working cycle (from opening time to closing time) because the installation position of the flowmeter was far from the NG injection device. Therefore, the converted measured cumulative flow rate of a single cycle was taken to compare to the simulated results of a transient CFD injection model. This converted value comes from the measured cumulative flow rate during several injection cycles (for example, 5 min). To eliminate the effects of the selected number of injection cycles, both the 5 min and 12 h cases were discussed. For the sake of safety and convenience, compressed air rather than NG was used as the fluid. The gas supply pressure of the bench was adjusted to 0.1, 0.12, 0.14, 0.16 and 0.2 MPa respectively. The flow measurement was carried out at about 293 K, and the compressed air could be guaranteed to be gaseous under pressures lower than 0.2 MPa. The working-cycle durations of 40 ms, 60 ms and 120 ms were taken into consideration, corresponding to the engine speeds of 3000 RPM, 2000 RPM and 1000 RPM, respectively. The injection duration time was set as one quarter of the working-cycle duration.

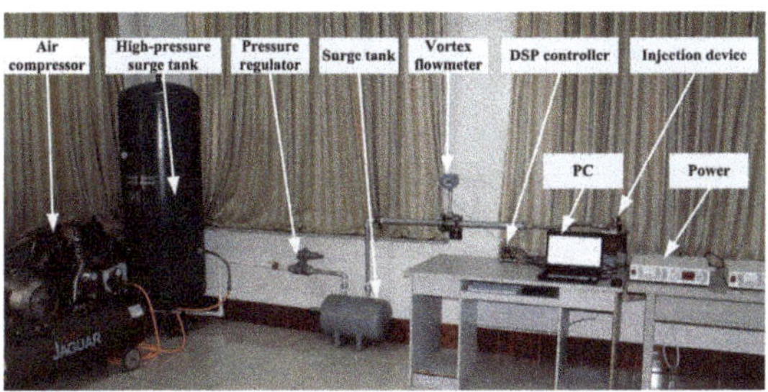

Figure 4. Flow measurement bench of the injector.

The simulated single-cycle flow rate was compared with the converted experimental one, as shown in Figure 5. It was easily found that the difference between the long-term (12 h) converted experimental flow rate and the simulation results was less than that between the short-term (5 min) rate and the simulated one. The authors' expectation was that the difference (caused by some random factors, such as upstream pressure fluctuation and data acquisition errors) between the converted experimental flow rate and the simulation results would be reduced as the cumulative injection time was extended. There was no obvious rule in the relationship between calculation errors and the upstream pressure of the injector. The maximum error, which was about 6.1%, appeared at an engine speed of 1000 RPM and an inlet pressure of 1.0 bar. In summary, the flow rate characteristics of the transient injection CFD model of the NG injector were verified from the perspective of cumulative single-cycle flow.

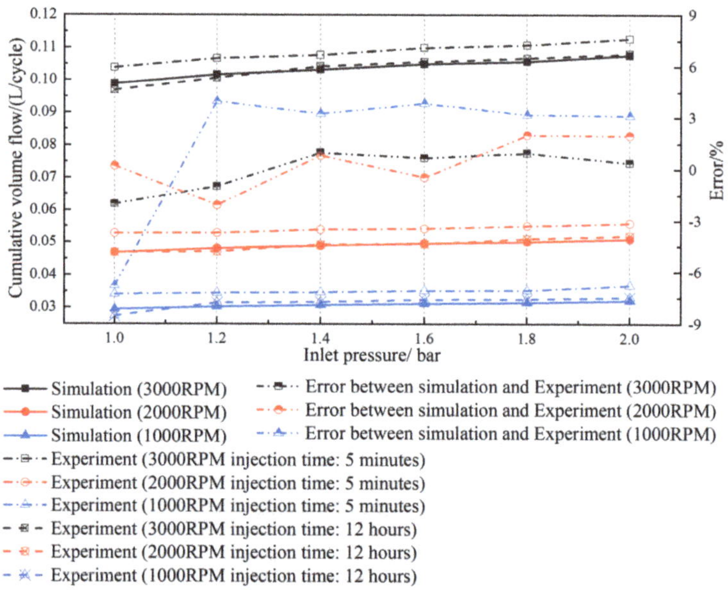

Figure 5. Comparison of single conversion and single-cycle transient volume flow.

2.2.2. Verification with Gas Jet Pattern

For the CFD simulation of the NG injector's transient injection process, it was most important to verify the calculation accuracy of the gas jet development process downstream of the injector's throat, although the accuracy of the jet flow rate was also rather important. To verify the ability to calculate the gas jet pattern downstream, the gas jet imaging results from Yosri et al. [22] were compared with the results of the corresponding transient CFD model in this paper, as shown in Figure 6. An NG injector prototype provided by the Continental company was used in the literature. The simulation results were postprocessed with the density gradient method. The valve lift curve took the measured valve lift when the pressure of the constant volume cavity (CVC) was 0.1 MPa [22]. The injector valve's vibration amplitude was about 25% of the maximum lift (0.4 mm) at the maximum lift position. In addition, the vibration amplitude was about 10% at the zero-lift position. The inlet pressure of the injector was set to 2.0 MPa, and the inlet temperature was 298 K. Methane (a surrogate for NG) was injected into the CVC with quiescent, nonreacting nitrogen. The CVC had an initial pressure of 0.1 MPa and an initial temperature of 298 K. Considering that accurate calculation of the jet development process requires the arrangement of 10–15 layers of grids on the cross-section of the throat [12], the mesh size was as small as 0.03 mm near the valve throat and up to 1 mm in other areas. It was found that the simulation results were rather consistent with the experimental gas jet imaging results both during the lifting process (222 µs after start of injection (ASOI)) and during the landing process (370 µs ASOI and 481 µs ASOI).

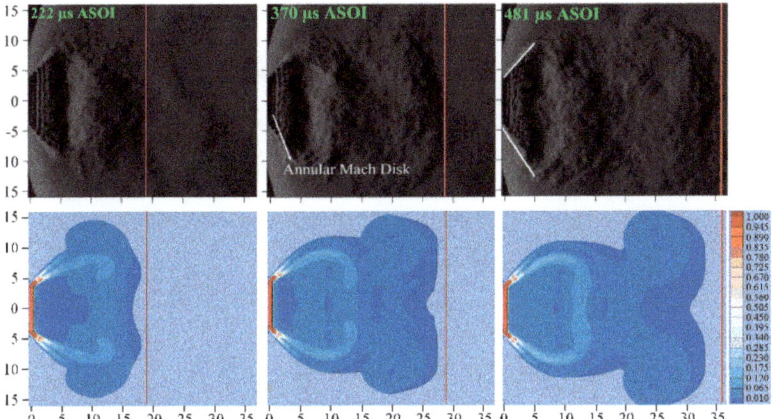

Figure 6. Instantaneous injection flow features during the injection, shown with Schlieren imaging [22] (first row) and simulation results using the density gradient method (second row).

3. Results and Discussion

The flow fluctuation after the end of the valve lift vibration (called delayed flow fluctuation (DFF), as shown in Figure 7a) is the most important focus of this study. DFF caused by lift vibration would have a serious impact on the calibration quantity of gas fuel under certain engine working conditions because DFF is usually uncontrollable. In order to facilitate a comparison, the ending times of the valve vibration processes for all discussed cases (as shown in Table 3 and Figure 2) were normalized to the ending times (1.4 ms) of the valve vibrations for cases 1–4. Only the DFF near the maximum valve lift will be discussed in this study, because there was no DFF near the zero-lift position.

Firstly, the effect of vibration amplitude was discussed as cases 1–4 were compared, and the NG flow-rate fluctuation processes through the injection device's throat in these four cases are presented in Figure 7. It was found that there was an almost linear correlation between the transient flow rate and instantaneous lift vibration during the lift-vibration process (from 1.0 ms to 1.4 ms), as shown in Figure 7a. Case 4 had the strongest DFF, and the strengths of the DFFs for case 3, case 2 and case 1 showed a gradual downward trend, as shown in Figure 7b. The cycle periods of the DFF for cases 1–4 were all the same. For a more comprehensive comparison, a case with no valve lift vibration (called no vibration in figure) was also taken into consideration. It was easily found that the case with no vibration had the weakest DFF and the shortest cycle period when compared with cases 1–4. It is obvious that higher vibration amplitude brings stronger DFF.

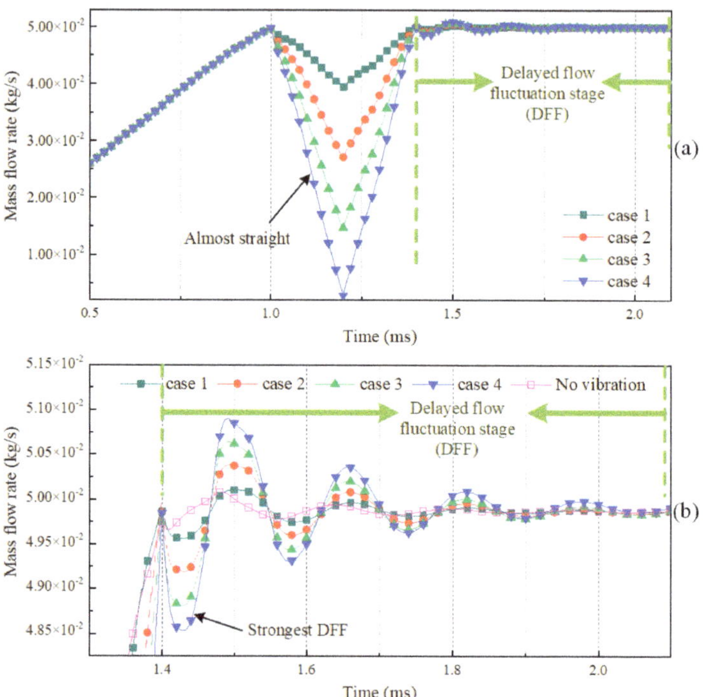

Figure 7. Effect of valve lift-vibration amplitude on the mass flow rate of the NG injector ((**b**) is the partial enlargement of (**a**)).

Secondly, the effect of the vibration period was discussed as cases 2 and 5–7 were compared, and the NG flow-rate fluctuation processes of these four cases are presented in Figure 8. The raw data during the lift-vibration process (for example, case 7: from 0.6 ms to 1.4 ms (normalized)) were almost the same as those when the vibration-amplitude factor was concerned. The linear relationships between the transient flow rate and instantaneous lift vibration still existed for case 6 and case 7. However, for case 5, which had the shortest vibration period, this linear relationship was somewhat broken, as shown in Figure 8a. It is obvious that the transient developing process of the supersonic gas jet was influenced by the faster velocity of valve vibration in case 5, and the transient mass flow rate was affected in turn.

When the vibration-period factor was considered, case 5 (period: 0.2 ms) had the strongest DFF, and the strengths of the DFFs for case 2 (period: 0.4 ms), case 7 (period: 0.8 ms) and case 6 (period: 0.6 ms) showed a gradual downward trend, as shown in Figure 8b. Case 5 had the longest DFF period, case 2 took the second place, case 7 took the third place and case 6 had the shortest value, as shown in Figure 8c. There seems to be some correlation among them: cases 2 and 5–7, which had shorter lift-vibration periods, had stronger DFFs and longer DFF periods. This rule is not applicable to case 6 or case 7, however. Actually, case 6, case 7 and the case with no vibration had some crossing areas when the case with no vibration was taken into consideration, as shown in Figure 8c. Their DFF strengths and periods were almost the same. Therefore, it can be concluded that the effect of valve vibration period on DFF will be rather little if the vibration period is long enough (the boundary was about 0.6 ms (case 6) in the example presented in this paper).

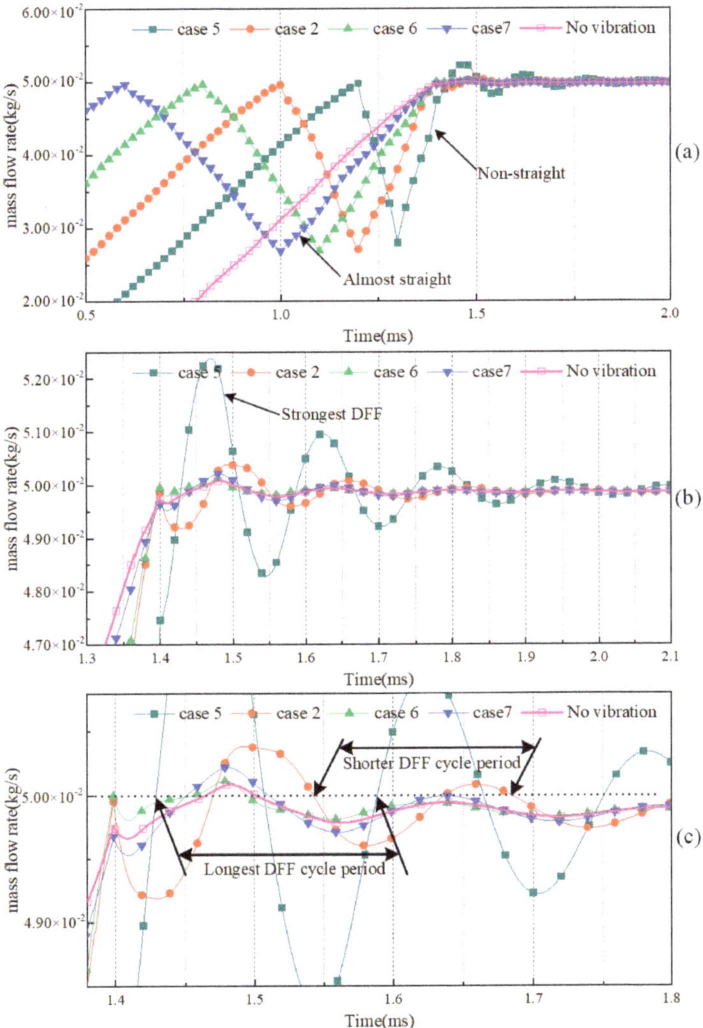

Figure 8. Effect of the valve lift-vibration period on the mass flow rate of the NG injector ((**b**,**c**) are the partial enlargements of (**a**)).

At last, the effect of vibration-cycle time was discussed as cases 2 and 8–10 were compared, as shown in Figure 9. It was found that the DFF of case 2 (cycle time: 1) was the strongest, and the DFFs of cases 8–10 were almost the same. What is more important is that the DFF strengths of the last three cases were slightly weaker (about 1.2‰ at the peak point of the flow rate) than that of case 2, as shown in Figure 9b. It can be concluded that more vibration time tends to bring weaker DFF, although the influence is quite little.

By now, the influences of injector-valve lift-vibration amplitude, period and cycle time on transient gas jet have been discussed in cases 1–10, above. It was interesting to find that case 4 had the same valve vibration speed (slope of valve lift-vibration curve) as case 5, as shown in Figure 2. This situation also existed for case 1 and case 7. Therefore, it is necessary to discuss these two cases further. Their valve lifts and corresponding mass flow rates are presented in Figures 10 and 11, respectively. It can be seen that case 5, which had a smaller lift-vibration amplitude, shows a stronger DFF than case 4, and that case 1, which had a smaller lift-vibration amplitude, shows a weaker DFF than case 7. These seem to lead to quite different conclusions. However, it is worth noting that the DFF strengths of case 1 and case 7 are actually almost the same as each other. The biggest difference between case 1 and case 7 is smaller than 4‰, and these two cases' DFFs were almost the same as that of the case with no vibration, as shown in Figures 7 and 8, because their lift-vibration speeds were rather slow (that is: their vibration periods were long enough). Shorter vibration tends to cause greater DFF when the vibration speed is the same.

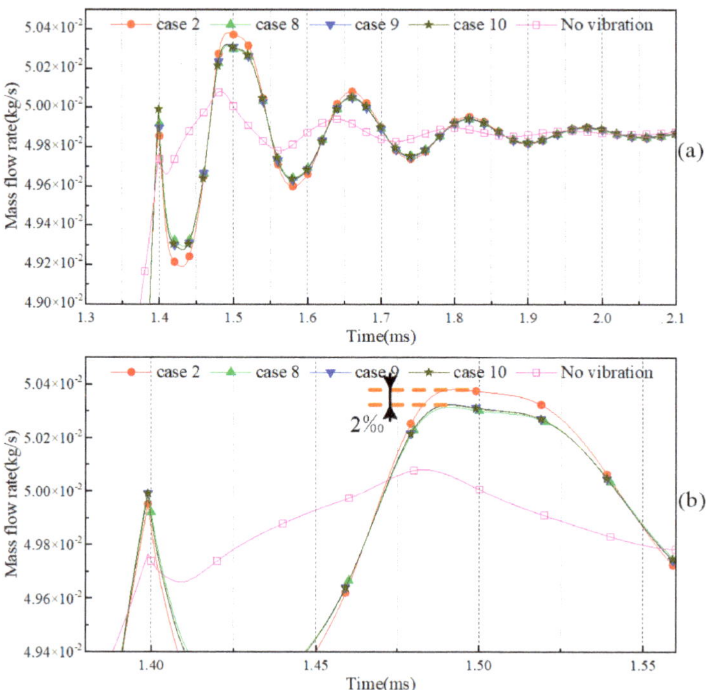

Figure 9. Effect of cycle time of valve lift vibration on the mass flow rate of the NG injector ((**b**) is the partial enlargement of (**a**)).

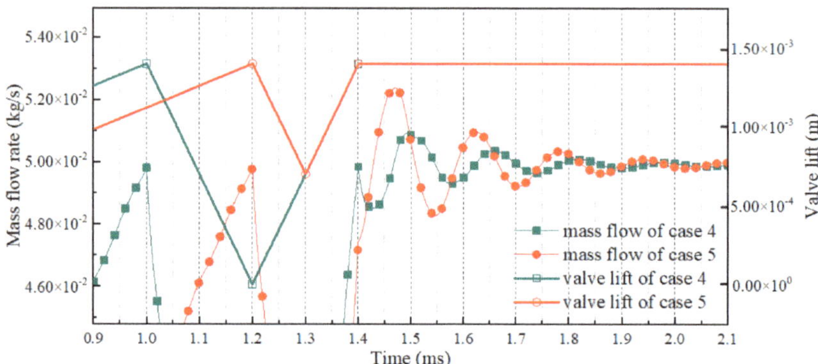

Figure 10. Valve lift vibrations and mass flow rates of case 4 and case 5 (they had the same valve vibration speed).

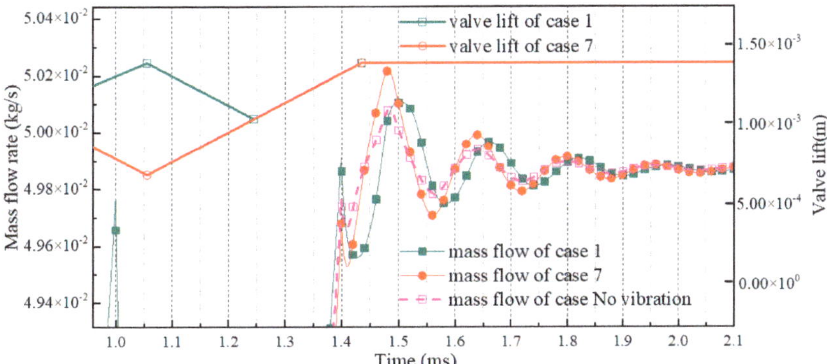

Figure 11. Valve lift vibrations and mass flow rates of case 1 and case 7 (they had the same valve vibration speed).

According to the mass-flow-rate formula of one-dimensional isentropic flow [17], the flow fluctuation of a jet nozzle generally results from pressure-ratio (ratio of pressure downstream of the nozzle throat and pressure upstream of the nozzle throat) fluctuation. Therefore, in order to reveal the reason for DFF in this paper, the pressure fluctuation situations downstream and upstream of the NG injection device's throat are presented in Figure 12. Based on the mass flow-rate fluctuation curves (Figures 7–9) after the end the of valve lift vibration, three peak points and three valley points of each fluctuation curve were selected as the inner pressure analyzing points. Firstly, the throat near-field pressure of the case with no vibration was presented because the DFF of this case could not come from valve lift vibration. The analyzing points were 1.009 ms, 1.082 ms, 1.162 ms, 1.239 ms, 1.320 ms and 1.400 ms. Secondly, as discussed above, case 2 was used as a base case for comparison, and its near-field pressure needed to be presented. The analyzing points were 1.430 ms, 1.492 ms, 1.580 ms, 1.657 ms, 1.741 ms and 1.820 ms.

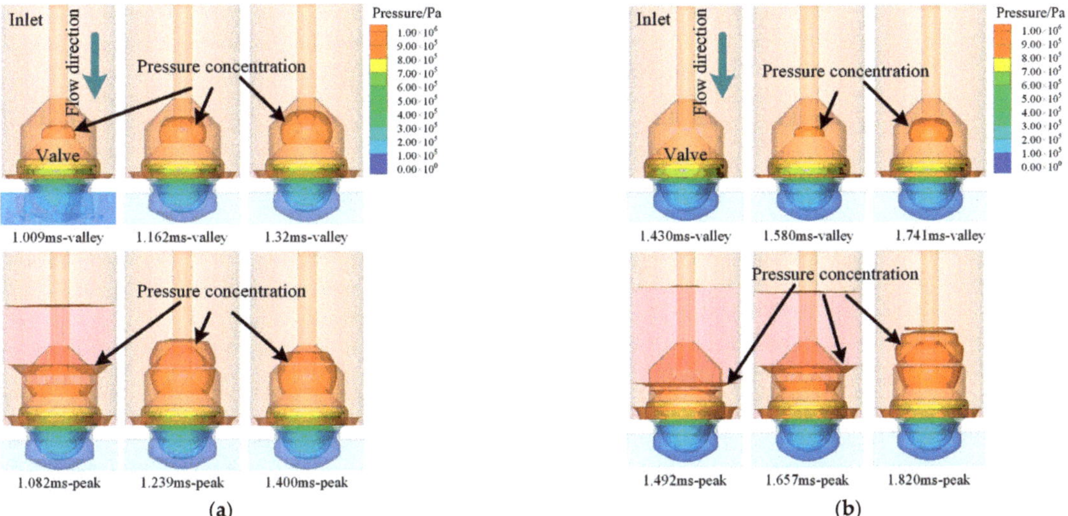

Figure 12. (**a**) Throat near-field pressure distribution of the case with no vibration (left). (**b**) Throat near-field pressure distribution of cases with vibration (case 2, for example; right).

It could be intuitively obtained from the near-field pressure distribution that a smaller pressure-concentration zone is formed on the back face of an injector's valve for the valley point (for example: case with no vibration, 1.162 ms-valley) when compared with that for the nearby peak points (for example: case with no vibration, 1.082 ms-peak and 1.239 ms-peak). In addition, the volume of the pressure concentration area tends to increase as injection time goes by. On the other hand, for the peak point (for example: case 2, 1.657 ms-peak), a larger-pressure concentration area is formed when compared with that for nearby valley points (for example: case 2, 1.580 ms-valley and 1.741 ms-valley), and the volume tends to decrease. In addition, the differences of pressure distribution between nearby valley and peak points tend to be gradually smaller. For example, the difference between the 1.320 ms point and the 1.400 ms point was much smaller than that between the 1.162 ms point and the 1.239 ms point. This rule existed for both the case with no vibration and case 2. This gradually decreased pressure difference leads to a smaller amplitude of DFF, and so the gas jet would finally be stabilized. It was also found that the difference between the nearby valley and peak points of the case with no vibration was smaller than that of case 2, resulting in a lower amplitude of DFF in the former case, as shown in Figure 7b.

4. Conclusions

The transient CFD model of a DI NG injector's transient injection process was established in this paper. Based on this CFD model, the influences of injector valve vibration amplitude, period, cycle time and velocity on transient injection characteristics were investigated. Flow fluctuation after the end of valve lift vibration was a focus. The main conclusions are as follows:

There is an almost linear correlation between the transient flow rate and instantaneous lift vibration during the lift-vibration process (from 1.0 ms to 1.4 ms for cases 1–4), and this linear correlation will be somewhat broken if the valve's vibration velocity is high enough or the vibration period is short enough.

A higher value of vibration amplitude tends to bring stronger DFF. In studying the seating control of gas fuel injectors, efforts should be made to reduce valve lift-vibration amplitude because it has a significant impact on flow during and after the injection period.

Under the condition of the same amplitude, a shorter valve vibration period means stronger DFF and a longer period of DFF. In addition, the effect of the valve vibration period on DFF is rather little if the vibration period is long enough (the boundary was about 0.6 ms in the example presented in this paper).

The influence of vibration-cycle time on DFF is quite little. More vibration time tends to bring weaker DFF. When studying the seating control of gas fuel injectors, we should try to avoid too-fast valve lift-vibration speed. However, vibration-cycle time does not need too much attention when only DFF is considered.

The throat near-field pressure fluctuation caused by valve lift vibration results in DFF.

Author Contributions: Methodology and software, T.W. and Y.Z.; verification, H.W. and L.Z.; resources, L.Z. and J.C.; writing—original draft preparation, W.G., L.L. and H.W. All authors have read and agreed to the published version of the manuscript.

Funding: This work was supported by the National Natural Science Foundation of China (Grants No. 52105260 and. 11802108), the Changzhou Sci & Tech Program (Grant No. CE20225049), the Natural Science Research Project of Higher Education Institutions in Jiangsu Province (Grants No. 21KJB460008 and 22KJA580002) and the Qinglan Engineering Project of Jiangsu Universities. The APC was funded by the National Natural Science Foundation of China, the Changzhou Sci & Tech Program and the Natural Science Research Project of Higher Education Institutions in Jiangsu Province (Grant No. 21KJB460008).

Data Availability Statement: Not applicable.

Conflicts of Interest: The authors declare no conflict of interest.

Nomenclature

ASOI	After start of injection
CFD	Computational fluid dynamics
CNG	Compressed natural gas
CVC	Constant volume cavity
DFF	Delayed flow fluctuation
DI	Direct injection
ICE	Internal combustion engine
IVC	Intake-valve closing time
LES	Large eddy simulation
MCELA	Moving-coil electromagnetic linear actuator
NG	Natural gas
NGV	Natural-gas vehicle
PFI	Port fuel injection
PLIF	Planar laser-induced fluorescence

References

1. Moon, S. Potential of direct-injection for the improvement of homogeneous-charge combustion in spark-ignition natural gas engines. *Therm. Eng.* **2018**, *136*, 41–48. [CrossRef]
2. IEA. The Contribution of Natural Gas Vehicles to Sustainable Transport. Available online: https://www.iea.org/reports/the-contribution-of-natural-gas-vehicles-to-sustainable-transport (accessed on 14 August 2022).
3. IEA. Gas. Available online: https://www.iea.org/fuels-and-technologies/gas (accessed on 14 August 2022).
4. Fadiran, G.; Sharma, T.; Rogan, F. Exploring a case transition to low carbon fuel: Scenarios for natural gas vehicles in Irish road freight. *SSRN Electron. J.* **2021**, 33. [CrossRef]
5. Honda. North American Environmental Report. 2015. Available online: https://csr.honda.com/environment/na-environmental-report/ (accessed on 14 August 2022).
6. Teoh, L.E.; Khoo, H.L. Analysis of natural gas vehicle acceptance behavior for Klang Valley, Malaysia. *Int. J. Sustain. Transp.* **2020**, *15*, 11–29. [CrossRef]
7. Bircann, R.; Kazour, Y.; Dauer, K.; Fujita, M. *Cold Performance Challenges with CNG PFI Injectors*; SAE Technical Paper 2013-01-0863; SAE: Warrendale, PA, USA, August 2013.
8. NICE Report Summary. Available online: http://cordis.europa.eu/publication/rcn/12052_en.html. (accessed on 31 August 2016).
9. Sankesh, D.; Lappas, P. *Natural-Gas Direct-Injection for Spark-Ignition Engines—A Review on Late-Injection Studies*; SAE Technical Paper 2017-26-0067; SAE: Warrendale, PA, USA, October 2017.
10. Baratta, M.; Andrea, E.; Pesce, F.C. Multidimensional modelling of natural gas jet and mixture formation in direct injection spark ignition engines—Development and validation of a virtual injector model. *J. Fluids Eng.* **2011**, *133*, 041304. [CrossRef]
11. Tan, C.; Ren, H.; Li, B.; Lu, J.; Li, D.; Tao, W. Design and analysis of a novel cascade control algorithm for braking-by-wire system based on electromagnetic direct-drive valves. *J. Franklin Inst.* **2022**, *359*, 8497–8521. [CrossRef]
12. Glasmachers, H.; Joachim, M.; Koch, A. *Sensorless Movement Control of Solenoid Fuel Injectors*; SAE Technical Paper 2006-01-0407; SAE: Warrendale, PA, USA, March 2006.
13. Swantek, A.B.; Duke, D.J. An experimental investigation of gas fuel injection with X-ray radiography. *Exp Therm. Fluid Sci.* **2017**, *87*, 15–29. [CrossRef]
14. Kim, G.H.; Allan, K.; Michell, C. Improvement of Poppet Valve Injection Performance in Large-Bore Natural Gas Engines. In Proceedings of the ASME 2004 Internal Combustion Engine Division Fall Technical Conference, Long Beach, CA, USA, 24–27 October 2004; Volume 37467.
15. Deshmukh, A.Y.; Bode, M. Simulation and Modeling of Direct Gas Injection through Poppet-type Outwardly opening Injectors in Internal Combustion Engines. In *Natural Gas Engines*; Srinivasan, K.K., Agarwal, A.K., Krishnan, S.R., Mulone, V., Eds.; Springer: Berlin/Heidelberg, Germany, 2019; pp. 65–115.
16. Deshmukh, A.Y.; Vishwanathan, G. Characterization of hollow cone gas jets in the context of direct gas injection in internal combustion engines. *SAE Int. J. Fuels Lubr.* **2018**, *11*, 353–377. [CrossRef]
17. Rogers, T.; Petersen, P. *Flow Characteristics of Compressed Natural Gas Delivery for Direct Injection Spark Ignition Engines*; SAE Technical Papers 2015-01-0002; SAE: Warrendale, PA, USA, March 2015.
18. Wang, T.B.; Zhang, L.C.; Chen, Q. Effect of Valve Opening Manner and Sealing Method on the Steady Injection Characteristic of Gas Fuel Injector. *Energies* **2020**, *13*, 1479. [CrossRef]
19. Tan, C.; Ge, W.; Li, B. A Novel Large-flow-rate Gas Fuel Injection Device with Sensorless Control. *Eng. Lett.* **2019**, *27*, 20.
20. Chiodi, M.; Berner, H.J. *Investigation on different Injection Strategies in a Direct-Injected Turbocharged CNG-Engine*; SAE Technical Paper 2006-01-3000; SAE: Warrendale, PA, USA, September 2006.

21. Wang, T.B.; Zhang, L.C.; Bei, S. Influence of injection valve opening manner and injection timing on mixing effect of direct injection compressed natural gas–fueled engine. *Int. J. Engine Res.* **2020**, *22*, 2244–2253. [CrossRef]
22. Yosri, M.R.; Ho, J.Z. Large-eddy simulation of methane direct injection using the full injector geometry. *Fuel* **2021**, *290*, 120019. [CrossRef]

actuators

Article

Design and Analysis of Electromagnetic Linear Actuation-Energy-Reclaiming Device Applied to a New-Type Energy-Reclaiming Suspension

Jianguo Dai, Lv Chang *, Youning Qin, Cheng Wang, Jianhui Zhu, Jun Zhu and Jingxuan Zhu

Faculty of Transportation Engineering, Huaiyin Institute of Technology, Huai'an 223003, China; djg619265809@hyit.edu.cn (J.D.)
* Correspondence: changlv@hyit.edu.cn

Abstract: In order to meet the increasing demand for high-performance and high-efficiency vehicles, this paper proposes a novel electromagnetic linear energy-reclaiming suspension technology based on the McPherson independent suspension, and analyzes its core component—ELA-ERD (Electromagnetic Linear Actuation Energy-Reclaiming Device). ELA-ERD, taking a shock absorber piston rod as the inner yoke, has a compact structure and reasonable layout by integrating the structural features of the suspension. In this paper, the design process of ELA-ERD is elaborated in detail. Aiming at the problem of over-saturation of the inner yoke magnetic density, this paper proposes a method to optimize the magnetic circuit by increasing the size of the inner yoke within the effective working area of the moving coil, thus effectively improving the electromagnetic characteristics of ELA-ERD. Moreover, the effect and potential of energy reclaiming on ELA-ERD were studied by using finite element software. The study on the energy-reclaiming law of ELA-ERD was carried out from the perspective of the changes in vibration frequency and amplitude. In addition, the internal relationship between the energy-reclaiming voltage and the vibration velocity was revealed in this work, and the energy-reclaiming voltage coefficient K_e was defined. Through calculation of a large amount of model data, the K_e value applicable to the designed ELA-ERD in this paper was approximately set to 4.5. This study lays an important theoretical foundation for the follow-up studies.

Keywords: energy reclaiming; ELA-ERD; McPherson independent suspension; electromagnetism; finite element

Citation: Dai, J.; Chang, L.; Qin, Y.; Wang, C.; Zhu, J.; Zhu, J.; Zhu, J. Design and Analysis of Electromagnetic Linear Actuation-Energy-Reclaiming Device Applied to a New-Type Energy-Reclaiming Suspension. *Actuators* **2023**, *12*, 142. https://doi.org/10.3390/act12040142

Academic Editor: Takeshi Mizuno

Received: 21 February 2023
Revised: 19 March 2023
Accepted: 23 March 2023
Published: 27 March 2023

1. Introduction

The suspension system is the key part of the vehicle as well as an important device to ensure smooth running and stable operation of the vehicle. However, traditional suspension can only passively reduce vibration, which is far from meeting the increasing demand of high performance and high energy efficiency in the vehicle industry with rapid development momentum. To this end, active suspension and energy-reclaiming suspension technology is gradually becoming a research hotspot [1–3].

Active suspension has the advantages of controlling the height of the vehicle body, improving the vehicle passability, and guaranteeing the operation smoothness and stability of the vehicle [4,5]. However, active suspension requires extra energy input, which brings additional energy consumption to the vehicle and is not conducive to the improvement of vehicle energy efficiency. Under the current development of the electric-driven vehicle industry, the improvement of battery and charge-discharge technologies, as well as the vehicle energy efficiency, is of great importance. Professor Yu fan from Shanghai Jiao Tong University and his team conducted an in-depth study on the energy consumption of passive suspension and active suspension, which confirmed the high energy consumption of active suspension, and indicated the necessity of the research on energy-reclaiming suspension technology [6].

Energy conservation and environmental protection have become important themes in the development of vehicle technology. In this context, the energy-reclaiming technology of suspension has attracted more and more attention [7,8]. An energy-reclaiming suspension collects and stores kinetic energy during suspension vibration so as to improve the efficiency of vehicles. In addition, combining with active suspension, energy-reclaiming suspension can reduce the energy loss of active suspension while satisfying active control, thus further improving the comprehensive performance of the vehicle [9–11].

Since the 1970s, energy-reclaiming suspension technology has been developed from theoretical research to product application. This technology is becoming increasingly mature and the product has become diversified, but there is still a long way to go before realizing actual popularization and application of this technology [12–14]. Currently, there are several different types of energy-reclaiming suspension, including the piezoelectric energy storage type, the hydraulic energy storage type and the electromagnetic energy storage type. With the continuous improvement of the electromagnetic theory and the improvement of the performance of permanent magnet materials and high-power electronic devices, the electromagnetic energy storage type has become the most promising one. Dr. Amara G Bose, the founder of Bose Corp, MA, USA., studied the optimization of the suspension system as early as 1980, and demonstrated by theoretical analysis that electromagnetism was an effective way to achieve the desired suspension performance [15].

For the electromagnetic energy storage type, although the linear motor has less satisfactory energy-reclaiming power than rotary motor [16], it has compact structure and great advantages in motion transformation [17]. At present, the research of electromagnetic linear energy-reclaiming suspension mainly focuses on the energy-reclaiming device, controller, energy-reclaiming circuit and structure arrangement, etc. The research goal is to improve the energy efficiency and comprehensive performance of the suspension system as much as possible. Among them, the research on the energy-reclaiming device is the core, and the quality of the energy-reclaiming device will directly affect the performance of the electromagnetic linear energy-reclaiming suspension. Some scholars have put forward various working modes for electromagnetic linear energy-reclaiming device and achieved outstanding results, such as professor Gysen from Eindhoven University of Technology in the Netherlands [18–20], professor Vijayakumara, P B from India [21], professor Zhaoxiang Deng from Chongqing University [22,23], etc. However, there are actually few studies on the electromagnetic linear energy-reclaiming device. Even for the Bose suspension system, a successful model of electromagnetic linear energy-reclaiming suspension, the data related to its actuator have not been published yet.

It can be seen that the technical research of the electromagnetic linear energy-reclaiming suspension is still in the preliminary exploration stage, and there is still a certain distance from the actual application. Nevertheless, it still has high research value and a broad application prospect owing to its high-efficiency motion conversion, compact structure arrangement and the relatively light weight of the electromagnetic linear motor.

In view of the above studies, this paper proposes a new-type electromagnetic linear energy-reclaiming suspension technology scheme based on McPherson independent suspension. It embeds the electromagnetic linear actuation—energy-reclaiming device (ELA-ERD) device to realize active control and passive energy feed of suspension, showing advantages of compact structure, easy modification and high reliability. This paper focuses on the design, simulation and optimization of ELA-ERD. Through analysis and comparison of simulation data, this study deeply explores the energy-feeding characteristics and laws of ELA-ERD, providing an important theoretical and practical basis for the development of electromagnetic suspension technology.

2. Overall Design of the New Suspension

The new electromagnetic linear energy-reclaiming suspension proposed in this paper is based on the McPherson independent suspension structure, which has the advantages of simple structure, high reliability and low cost [24]. The design based on the McPherson independent suspension structure is beneficial to the application and promotion of the new suspension scheme.

2.1. Suspension Scheme

Figure 1 shows the structure of the new energy-reclaiming suspension. Unlike the traditional passive suspension, the electromagnetic linear energy-reclaiming suspension is equipped with an ELA-ERD at the upper end of the shock absorber to achieve active control and passive energy reclaiming. Compared with other similar active or energy-reclaiming suspension [25,26], the new suspension scheme proposed in this paper retains the shock absorber, and realizes the excellent damping effect of the whole suspension by controlling the output size and direction of the electromagnetic force of the ELA-ERD. In this way, the features and advantages of the McPherson independent suspension can be maintained. Because the shock absorber is retained, the suspension system can operate as a traditional suspension device even if the ELA-ERD does not work or fails, thereby ensuring the reliability of the suspension.

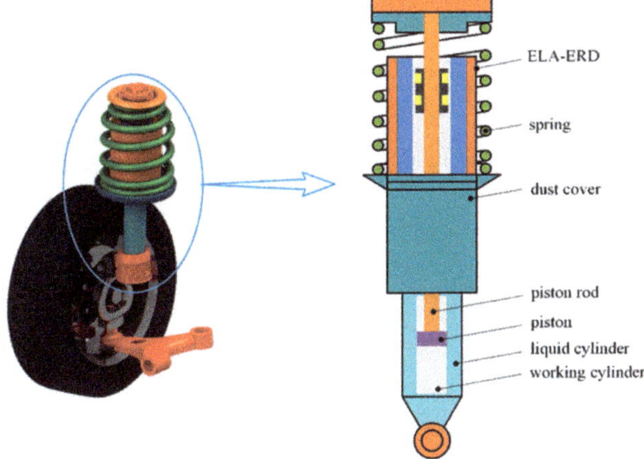

Figure 1. The new energy-reclaiming suspension structure.

The electromagnetic linear energy-reclaiming suspension can realize passive energy reclaiming and active control while maintaining the advantages of the McPherson independent suspension, which is mainly attributed to the ELA-ERD. It can be seen from Figure 1 that ELA-ERD is mounted on a suspension shock absorber, in series with the shock absorber, and in parallel with the spring. The piston rod of the damper is fixedly connected with the coil skeleton of the ELA-ERD. The piston rod also plays a role of inner yoke. When the shock absorber moves, the coils of ELA-ERD will move synchronously with the shock absorber piston rod. Such layout realizes organic integration of ELA-ERD and suspension structure, which is more compact and reasonable.

For existing electromagnetic energy-reclaiming suspension technology [27], the electromagnetic energy-reclaiming actuator is usually arranged inside the shock absorber, which increases the difficulty in assembly and later maintenance repair and leads to shock absorber jamming, as the permanent magnet easily falls off during vibration. The technical scheme proposed in this paper can effectively solve this problem without hollowing out the shock absorber piston rod, thus avoiding the reduction in piston rod strength.

2.2. Parameters Determination

The design of suspension should be based on vehicle parameters and application requirements. An existing commercial available model is adopted in this paper, whose basic parameters are shown in Table 1.

Table 1. Basic parameters of selected vehicle model.

Parameters	Value	Parameters	Value
Unladen mass	1220 kg	The load distributed by the front axle under no-load condition	60%
Wheelbase	2610 mm	Wheel specification	205/55 R16
Max power	81 kw	Max speed	188 km/h
Max torque	155 N·m	Displacement	1598 mL

After determining the vehicle parameters, the design and calculation of the suspension parameters is the basis for "Vehicle Design", which will not be presented in this paper.

The two main components of the McPherson independent suspension are the shock absorber and the spring. On the basis of determining the basic parameters of the vehicle, the dimensions and materials of the two components can be defined through calculation, model selection and verification. Figure 2 shows the size marking of the shock absorber and spring in suspension. The specific dimensions of the shock absorber and spring are shown in in Tables 2 and 3, respectively.

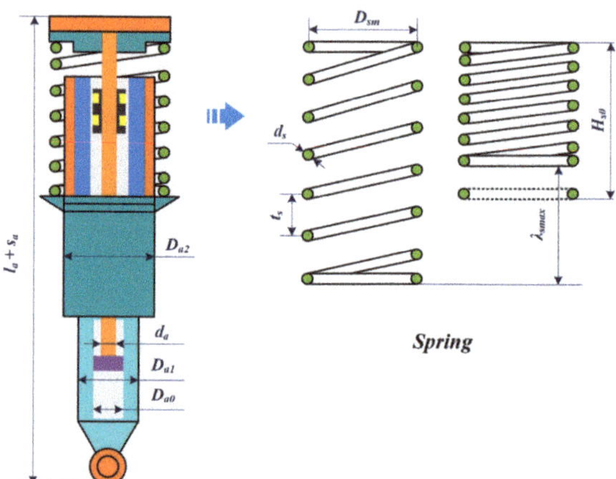

Figure 2. Size marking of the shock absorber and spring in suspension.

Table 2. Parameters of shock absorber.

Parameters	Value
Diameter of working Cylinder D_{a0}	20 mm
Base length l_a	80 mm
Oil tank diameter D_{a1}	34 mm
Piston stroke s_a	100 mm
Outer diameter of dust Cover D_{a2}	40 mm
Diameter of piston rod d_a	10 mm

Table 3. Parameters of spring.

Parameters	Value
Pitch diameter D_{sm}	112 mm
Steel wire diameter d_s	14 mm
Number of active coils n_s	10
Unsupported height H_{s0}	420 mm
Pitch of teeth t_s	39.2 mm
Max deflection λ_{smax}	201.6 mm

The working cylinder diameter D_{a0} of the shock absorber can be determined by Equation (1), prior to which the max unloading force F_{a0} must be calculated by Equation (2). From Equations (1) and (2), it can be known that the selection of the working cylinder diameter of the shock absorber is closely related to the vibration status and sprung mass of the vehicle.

$$D_{a0} = \sqrt{\frac{4F_{a0}}{\pi[p](1 \times \lambda_a^2)}} \tag{1}$$

where $[p]$ is the maximum allowable pressure of the working cylinder, which is set to 3.5 MPa; λ_a is the ratio of the connecting rod diameter to the cylinder diameter, which is set to $\lambda_a = 0.4$ for telescopic shock absorber.

$$F_{a0} = \delta v_{a0} = \delta A \sqrt{c/m_s} \cos \alpha \tag{2}$$

where v_{a0} is the unloading velocity, δ is the damping coefficient of shock absorber, A is the vibration amplitude of vehicle body, which is set to ± 40 mm, c is the suspension stiffness, m_s is the sprung mass, and α is the arrangement angle of shock absorber.

In order to meet the actual strength demand, the spring steel wire diameter d_s should meet:

$$d_s \geq 1.6 \sqrt{\frac{F_{s1} K C}{\tau_s}} \tag{3}$$

where τ_s is the allowable stress of spring, $\tau_s = 471$MPa. F_{s1} is the axial load on the spring; C is the spring index, which is set to $C = 8$ in this paper; K is the curvature correction factor, and the relation between K and C is shown as follow:

$$K = \frac{4C - 1}{4C - 4} + \frac{0.615}{C} \tag{4}$$

After calculation and verification, the dimension parameters of the suspension determined in this paper meet the requirements of vehicle suspension design.

3. Design of ELA-ERD

ELA-ERD is a key component to realize passive energy reclaiming and active control. The design of ELA-ERD is the core of electromagnetic linear energy-reclaiming suspension.

3.1. Basic Structure and Working Principle

Based on the previous study on ELA [28,29], this paper proposes ELA-ERD technology scheme combining the characteristics of suspension structure and long stroke, as shown in Figure 3. Considering that the moving coil type is easier to be controlled and the mover mass is smaller than the moving iron type, the ELA-ERD in this paper adopts the moving coil type.

As shown in Figure 3, ELA-ERD is mainly composed of external magnetic yoke, permanent magnets, moving coils and inner core. The piston rod of the shock absorber is used as the inner yoke of the ELA-ERD. The coil skeleton is fixedly connected with the piston rod. When the suspension vibrates and the shock absorber piston reciprocates, the coil of ELA-ERD moves synchronously, cutting the magnetic induction line to generate induction current.

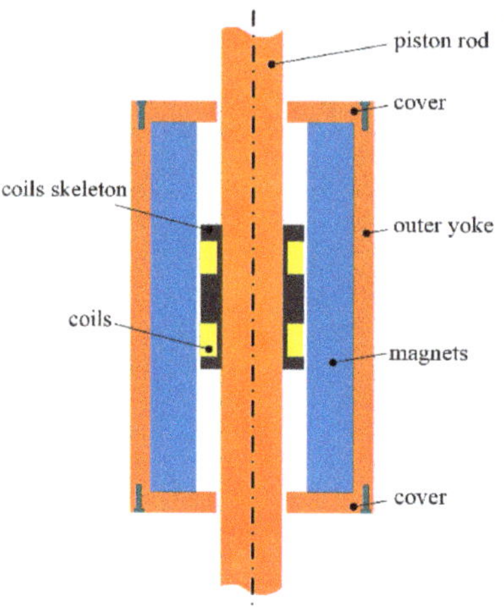

Figure 3. Schematic diagram of ELA-ERD structure.

The mechanism of generating magnetic induction voltage by moving coils cutting the magnetic induction line can be expressed by Equation (5):

$$U = BLv \tag{5}$$

where U is the induced voltage, B is the magnetic induction intensity, L is the effective coil length for cutting magnetic line, and v is the cutting speed of the coil. To meet Equation (5), B, L, v must be perpendicular to each other, which is an ideal state. However, angle deviations will be inevitably caused in the actual process. The mechanism of energy reclaiming is analyzed according to Equation (5).

As can be seen from Equation (5), higher induction voltage requires stronger magnetic induction intensity. For ELA-ERD, it is necessary to increase the magnetic induction intensity in the radial direction of the working domain of the mover as much as possible. Therefore, the Halbach array pattern is adopted for the permanent magnet of ELA-ERD, which is conducive to enhancing one side's magnetic field while weakening the other side's magnetic field [30]. The working principle of the Halbach array pattern is shown in Figure 4.

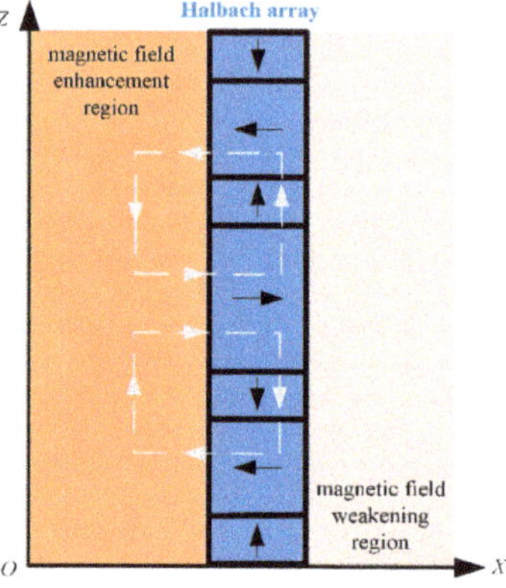

Figure 4. Working principle of Halbach array pattern.

3.2. Determination of Dimension and Material

In order to achieve reasonable layout and avoid motion interference, the overall dimension and motion stroke of the ELA-ERD are limited by the size of the McPherson independent suspension.

In order to avoid the impact failure in the compression process, the height H_e of ELA-ERD should be lower than the height of the suspension spring when compressed to the bottom, i.e.,

$$H_e \leq H_{s0} - \lambda_{smax} \tag{6}$$

Through calculation, it can be known that $H_e \leq 218.4$ mm, in this paper $H_e = 200$ mm.

The outer diameter D_e of ELA-ERD is identical to that of the shock absorber dust cover, both of which are set to 40 mm in this study. In order not to interfere with the normal operation of the suspension shock absorber, the motion travel of the ELA-ERD should be slightly larger than that of the shock absorber. As the shock absorber travel is 100 mm, so the motion travel of ELA-ERD s_e is set to 110 mm.

Air gap thickness δ is an important parameter affecting the electromagnetic performance of the ELA-ERD, a too big or too small value of δ will be inconducive to the full utilization of magnetic energy and performance improvement of ELA-ERD. The determination of air gap thickness is based on the magnetic circuit design of ELA-ERD, which is closely related to the magnetic circuit structure, permanent magnets size and working point of the magnets, as shown in Equation (7).

$$B_g{}^2 = (-H_m B_m) \frac{V_m}{V_g} \frac{\mu_0}{K_f K_r} \tag{7}$$

Where B_g is the gap flux density. $H_m B_m$ is the magnetic energy product at working point of permanent magnet, V_m is the permanent magnet volume, V_g is the air gap volume, μ_0 is the permeability of vacuum, K_f is the magnetic leakage coefficient, and K_r is the magnetic reluctivity.

In the design of magnetic circuit, the air gap thickness is determined through multiple calculations. According to previous design experience on the linear motor, current structural features of ELA-ERD, the layout pattern of permanent magnets and materials, the air gap thickness δ is preliminarily set to 0.15 mm in this study. After the air gap thickness δ is determined, the permanent magnets thickness d_m and the external magnetic yoke thickness d_y in the radial direction of ELA-ERD can be determined successively.

In the axial direction, ELA-ERD is in a symmetrical arrangement, so there is

$$H_e = h_m + 2h_c \tag{8}$$

where h_m is the overall axial height of the permanent magnet, h_c is the axial thickness of end cover.

At this point, the structural dimension of ELA-ERD has been basically determined, and the specific parameters are shown in Table 4.

Table 4. Basic dimensional parameters of ELA-ERD.

Parameters	Value	Parameters	Value
Height H_e	200 mm	Air gap thickness δ	0.15 mm
Outer diameter D_e	40 mm	Radial thickness of permanent magnet d_m	5.5 mm
Stroke s_e	110 mm	Radial thickness of the outer yoke d_y	1.5 mm
Inner yoke diameter (diameter of piston rod d_a)	10 mm	Axial thickness of end cover h_c	9 mm

The material of each component of ELA-ERD is shown in Table 5. To obtain better magnetic effect, steel-1008 with high strength and good magnetic property was selected for the outer cover, end cover and inner yoke of ELA-ERD. The permanent magnets are made of the sintered NdFeB N45 H with high remanent magnetism and high temperature resistance, which can provide a powerful magnetic field for ELA-ERD. The coils are made of copper-core enameled wire, which can maximize the winding coverage.

Table 5. Material of each component of ELA-ERD.

Component	Material	Component	Material
Outer cover	Steel-1008	Permanent magnets	N45 H
End cover	Steel-1008	Coil skeleton	Teflon
Inner yoke (piston rod)	Steel-1008	Coils	Copper-core enameled wire

3.3. Winding Design

According to the ELA-ERD structure shown in Figure 3, the height h_{sk} of the moving coil skeleton is affected by the device stroke. In this paper, the height of the moving coil skeleton can be calculated by Equation (9).

$$h_{sk} = h_m - s_e \tag{9}$$

The arrangement of windings is closely related to the array and dimension of permanent magnets. As can be seen from Figure 5a, as the coil skeleton height span consists of three radial permanent magnets, three groups of windings were arranged in order to realize maximum energy-reclaiming effect. Further research can be carried out on the winding-permanent magnet arrangement, which will be of great significance to the performance improvement of ELA-ERD, and will not be discussed in this paper.

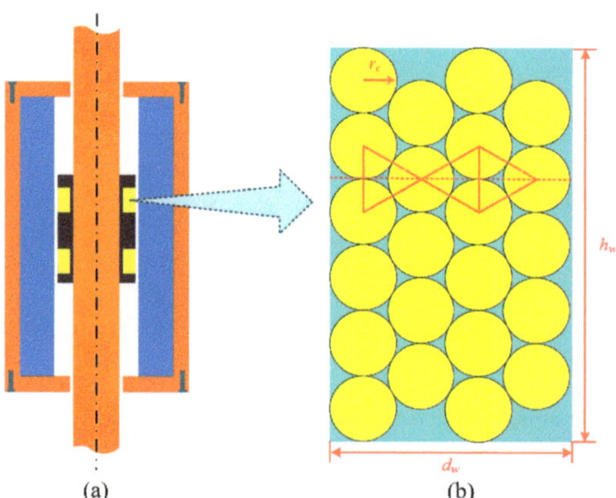

Figure 5. Winding arrangement and coil winding: (**a**) ELA-ERD; (**b**) Winding.

The moving coils are winded enameled wire. In order to improve the current density of the winding as much as possible, the coils should be winded as tightly as possible, and the winding state is shown in Figure 5b, from which we can see that the winding meets the following relationship:

$$d_w = \left(2 + \frac{\sqrt{3}(n-1)}{2}\right)r_c \tag{10}$$

$$N_c = \frac{h_w}{2r_c} \times n \tag{11}$$

According to Equations (10) and (11), given the height h_w and thickness d_w of the winding, enameled wire radius r_c and the winding number N_c can be obtained, where n is the number of enameled wire winding layers, which is set to $n = 4$ in this paper. After determining the enameled wire radius r_c, the maximum current or current density can be defined by referring to the characteristic parameter table of enameled wire, and reverse design of the ELA-ERD windings can be realized according to the maximum current limit.

After the winding parameters are determined, the resistance value R_c of the windings of the ELA-ERD can be obtained according to the resistance formula (12) of copper.

$$R_c = \frac{\rho l_c}{S_c} = \frac{6 g r_w N_c}{r_c{}^2} \tag{12}$$

where ρ is the copper resistivity, l_c is the total length of enameled wire in the windings, S_c is the cross-sectional area of a single-stranded copper-core enameled wire, and r_w is the effective median diameter of the windings.

3.4. Improvement of Magnetic Circuit

Magnetic circuit design directly determines the performance of the motor. In the magnetic circuit design process, the magnetic circuit needs to be modified and improved continuously. The magnetic circuit of the ELA-ERD designed in this paper needs to be analyzed and optimized as well. The traditional calculation method is very inefficient. Finite element simulation software is an important tool for rapid and accurate evaluation and feedback of motor magnetic circuit design, based on which we can simulate and calculate many complex problems.

In this paper, Ansoft Maxwell was used to model and simulate the ELA-ERD. Figure 6 shows the cloud diagram of magnetic induction intensity distribution obtained by simulation.

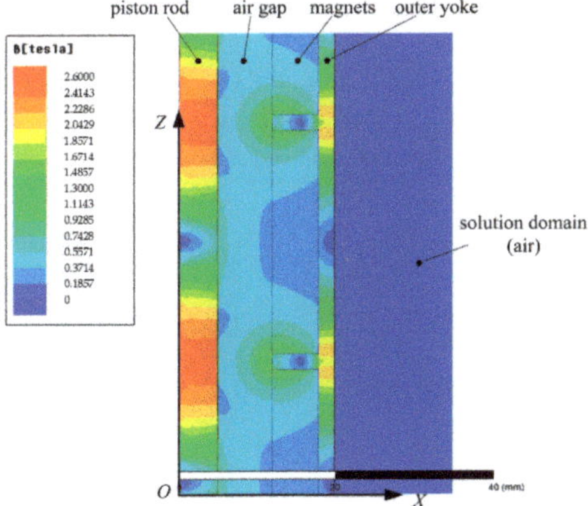

Figure 6. Finite element model and magnetic induction intensity distribution cloud map of ELA-ERD.

As can be seen from Figure 6, the magnetic induction intensity at the inner yoke (i.e., piston rod) position is seriously saturated, while the magnetic density in the working area of the moving coils is far from the ideal value. This indicates that the inner yoke is too small, which leads to the waste of a large part of magnetic energy, thus damping the output performance. Therefore, the magnetic circuit needs to be improved. Since the diameter of the piston rod has already been determined at the time when the selection design of shock absorber type is carried out, its size is not changeable. If the overall size of the ELA-ERD is reduced to match the small inner yoke, it is not conducive to maximizing the use of suspension layout space, and the overall performance of the ELA-ERD will be greatly reduced. Under the premise of keeping the current inner yoke size unchanged and not reducing the overall external diameter of ELA-ERD, the permanent magnet size is bound to be larger so as to achieve a more ideal working domain magnetic density, and the work domain thickness will be compressed. This will only aggravate the over-saturation of the inner yoke magnetic density, resulting in a large amount of magnetic energy waste.

Based on the above analysis, it can be known that a too small inner yoke is the root cause. Aiming at this problem, this paper adopts an improvement scheme, as shown in Figure 7. A part of the same material is added at the junction of the moving coils and the inner yoke, which can form a good magnetic circuit within the effective working range of the moving coils, increasing the magnetic density at the position of the moving coils, and effectively alleviating the over-saturation of the inner yoke magnetic density.

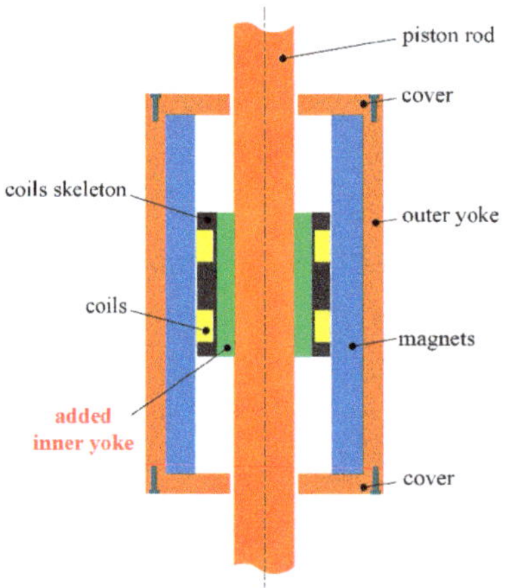

Figure 7. The structure diagram of improved ELA-ERD.

Similarly, with the help of Ansoft Maxwell software, the finite element model of the improved ELA-ERD was reconstructed and simulated. The magnetic induction intensity distribution of the improved scheme can be obtained, as shown in Figure 8.

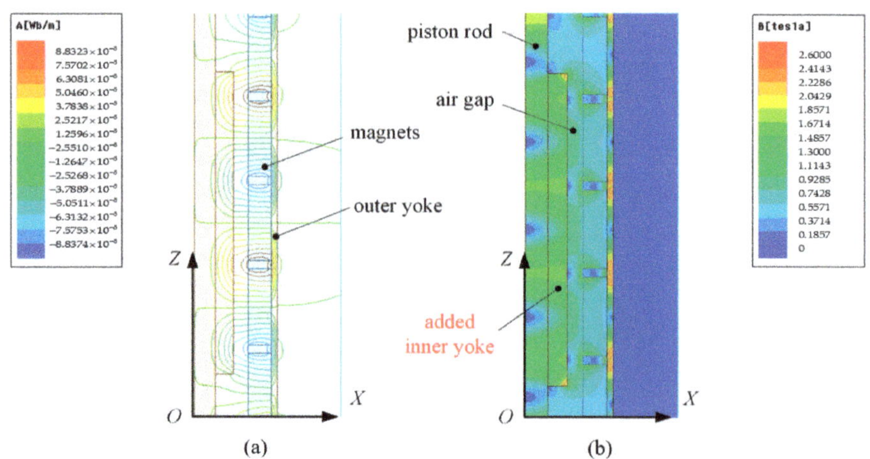

Figure 8. Distribution of magnetic force lines and magnetic induction intensity of ELA-ERD after improvement: (**a**) Distribution of magnetic force lines; (**b**) Distribution of magnetic induction intensity.

As shown in Figure 8, the magnetic induction intensity at the improved moving coils working area is significantly enhanced and in a more uniform distribution, and the magnetic density saturation of the inner yoke is significantly alleviated as well. In addition, as the inner yoke of ELA-ERD is replaced by the shock absorber piston rod, which runs through the whole device, certain magnetic leakage will be inevitably resulted. It can be seen from Figure 8 that the magnetic leakage situation at both ends of the inner yoke of the improved ELA-ERD has been significantly improved.

4. Analysis of Energy-Reclaiming Characteristics of ELA-ERD

After completing the design of ELA-ERD, it is necessary to carry out in-depth analysis of its energy-reclaiming characteristics. This paper mainly studies the ELA-ERD from energy-reclaiming effect and the law of energy reclaiming.

4.1. Analysis of Energy-Reclaiming Effect

First, the energy-reclaiming effect of the designed ELA-ERD needs to be verified. With the help of Ansoft Maxwell, modeling, mesh generation and excitation setting of ELA-ERD were carried out, in which the excitation was based on motion as input, and the simulation results of the model were obtained through post-processing.

The energy reclaiming of ELA-ERD was analyzed by taking the motion of the moving coils as input. However, the vibration is irregular in the actual operation of the vehicle. In order to facilitate the research and subsequent regular analysis, this paper takes the simple harmonic motion as the vibration input in the simulation process. The kinematic equation of the ELA-ERD moving coils is shown in Equation (13):

$$y(t) = A\sin(\omega t + a) \tag{13}$$

where A is the vibration amplitude, and the vibration frequency $f = \omega/2\pi$. The offset frequency of the front suspension is kept at 1–1.6 Hz when full load is required for passenger vehicles. Under actual working conditions, slight jolt will increase the vibration of the vehicle suspension. In this paper, the simple harmonic vibration with vibration frequency of 2 Hz and vibration amplitude of 50 mm was preliminarily selected as the motion inputs of the model. The output characteristic curve can be obtained through simulation, as shown in Figure 9.

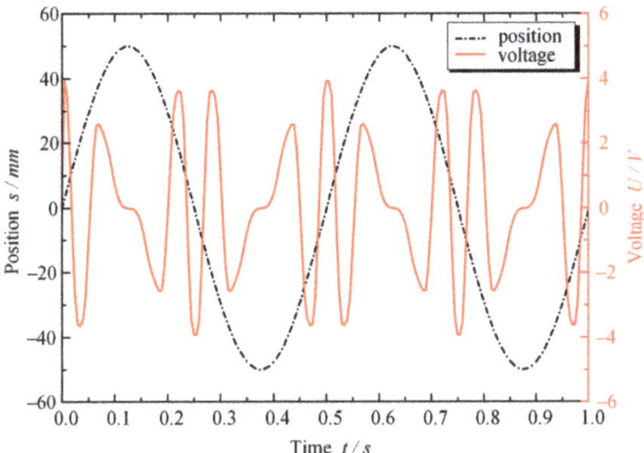

Figure 9. The vibration and energy-reclaiming voltage curve of ELA-ERD under vibration frequency of 2 Hz and vibration amplitude of 50 mm.

As can be seen from Figure 9, the output curve of energy-reclaiming voltage of ELA-ERD shows a periodical change. The peak value of the energy-reclaiming voltage reached 4 V under the simple harmonic input with vibration amplitude of 50 mm and vibration frequency of 2 Hz. Combining the resistance R_c of the windings of ELA-ERD, the energy-reclaiming power under the current vibration condition was calculated to be about 42 W, which is also the energy-reclaiming power of the single side suspension. In the actual operation process, slight unevenness of road will increase the vibration of vehicle, and four wheels will all vibrate, so the electromagnetic linear energy-reclaiming suspension has a considerable potential for energy reclaiming.

4.2. Study of Energy-Reclaiming Law

The change of vibration input is a key issue that should be taken into consideration when studying the energy-reclaiming law of ELA-ERD. According to Equation (13), amplitude A and frequency f are the two most important parameters affecting the state of simple harmonic vibration. The parametric analysis of the motion input of ELA-ERD was carried out, and the change of energy-reclaiming voltage under two different vibration frequencies and vibration amplitudes was observed, as shown in Figures 10 and 11.

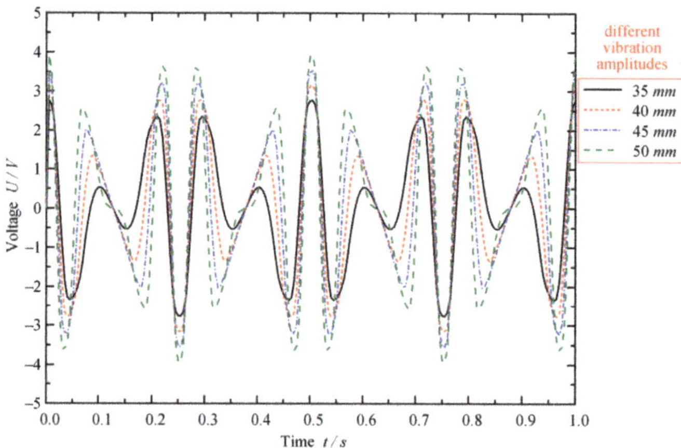

Figure 10. Vibration and energy-reclaiming voltage curve of ELA-ERD under different vibration amplitudes.

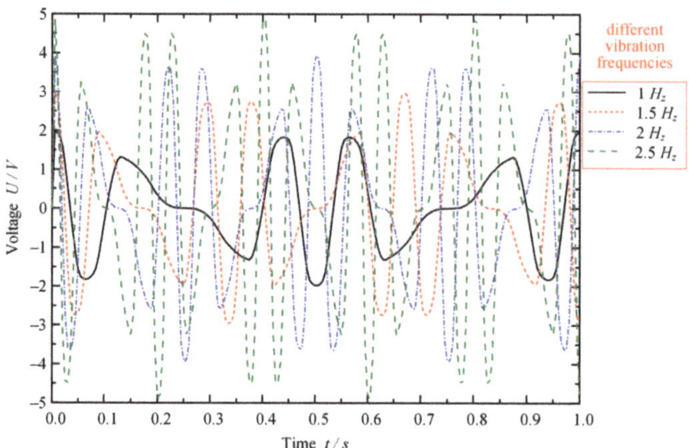

Figure 11. Vibration and energy-reclaiming voltage curve of ELA-ERD under different vibration frequencies.

As can be seen from Figures 10 and 11, the energy-reclaiming voltage of ELA-ERD changes periodically, and the curve shapes are almost identical, indicating stable output of the energy-reclaiming voltage. As can be seen from Figure 10, the energy-reclaiming voltage increases significantly with the increase in vibration amplitude. As can be seen from Figure 11, the increase in vibration frequency can increase the energy-reclaiming voltage in terms of both intensity and amplitude. In conclusion, the vibration frequency increase or vibration amplitude increase leads to the increase in energy-reclaiming voltage, as evidenced by the relationship between the absolute average value of the energy-reclaiming voltage and vibration frequency and vibration amplitude as shown in Figure 12.

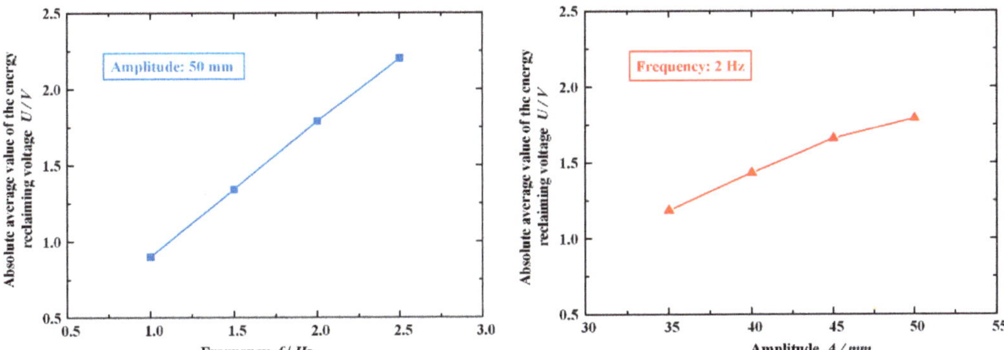

Figure 12. Relationship between the absolute average value of the energy-reclaiming voltage and vibration frequency and vibration amplitude.

The change of vibration frequency and amplitude eventually leads to the change of vibration velocity. However, the magnitude of vibration velocity is the real factor affecting the energy-reclaiming voltage according to ELA-ERD's energy-reclaiming principle Equation (5) that is established under an ideal state. In ELA-ERD, the magnitude B and direction of the magnetic induction at the moving coils working area vary from position to position. In the process of reciprocating motion of the moving coils, both B and L are difficult to determine. Therefore, it is necessary to carry out studies from other perspectives.

Under working conditions, the vibration velocity curve can be obtained by taking the derivative of the vibration displacement curve. The curves of vibration velocity and energy-reclaiming voltage are compared, as shown in Figure 13a. Due to the varying direction of the moving coil when cutting the magnetic induction line as well as the changing direction of the magnetic field, the direction of energy-reclaiming voltage is also changing, which cannot be observed from Figure 13a. Figure 13b shows the curve after taking the absolute value of vibration velocity and energy-reclaiming voltage with consideration on the influence of direction. It is obvious from Figure 13b that the overall curve trend of energy-reclaiming voltage is well consistent with that of vibration velocity.

Figure 13 shows the working condition where the vibration frequency is 2.5 Hz and the vibration amplitude is 50 mm, which is of no universal significance. Through conducting the above process under other working conditions, the relationship shown in Figure 14 can be obtained. Figure 14 mainly lists four working conditions, and it can be clearly seen that the above regular features are presented regardless of the vibration frequency and vibration amplitude.

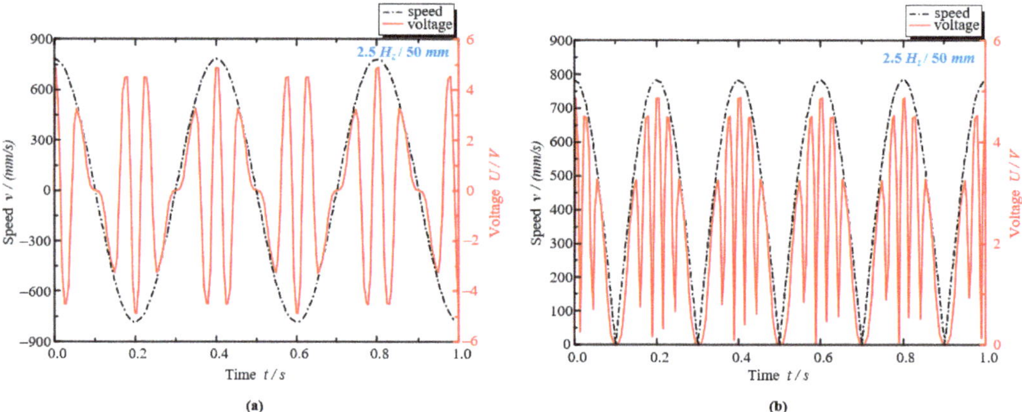

Figure 13. The relation between the curve of energy-reclaiming voltage and the curve of vibration velocity under vibration frequency of 2.5 Hz and vibration amplitude of 50 mm: (**a**) The curves of vibration velocity and energy-reclaiming voltage; (**b**) The curve after taking the absolute value of vibration velocity and energy-reclaiming voltage.

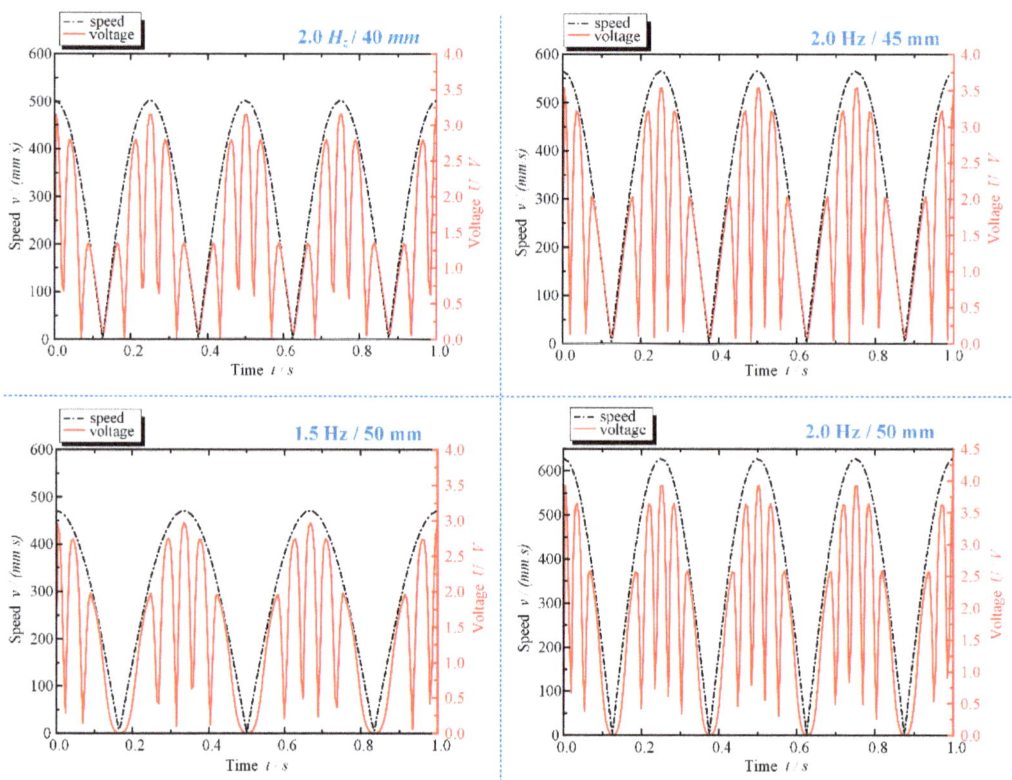

Figure 14. The relation between the curve of energy-reclaiming voltage and the curve of vibration velocity under four working conditions.

As can be seen from the relation between the curve of energy-reclaiming voltage and the curve of vibration velocity, there is a certain proportional relation between the absolute value of the vibration velocity and the energy-reclaiming voltage. To explore the numerical relationship between them, we averaged the absolute value of the energy-reclaiming voltage and the vibration velocity, and calculated the ratio of the two, i.e., $\overline{|U|}/\overline{|V|}$. Table 6 shows the ratio $\overline{|U|}/\overline{|V|}$ under different vibration frequencies and vibration amplitudes.

Table 6. The ratio $\overline{|U|}/\overline{|V|}$ under different vibration frequencies and vibration amplitudes.

Vibration Frequency	Vibration Amplitude	$\overline{|U|}/\overline{|v|}$
1 Hz	35 mm	4.47
1 Hz	40 mm	4.56
1 Hz	45 mm	4.46
1 Hz	50 mm	4.49
1.5 Hz	35 mm	4.41
1.5 Hz	40 mm	4.53
1.5 Hz	45 mm	4.59
1.5 Hz	50 mm	4.47
2 Hz	35 mm	4.22
2 Hz	40 mm	4.45
2 Hz	45 mm	4.58
2 Hz	50 mm	4.38
2.5 Hz	35 mm	4.42
2.5 Hz	40 mm	4.40
2.5 Hz	45 mm	4.45
2.5 Hz	50 mm	4.30

To find the rule and avoid the accidental factors, Table 6 shows the values of $\overline{|U|}/\overline{|V|}$ under 16 different working conditions. It can be found that no matter how the vibration frequency and amplitude change, the value of $\overline{|U|}/\overline{|V|}$ under different conditions basically remains around 4.5. In this paper, K_e is defined as the energy-reclaiming voltage coefficient of ELA-ERD.

$$\overline{|U|} = K_e\overline{|V|} = K_e \times \overline{\left|\frac{dy(t)}{dt}\right|} \tag{14}$$

where $\overline{|U|}$ is the mean absolute value of the energy-reclaiming voltage, $\overline{|V|}$ is the mean absolute value of the vibration velocity, $y(t)$ is the displacement function of the vibration of ELA-ERD moving coils. From Table 6, it can be known that the energy voltage reclaiming coefficient of ELA-ERD designed in this paper is $K_e \approx 4.5$.

According to Equation (14), it can be known that the energy-reclaiming voltage of ELA-ERD is proportional to the vibration velocity. According to Equation (14), as long as the time-domain data $y(t)$ of the vibration amplitude of shock absorber piston of the electromagnetic linear energy-reclaiming suspension can be obtained, the magnitude of the energy-reclaiming voltage of ELA-ERD can be estimated, which lays a theoretical foundation for further study of ELA-ERD.

5. Conclusions

The following conclusions are drawn from this paper:

1. A novel electromagnetic linear energy-reclaiming suspension based on the McPherson independent suspension is proposed in this study. The suspension has the advantages of compact structure, easy modification, and high reliability. Even if ELA-ERD fails, the normal operation of the suspension will not be affected, showing obvious advantages over other electromagnetic energy-reclaiming suspension.
2. The ELA-ERD applied to the novel suspension was designed to realize passive energy reclaiming and active control. This device adopts the piston rod of the shock absorber as the inner yoke and organically integrates the structural characteristics of the sus-

pension. The permanent magnets are arranged in Halbach array pattern to enhance the magnetic density within the work field.

3. To solve the problem of magnetic density oversaturation of the inner yoke in the initial design phase, the magnetic circuit of ELA-ERD is optimized by increasing the size of the inner yoke in the effective working area of the moving coils. The simulation results show that this measure effectively improved the electromagnetic performance of ELA-ERD.

4. The simulation analysis showed that the energy-reclaiming power of ELA-ERD reached 42 W under vibration amplitude of 50 mm and vibration frequency of 2 Hz. Therefore, the electromagnetic linear energy-reclaiming suspension with ELA-ERD has considerable potential for energy reclaiming.

5. The factors influencing the energy-reclaiming law of ELA-ERD were analyzed from the perspectives of the changes of vibration amplitude and vibration frequency, and then the most fundamental influencing factor, vibration velocity, was further explored. According to Equation (14), the energy-reclaiming voltage coefficient was defined. Through comparison of a large number of data, the energy-reclaiming voltage coefficient $K_e \approx 4.5$ of the ELA-ERD designed in this paper was derived, which lays a theoretical foundation for the subsequent research of ELA-ERD.

Although achieving progress and conclusions, the research on electromagnetic linear energy-reclaiming suspension is still in the preliminary stage. Further research should be carried out on the improvement of suspension energy-reclaiming power and reclaiming efficiency as well as on the implementation of active control, so as to promote the application of electromagnetic linear energy-reclaiming suspension and enhance the development of vehicle suspension technology.

Author Contributions: Conceptualization, J.D. and L.C.; methodology, J.D., L.C. and Y.Q.; software, formal analysis and investigation, J.D., Y.Q., C.W. and J.Z. (Jianhui Zhu); resources, J.D., L.C. and C.W.; data curation, J.D., J.Z. (Jun Zhu) and J.Z. (Jingxuan Zhu); writing—original draft preparation, J.D., Y.Q., J.Z. (Jianhui Zhu) and J.Z. (Jun Zhu); writing—review and editing, J.D., L.C. and Y.Q. All authors have read and agreed to the published version of the manuscript.

Funding: This work was supported by the National Natural Science Foundation of China (Grant No. 51605183, 51975239), and the Natural Science Foundation for Jiangsu Colleges and Universities (Grant No. 21KJB480005).

Institutional Review Board Statement: Not applicable.

Informed Consent Statement: Not applicable.

Data Availability Statement: Not applicable.

Conflicts of Interest: The authors declare no conflict of interest.

References

1. Abdelkareem, M.A.; Xu, L.; Ali, M.K.A.; Elagouz, A.; Mi, J.; Guo, S.; Liu, Y.; Zuo, L. Vibration energy harvesting in automotive suspension system: A detailed review. *Appl. Energy* **2018**, *229*, 672–699. [CrossRef]
2. Zheng, P.; Gao., J.; Wang, R.; Dong, J.; Diao, J. Review on the Research of Regenerative Shock Absorber. In Proceedings of the 2018 24th International Conference on Automation and Computing (ICAC), Newcastle upon Tyne, UK, 6–7 September 2018; pp. 1–12.
3. Zhang, Y.; Guo, K.; Wang, D.; Chen, C.; Li, X. Energy conversion mechanism and regenerative potential of vehicle suspensions. *Energy* **2017**, *119*, 961–970. [CrossRef]
4. Čorić, M.; Deur, J.; Kasać, J.; Tseng, H.E.; Hrovat, D. Optimisation of active suspension control inputs for improved vehicle handling performance. *Veh. Syst. Dyn. Int. J. Veh. Mech. Mobil.* **2016**, *54*, 1574–1600. [CrossRef]
5. Gong, M.; Yan, X. Robust Control Strategy of Heavy Vehicle Active Suspension Based on Road Level Estimation. *Int. J. Automot. Technol.* **2021**, *22*, 141–153. [CrossRef]
6. Zheng, X.; Yu, F. Study on the potential benefits of an energy-regenerative active suspension for vehicles. *SAE Trans.* **2005**, *114*, 242–245.
7. Lato, T.; Zhao, H.; Lin, Z.; He, Y. An Energy-Regenerative Suspension System. In Proceedings of the ASME 2018 International Mechanical Engineering Congress and Exposition IMECE 2018, Pittsburg, PA, USA, 9–15 November 2018.

8.   Zheng, P.; Wang, R.; Gao, J. A Comprehensive Review on Regenerative Shock Absorber Systems. *J. Vib. Eng. Technol.* **2020**, *8*, 225–246. [CrossRef]
9.   Long, G.; Ding, F.; Zhang, N.; Zhang, J.; Qin, A. Regenerative active suspension system with residual energy for in-wheel motor driven electric vehicle. *Appl. Energy* **2020**, *260*, 114180. [CrossRef]
10.   Salman, W.; Qi, L.; Zhu, X.; Pan, H.; Zhang, X.; Bano, S.; Zhang, Z.; Yuan, Y. A high-efficiency energy regenerative shock absorber using helical gears for powering low-wattage electrical device of electric vehicles. *Energy* **2018**, *159*, 361–372. [CrossRef]
11.   Fu, C.; Lu, J.; Ge, W.; Tan, C.; Li, B. A Review of Electromagnetic Energy Regenerative Suspension System & Key Technologies. *Comput. Model. Eng. Sci.* **2022**, *135*. [CrossRef]
12.   Tvrdić, V.; Podrug, S.; Jelaska, D.; Perkušić, M. A Concept of the Novel Regenerative Hydraulic Suspension: The Prototype Description. In Proceedings of the 2018 3rd International Conference on Smart and Sustainable Technologies (SpliTech), Split, Croatia, 26–29 June 2018; pp. 1–6.
13.   He, X.; Xiao, G.; Hu, B.; Tan, L.; Tang, H.; He, S.; He, Z. The applications of energy regeneration and conversion technologies based on hydraulic transmission systems: A review. *Energy Convers. Manag.* **2020**, *205*, 0196–8904. [CrossRef]
14.   Dai, J.G.; Wang, C.; Liu, Z.F.; Zhu, J.H.; Hu, X.M. Review of energy reclaiming suspension technology. *Sci. Technol. Eng.* **2018**, *18*, 131–139.
15.   Bose Automotive Systems (China). BOSE's innovative suspension system. *Auto Accessories* **2012**, *12*, 38–39.
16.   Boduroglu, A.; Gulec, M.; Demir, Y.; Yolacan, E.; Aydin, M. A New Asymmetric Planar V-Shaped Magnet Arrangement for A Linear PM Synchronous Motor. *IEEE Trans. Magn.* **2019**, *55*, 1–5. [CrossRef]
17.   Lv, X.; Ji, Y.; Zhao, H.; Zhang, J.; Zhang, G.; Zhang, L. Research Review of a Vehicle Energy-Regenerative Suspension System. *Energies* **2020**, *13*, 441. [CrossRef]
18.   Gysen, B.L.; Paulides, J.J.; Janssen, J.L.; Lomonova, E.A. Active electromagnetic suspension system for improved vehicle dynamics. *IEEE Trans. Veh. Technol.* **2010**, *59*, 1156–1163. [CrossRef]
19.   Gysen, B.L.; van der Sande, T.P.; Paulides, J.J.; Lomonova, E.A. Efficiency of a regenerative direct-drive electromagnetic active suspension. *IEEE Trans. Veh. Technol.* **2010**, *60*, 1384–1393. [CrossRef]
20.   Gysen, B.L.; Janssen, J.L.; Paulides, J.J.; Lomonova, E.A. Design aspects of an active electromagnetic suspension system for automotive applications. In Proceedings of the Industry Applications Society Meeting, Houston, TX, USA, 4–8 October 2009; IEEE: New York, NY, USA, 2009; pp. 1–8.
21.   Vijayakumara, P.; Mallikarjuna, D.C.; Suresh, R. Generating of power suspension shock absorber. *Int. J. Adv. Eng. Res. Dev.* **2017**, *4*, 713–721.
22.   Deng, Z.; Lai, F. Vehicle Active suspension Electromagnetic linear actuator Finite element analysis. *Chin. J. Mech. Eng. Engl. Ed.* **2011**, *47*, 121–128. [CrossRef]
23.   Chen, X.; Luo, H.; Deng, Z. Design of an energy-regenerative suspension control systemusing linear motor and energy recovery analysis. *J. Vib. Shock.* **2012**, *31*, 124–129.
24.   Deubel, C.; Prokop, G. Friction of a MacPherson suspension system at various load cases. *J. Vib. Control.* **2022**, *2022*, 10775463221140324. [CrossRef]
25.   Xiao-dong, S.U.N.; Feng, C.A.I.; Ying-feng, C.A.I.; Long, C.H.E.N. Improved Model Predictive Thrust-force Control of Linear Motors for Active Suspensions. *China J. Highw. Transp.* **2021**, *34*, 85–100.
26.   Yang, M.; Yu, J.; Guo, Y. Present Situation and Development Tendency of Shock Absorbers in Vehicle Suspensions. *Eng. Test* **2019**, *59*, 97–100 + 103.
27.   Yu, L.; Zhang, M.; Xue, W.; Luo, W.; He, J.; Tian, H. A Comprehensive Review of Permanent Magnet Synchronous Linear Motors in automotive electromagnetic suspension system. In Proceedings of the 2021 IEEE 4th Advanced Information Management Communicates, Electronic and Automation Control Conference (IMCEC), Chongqing, China, 18–20 June 2021; pp. 792–796.
28.   Dai, J.; Zhao, Z.; Xu, S.; Wang, C.; Zhu, J.; Fan, X. Inhibition of iron loss of the inner yoke in electromagnetic linear actuator. *IET Electr. Power Appl.* **2019**, *13*, 419–425. [CrossRef]
29.   Dai, J.; Xia, J.; Wang, C.; Xu, S. Thermal analysis of an electromagnetic linear actuator. *Adv. Mech. Eng.* **2017**, *9*, 1687814017745387. [CrossRef]
30.   Li, Z.; Wu, Q.; Liu, B.; Gong, Z. Optimal Design of Magneto-Force-Thermal Parameters for Electromagnetic Actuators with Halbach Array. *Actuators* **2021**, *10*, 231. [CrossRef]

Article

Simulation Study on Direct-Drive Compressor with Electromagnetic Linear Actuator

Jianhui Zhu *, Mengmeng Xue, Jianguo Dai, Zongzheng Yang and Jingnan Yang

Faculty of Transportation Engineering, Huaiyin Institute of Technology, Huai'an 223003, China; 15861787982@163.com (M.X.); djg619265809@hyit.edu.cn (J.D.); zz19805088799@163.com (Z.Y.); mm13338134521@163.com (J.Y.)
* Correspondence: zhujianhui@hyit.edu.cn; Tel.: +86-15061212672

Abstract: In order to enhance compressor efficiency and meet the requirements of energy conservation and environmental protection, this study designed a direct-drive compressor with an electromagnetic linear actuator. Starting with the structural design and performance analysis of the linear compressor, the working process of the moving-coil linear compressor was analyzed. The basic performance of the designed moving-coil linear motor was simulated and analyzed using the Maxwell software, while the overall performance of the linear compressor was simulated and analyzed using the Simulink software to verify the feasibility of controlling a linear compressor via the direct drive of a linear motor.

Keywords: linear compressor; simulation analysis; linear motor; moving coil

1. Introduction

Compressors play an important role in modern machinery. Currently, compressors extensively used in production and daily life include crankshaft connecting rod type, reciprocating type, and rotary type, among others. Despite the mature technology and a complete industrial chain, such compressors have a low mechanical efficiency [1]. By using an electromagnetic actuator to directly drive the piston, a linear compressor conducts reciprocating linear motion in the cylinder, with the mechanical efficiency increasing by 15~25%, thus greatly contributing to energy conservation [2]. In addition, a linear compressor has the advantages of small volume, long service life, low noise, controllable piston stroke, and almost zero lateral force [3]. Nonetheless, research on linear compressors is mainly carried out by large technology companies and scientific research institutes in Japan, South Korea, the United States, etc. Sunpower of the United States, LG of South Korea, and Panasonic of Japan have attained remarkable achievements in this respect [4].

Compared to foreign countries, domestic research on linear compressors in China is disadvantageous in both theory and technology. With the advancement of science and technology, domestic scholars have started to study linear compressors after being aware of their importance. Since the 1980s, China has introduced foreign linear compressor technology and carried out extensive experimental research and product development on this basis. However, due to the weak theoretical foundation and backward manufacturing technology, domestic production of compressors has low mechanical efficiency and high cost, and cannot replace traditional compressors [5].

Based on the Redlich moving-magnet linear motor structure, Gao Yao et al. designed a linear compressor using the vector analysis method, which has the advantages of low weight and high efficiency [6]. Fang Xueliang et al. studied a new oil-free linear compressor system and proposed a thermodynamic compressor model [7]. Zhang Kai et al. from the Kunming Institute of Physics studied flexible springs in linear compressors, which provided guiding significance for the design of stacked components for flexible springs [8]. Chen Hongyue et al. from the Liaoning Technical University studied E-type electromagnetic

Citation: Zhu, J.; Xue, M.; Dai, J.; Yang, Z.; Yang, J. Simulation Study on Direct-Drive Compressor with Electromagnetic Linear Actuator. *Actuators* **2023**, *12*, 185. https://doi.org/10.3390/act12050185

Academic Editors: Eihab M. Abdel-Rahman and Ioan Ursu

Received: 25 March 2023
Revised: 19 April 2023
Accepted: 21 April 2023
Published: 25 April 2023

structure of moving-magnet linear compressors and found a way to improve the efficiency of these linear compressors [9]. Chen Xinwen et al. studied the moving-magnet linear motor of linear compressors, with certain results achieved [4].

China has continued its research and exploration of linear compressors. Various research institutes and universities have carried out related research, and experts are engaged in continuous innovation and optimization. Nonetheless, there is insufficient basic theory and practical application of linear compressors, so further exploration and practice are needed before the commercialization stage [10].

Compared with general-purpose compressors, small direct-drive compressors with an electromagnetic linear actuator are more demanding in structure and performance. Thus, a direct-drive linear compressor should meet the following requirements:

(1)   Good refrigeration performance to ensure good refrigeration effect under different working conditions.
(2)   Greater mechanical efficiency on the basis of an existing compressor to meet the requirements of energy conservation and environmental protection.
(3)   Compact structure, small size, and low weight, so that installation and fixture in the limited space of the engine room are possible.
(4)   The compressor should run smoothly, especially in the working conditions of startup and stop, to reduce noise and vibration.

2. Structure of Direct-Drive Compressor with Electromagnetic Linear Actuator

According to different types of driving parts, linear compressors can be divided into the following three types: the first type is electromagnetic vibration type of compressors, the second type is linear electric type of compressors; and the third type is linear stepping motor-driven compressors. According to different electromagnetic drive patterns, electromagnetic vibration type of compressors can be further divided into moving-coil type, moving-iron type, and moving-magnet type [3]. Compared to a moving-coil linear compressor, a moving-iron linear compressor easily generates radial force, so the piston significantly rubs the cylinder wall, while a moving-magnet linear compressor has a heavy promoter, which requires a greater elasticity of the resonant spring and makes the resonance difficult to control. Hence, a moving-coil linear compressor is preferred [11].

A moving-coil linear compressor is mainly composed of a power unit (moving-coil linear motor), an engine body, a connecting rod, cylinder block parts, piston parts, intake and exhaust pipelines, bolt screw holes, and other accessories, as shown in Figure 1. The power unit mainly includes the motor and related support parts. The main power source of the linear compressor is the internal moving-coil linear motor, which directly drives the piston to reciprocate in the cylinder and can maintain the stability of the direction of movement to complete subsequent cycle processes. The body mainly plays a bearing role in the structure of the linear compressor, including the two major structures of the fuselage and the base. There are various moving parts inside the fuselage that provide positioning and guiding for each transmission part. A cylinder formed by the base is installed on the base, and the linear motor is fixed to the base by fasteners [12].

In a linear compressor, the cylinder block is an important structure for compressing refrigerants. The main components include the cylinder block, the cylinder head, etc. The cylinder parts have a more complex structure than other parts because they bear a heavy air pressure. The cylinder is provided with an intake and exhaust chamber at one end, with an air inlet and exhaust port arranged on both sides of the cylinder. The two are connected to the intake and exhaust chamber through the ventilation pipe, and a piston is placed in between. The piston parts mainly include the piston, the piston ring, etc. The piston is connected to the piston rod, the other end of which is connected with the promoter of the linear motor by fasteners. A 3D model is shown in Figure 2 [13].

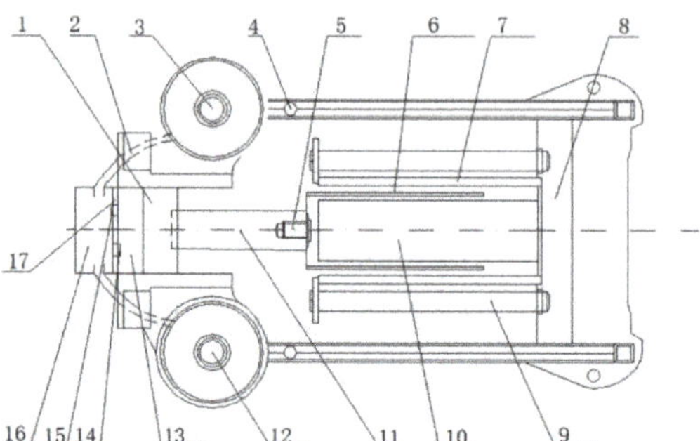

Figure 1. Structure diagram of a moving-coil linear compressor: 1. piston; 2. air pipe; 3. exhaust port; 4. open hole; 5. screw; 6. promoter; 7. permanent magnet; 8. boss; 9. Fixing bolt; 10. iron core; 11. connecting rod; 12. air inlet; 13. cylinder; 14. intake valve; 15. exhaust valve; 16. intake and exhaust chamber; and 17. cylinder head.

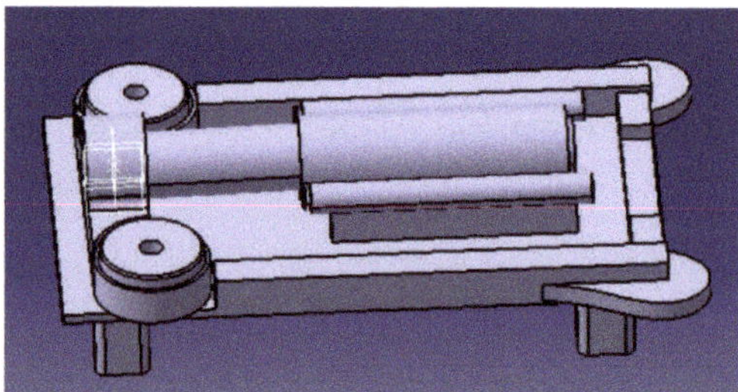

Figure 2. Three-dimensional model of the moving-coil linear compressor.

3. Modeling of Direct-Drive Compressor with Electromagnetic Linear Actuator

In the design of a moving-coil linear motor, the priority should be size design and material selection according to its intended use. The next stage is the stator and promoter structure design, magnetic circuit analysis, and coil structure design. Finally, the structure and function of the moving-coil linear motor should be determined as a whole to achieve the best results through parameter adjustment and structure optimization.

3.1. Mathematical Model of the Moving-Coil Linear Motor

The running equation of a linear motor is composed of a balance among the circuit system, the magnetic field system, and the mechanical system [14].

The circuit system converts voltage in the moving coil into current, which can be equivalent to a resistor and an inductor in series, and the moving coil moves in a magnetic field, generating a back EMF. The coil voltage balance equation [15] is as follows:

$$U_{in} = Ri + L\frac{di}{dt} + E_{emf} \tag{1}$$

The counter electromotive force E_{emf} can be expressed as follows:

$$E_{emf} = BlNv = K_e v \tag{2}$$

where U_{in} is the supply voltage; i is the coil current; R is the coil resistance; L is the coil inductance; E_{emf} is the counter electromotive force during moving coil motion in the magnetic field; K_e is the counter electromotive force constant; v is the promoter motion velocity; B is the magnetic flux density; and N is the turns per coil.

The magnetic field system enables the linear motion of the promoter through the electromagnetic force generated in the interaction between the circuit system and the magnetic field. According to the Lorentz force on the energized coil in the magnetic field, the electromagnetic force balance equation is as follows:

$$F_e = k_b BlNi = k_e i \qquad (3)$$

where k_b is the structural coefficient; B is the magnetic flux density of the coil; l is the effective length of each coil in the magnetic field; and k_e is the driving force constant of the linear motor.

During operation, the moving-coil linear motor needs to overcome friction force and inertia force in motion. In addition, there is a load force; thus, the mechanical balance equation [15] is as follows:

$$F_e = m\frac{d^2x}{dt^2} + k_v\frac{dx}{dt} + F_L \qquad (4)$$

where F_m is the electromagnetic force; m is the mass of the coil assembly; x is the promoter displacement; k_v is the damping force coefficient of the moving component in the magnetic field; and F_L is the load force.

3.2. Parameter Design and Material Selection of the Moving-Coil Linear Motor

In the design of a moving-coil linear motor used in a direct-drive compressor with an electromagnetic linear actuator, there are the following requirements:

(1) The driving force and the working stroke of the linear motor should be increased within the limited installation range;
(2) The thrust performance of the linear motor should be increased to reduce fluctuation during motor operation;
(3) The electromagnetic density and the magnetic circuit should be properly designed to reduce temperature rise during motor operation;
(4) The mechanical efficiency and energy utilization efficiency of the motor should be increased to reduce the working energy consumption as much as possible;
(5) It is necessary to meet the requirements of small size, compact structure, reliable operation, and high space utilization rate.

Therefore, the basic parameters and sizes of the moving-coil linear motor are designed as shown in Tables 1 and 2 below:

Table 1. Basic parameters of a moving-coil linear motor.

Item Name	Parameter Requirement
Rated voltage/V	12 (V)
Maximum current/A	10 (A)
Rated power/kw	3 (kw)
Maximum thrust/N	450 (N)
Maximum stroke/mm	20 (mm)

Table 2. Basic size of a moving-coil linear motor.

Structure Name	Size
Stator outer diameter (mm)	84
Stator inner diameter (mm)	24
Permanent magnet internal diameter (mm)	38
Motor length (mm)	194
Moving-coil length (mm)	185

3.3. Mathematical Model of the Moving-Coil Linear Compressor

The mechanical motion system of a moving-coil linear compressor is dominated by the reciprocating motion of the promoter, so it can be simplified to a model of a forced damping vibration system with a single degree of freedom and a single mass [16]. For a sample machine of the moving-coil linear compressor, the specific parameters are shown in Table 3. By studying the principle of the linear compressor, it is possible to simplify the mechanical subsystem into a damping vibration system with a single degree of freedom if thermodynamics factors are ignored.

Table 3. Basic parameters of the linear compressor.

Parameter Name	Value
Promoter mass	0.74 kg
Piston diameter	0.025 m
Linear motor force constant	40 N/A
Intake pressure	0.84378 MPa
Exhaust pressure	1.0166 MPa
Piston stroke	20 mm
Self-inductance coefficient	1.1 mH
Coil loop resistance	0.66 Ω
Damping factor	20 N·s/m
Spring rate	22,950 N/m

The mechanical subsystem of the moving-coil linear compressor is dominated by the reciprocating linear motion of the promoter of the linear motor, which is now simplified as forced damped vibration with a single degree of freedom and a single mass. The promoter is taken as the research object, which is subjected to an electromagnetic force $F_e(t)$, a gas load force $F_g(t)$, a friction force $F_m(t)$, a spring force $F_s(t)$, an inertia force, and a damping force. These forces form the force balance equation of the mechanical subsystem of the moving-coil linear compressor. According to the mechanical subsystem model of the permanent-magnet moving-coil linear compressor, its force balance equation can be obtained as follows [17]:

$$m\frac{d^2x(t)}{dt^2} + c\frac{dx(t)}{dt} + K[x(t) - X_s] = F_e(t) + F_g(t) \tag{5}$$

where m is the promoter mass; $x(t)$ is the promoter motion displacement; t is the working time; X_s is the static equilibrium position of the promoter; K is the spring stiffness coefficient; c is the damping coefficient of the mechanical subsystem; $F_e(t)$ is the electromagnetic driving force; and $F_g(t)$ is the gas load force.

The gas load force $F_g(t)$ is the force acting on the piston under the pressure difference between the interior and exterior of the cylinder, which size is proportional to the area and the pressure difference of the piston:

$$F_g(t) = A[P_c(t) - P_b] \tag{6}$$

where A is the cross-sectional area of the piston; $P_c(t)$ is the gas pressure in the cylinder; and P_b is the exhaust back pressure outside the cylinder, which is the atmospheric pressure under normal situations.

In the calculation of the electromagnetic force $F_e(t)$ generated by the driving linear motor, the turns per coil N, the air-gap magnetic induction intensity B, and the effective length l of the single-turn coil are all constant values; only current i is a variable. Thus, the electromagnetic driving force $F_e(t)$ can be expressed as follows:

$$F_e(t) = K_e i \tag{7}$$

Because of the constant reciprocating motion of the piston in the cylinder, the gas pressure $P_c(t)$ in the cylinder changes constantly. A reciprocating linear motion of the piston involves four processes: intake, compression, exhaust, and expansion. The gas pressure in the cylinder can be expressed using a piecewise function [18]:

$$P_c(t) = \begin{cases} P_s & \text{Air intake process} \\ P_s\left[\frac{S+S_0}{x(t)}\right]^n & \text{Compression process} \\ P_d & \text{Exhaust process} \\ P_d\left[\frac{S_0}{x(t)}\right]^n & \text{Expansion process} \end{cases} \tag{8}$$

where P_s is the intake pressure of the air inlet, which is the atmospheric pressure under normal circumstances; P_d is the exhaust pressure of the exhaust port; S is the stroke of the linear compressor piston; S_0 is the distance from the piston apex to the valve assembly at the end of the exhaust process, i.e., the clearance; and n is the polytropic exponent.

The movement pressure diagram of the compressor [19] is shown in Figure 3. The figure shows the following: one to two is the compression process; two to three is the exhaust process; three to four is the expansion process; and four to one is the intake process.

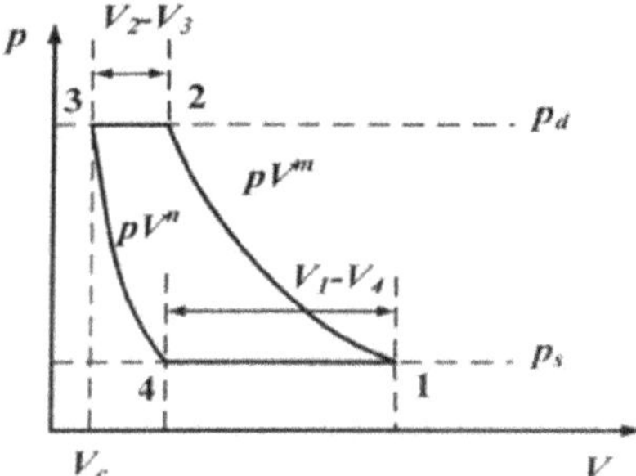

Figure 3. Compressor pressure diagram.

Due to the use of suction and exhaust valves to control the flow of working fluids, a linear compressor for refrigeration has an ideal working process, as shown in Figure 3, and the process is a nonlinear process.

Due to the existence of gap volume, the gas in the cylinder will exchange heat with the wall in contact during the expansion process; thus, the expansion process is not a straight line as shown by the theoretical cycle, but a curve, as shown in Figure 3.

During the actual operation, because of the influence of the spring force, flow resistance, heat exchange, etc., of the gas valve, the suction and exhaust valve is delayed in opening, and the pressure and temperature of the actual suction and exhaust process change, which is not according to the linear process shown by the theoretical cycle, but a curve process with fluctuations that slightly convex up and down; however, its fluctuations and changes are small, and it is still regarded as the linear process shown by the theoretical cycle when analyzed.

The intake and exhaust valves are based on the LG compressor which model is LA95LAEM. The exhaust mechanism collects compressed gas in the cylinder to reduce exhaust noise and introduce the gas to the next place of use, as shown in Figures 4 and 5.

The intake mechanism can suck low-pressure gas into the body, which is screened by the intake valve and then fed into the working chamber to compress air in the piston to complete the compression of the gas. This is shown in Figures 6 and 7.

Figure 4. Vent valve disc and seat.

Figure 5. Connection diagram of each exhaust part.

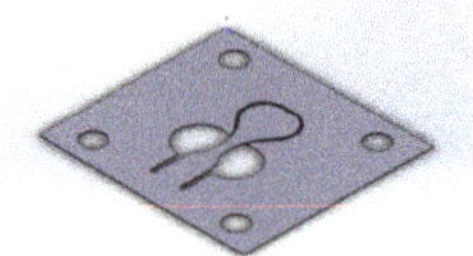

Figure 6. Inlet valve plate.

Figure 7. Connection diagram of each component of the intake valve.

R134a is selected as the compressor refrigerant. According to the pressure enthalpy diagram, the exhaust pressure is 1.016 MPa and the intake pressure is 0.08437 MPa. The dividing points of the linear compressor during the compression, exhaust, expansion, and intake processes are obtained from Equation (8).

According to the equivalent circuit diagram of the electromagnetic subsystem of the moving-coil linear compressor, the voltage balance equation can be obtained as follows [20]:

$$R_e i(t) + B_e l_e \frac{dx(t)}{dt} + L \frac{di(t)}{dt} = U(t) \tag{9}$$

where R_e is the effective resistance of the moving coil; i is the current in the moving coil; B_e is the magnetic field intensity in the air gap; l_e is the effective total length of the moving coil; L is the self-inductance coefficient of the moving coil; and $U(t)$ is the voltage at both ends of the moving coil.

The linear motor efficiency η of the permanent-magnet moving-coil linear compressor can be expressed as follows:

$$\eta = \frac{W_1}{W_0} = \frac{\int F_e v dt}{\int U i dt} \tag{10}$$

where η is the linear motor efficiency of the permanent-magnet moving-coil linear compressor; W_1 is external work performed by the permanent-magnet moving-coil linear compressor; and W_0 is the input energy of the permanent-magnet moving-coil linear compressor.

4. Results

4.1. Simulation Analysis of the Moving-Coil Linear Motor under Maxwell

For an axially symmetric moving-coil linear motor, only half of the two-dimensional motor model needs to be built, as shown in Figure 8. The two-dimensional model of the moving-coil linear motor mainly consists of an inner magnetic yoke, an outer magnetic yoke, a bottom plate, a coil, a coil former and a permanent magnet, among other components. Hence, after the establishment of the two-dimensional model for the motor, material selection and distribution is carried out for the above parts, with the materials of each part shown in Table 4 [21].

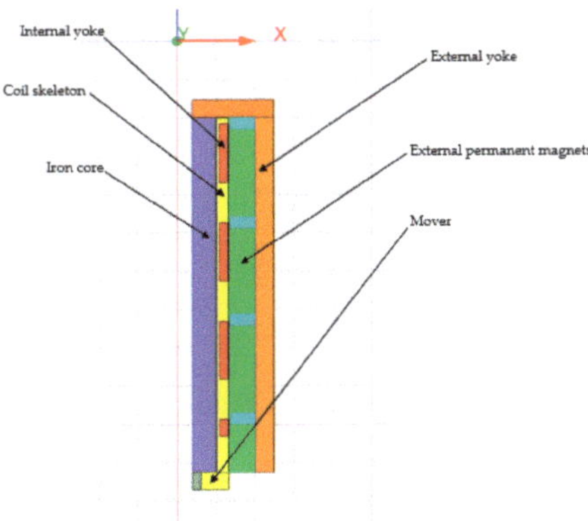

Figure 8. Two-dimensional model of the moving-coil linear motor under Maxwell.

Table 4. Materials for each part of the moving-coil linear motor.

Item Name	Material Name
(inner) inner magnetic yoke	Steel_1008
(outer) outer magnetic yoke	Steel_1008
(under) bottom plate	Steel_1008
(coil)	Cooper
(former) coil former	Teflon
(PM) Permanent magnet	NdFeB35

For axisymmetric moving-coil linear motors, only half of the 2D motor model needs to be created. Using the Maxwell 2D module, the coordinates are established so that the Z-axis is the axis of rotational symmetry, and a model is created, as shown in Figure 9, according to the specified motor parameters. Among them, Figure 9a–c show the magnetic contour distribution diagram, the magnetic vector distribution diagram, and the magnetic flux distribution diagram of the linear motor.

According to the simulation results, the maximum magnetic field line is 4.7422×10^{-4} Wb/m, and the maximum magnetic induction intensity is about 2.25 T. According to Gauss's law, each magnetic field line is closed, and the size of the magnetic field line gradually decreases from the inside to the outside due to the mutual repulsion of the magnetic field lines, resulting in each magnetic field line being independent and not intersecting with other magnetic field lines.

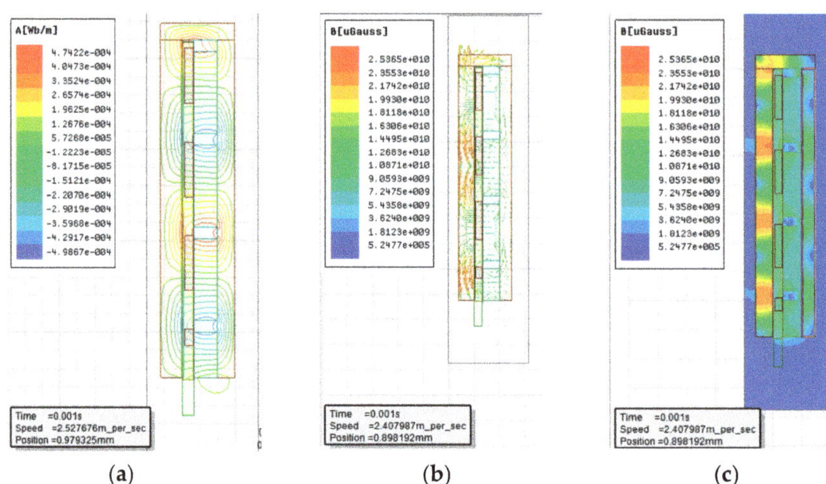

(a)   (b)   (c)

Figure 9. Magnetic distribution diagram of the linear motor. (**a**) Magnetic contour distribution diagram of the linear motor; (**b**) Magnetic vector distribution diagram of the linear motor; (**c**) Magnetic flux distribution diagram of the linear motor.

Through simulation, it is possible to know the thrust fluctuation, the promoter motion velocity variation, and the promoter position change of the moving-coil linear motor, with the specific simulation results shown in Figures 10–12.

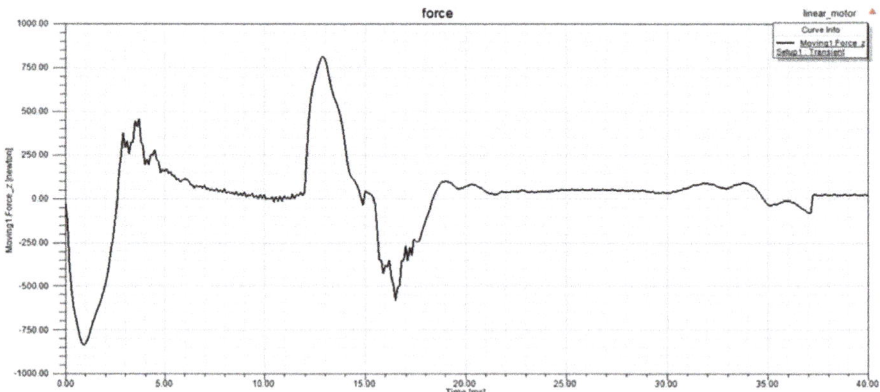

Figure 10. Positioning thrust fluctuation diagram of the moving-coil linear motor.

When the linear motor begins operation, its thrust is negative and increases rapidly. With the passage of time, the thrust gradually decreases, with the direction turning from negative to positive. At about 13 ms, the thrust reaches the maximum value of about 800 N, which satisfies the expected design requirements. The thrust then gradually decreases and hovers around 0 N.

The designed moving-coil linear motor has a maximum stroke of 20 mm. According to the variation curve of its motion position with time, the promoter of the moving-coil linear motor completes two reciprocating motions within 40 ms, reaching the maximum stroke of 20 mm at about 5 ms and 23 ms, respectively.

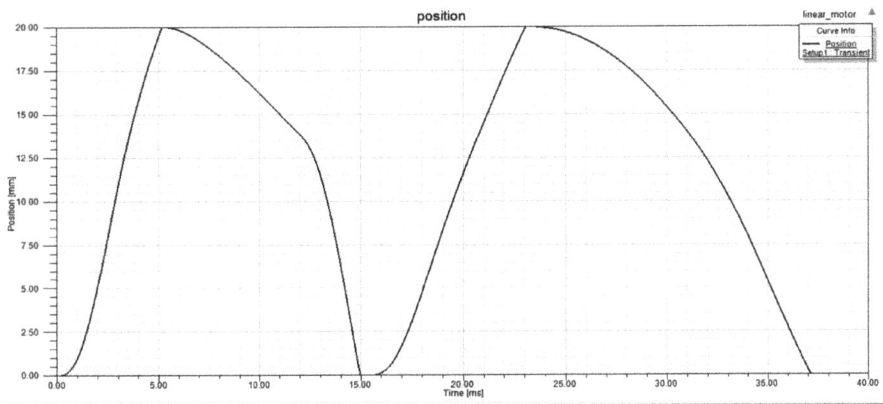

Figure 11. Linear motion position diagram of the moving-coil linear motor.

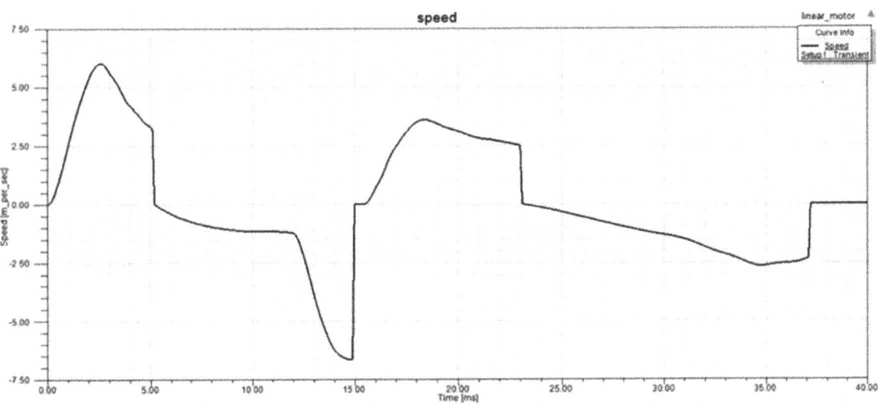

Figure 12. Velocity fluctuation diagram of the moving-coil linear motor.

At the beginning, the promoter velocity first increases and then decreases. When the maximum motion stroke is reached, that is, after about 5 ms, the velocity is 0. Then, the promoter motion velocity increases in the opposite direction. In the meantime, the promoter performs a counter motion and returns to its origin with a velocity of 0 at about 15 ms. Afterward, the motion velocity increases in the positive direction. When the maximum stroke is reached again, the velocity slows to 0 at about 23 s. Then, it accelerates in the reverse direction, and the promoter returns in the opposite direction, with its velocity decreases to zero at the end of the motion.

The above simulation results show that, when the linear motor starts operation, the thrust increases in a reverse direction under the system interference, and both the velocity and motion stroke of the moving coil are 0 mm. Over time, the thrust changes from negative to positive, and the motion velocity of the moving coil also increases, with the motion stroke curve presenting an upward trend. When the maximum stroke is nearly reached, the linear motor current provided by the system decreases, with the motion velocity and thrust of the moving coil also decreasing. The velocity slows to 0 m/s when the moving coil stroke is 20 mm. During the return of the moving coil, the velocity increases in the reverse direction, and the thrust decreases slowly with the decrease in the stroke until it reaches 0 mm. At about 15 ms, the moving coil restores the initial state in terms of thrust, velocity, and stroke.

4.2. Dynamic Simulation and Analysis of the Moving-Coil Linear Compressor under Simulink

4.2.1. Dynamic Modeling Simulation

According to the working principle of the moving-coil linear compressor, a model of air pressure under the four working conditions can be established in Simulink. When the piston moves, the piston movement needs to be exchanged between the four states, and the corresponding conditions need to be set to change the state. This can be achieved with the help of the chart module in Stateflow, as shown in Figure 13. In order to bring the simulation closer to reality, special cases must be considered, including when the electromagnetic force is too small, so that the displacement does not reach 0.003471 m during expansion and the speed is already less than 0 m/s, which causes the expansion state to jump directly into the compression state. Similarly, during the compression process, when the displacement does not reach 0.006050 m, the speed is already greater than 0 m/s, causing the compression state to jump directly into the expansion state. The Simulink model of the gas load force is built; the input signals are the displacement and velocity, and the output signals are the gas load force and the gas pressure in the cylinder, as shown in Figure 14.

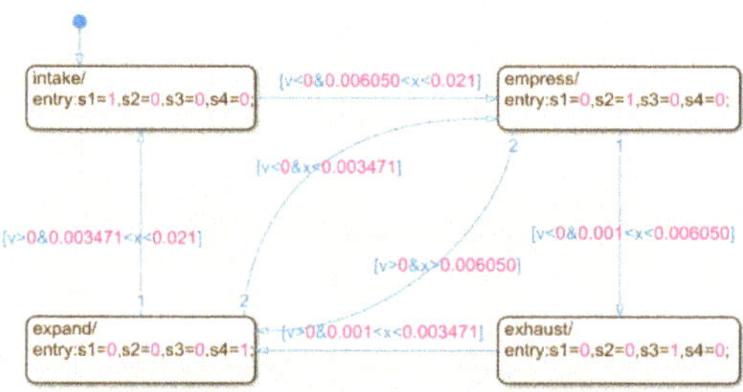

Figure 13. Stateflow model diagram of piston movement.

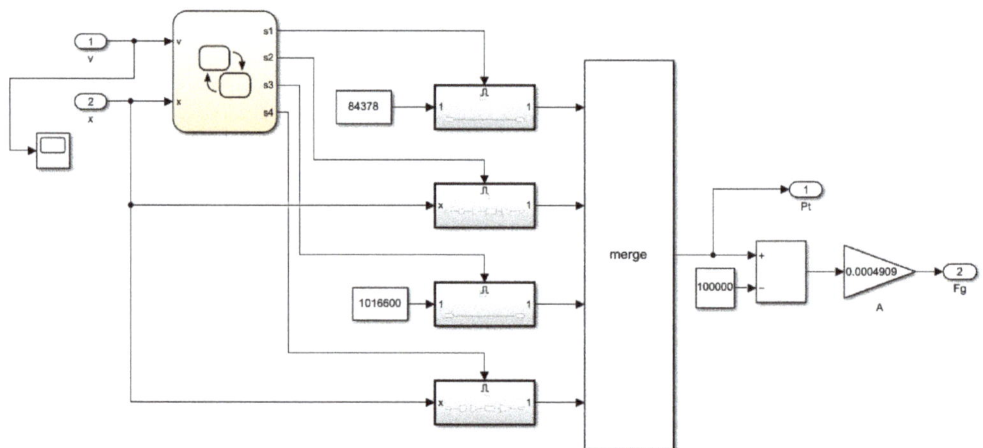

Figure 14. Simulink model diagram of gas load force.

After the gas load force Simulink model is built, according to Equation (5), the mechanical subsystem of the Simulink model could be obtained, and the output signals of this model are the displacement x, velocity v, current I, and electromagnetic driving force $F_e(t)$, as shown in Figure 15.

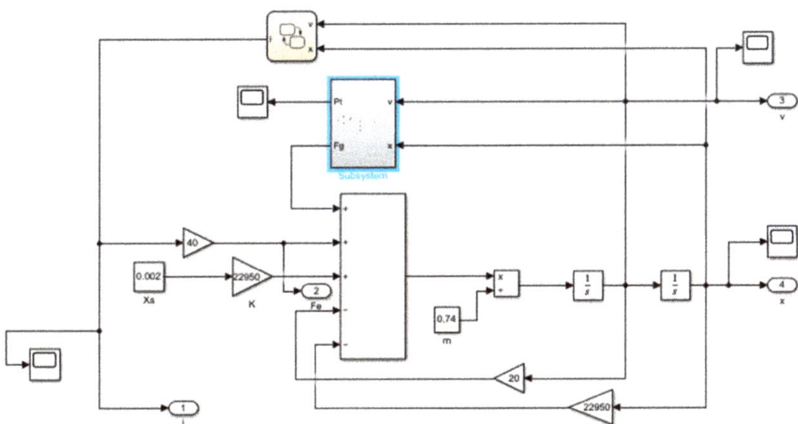

Figure 15. Power system model diagram of the moving-coil linear compressor.

According to Equation (9), the electromagnetic subsystem of the Simulink model could be obtained. The input signals of the electromagnetic subsystem of the Simulink model are the motion speed v of the piston and the current i in the moving coil, and the output signal is the voltage U in the moving coil, as shown in Figure 16.

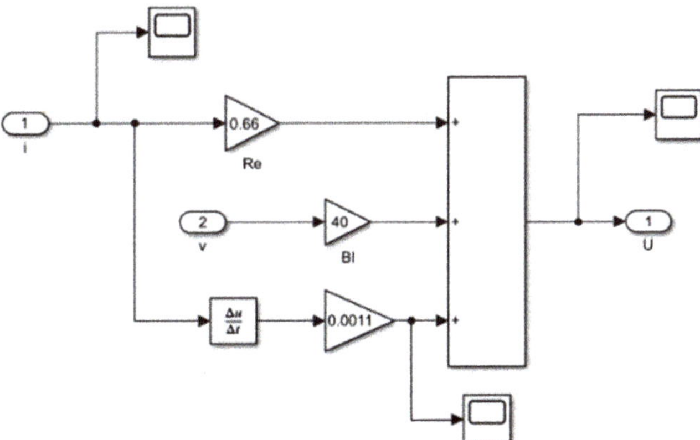

Figure 16. Model diagram of electromagnetic subsystem.

According to Equation (10), the calculation model of the efficiency of the direct-drive compressor [22] with an electromagnetic actuator in Simulink could be obtained, as shown in Figure 17. The input signals of this model are the electromagnetic driving force, the speed v of the moving coil, the voltage U across the moving coil, and the current i in the moving coil. The output signal is the efficiency of the permanent-magnet moving-coil linear compressor. As can be seen from Figure 18, the Simulink model for this efficiency continuously accumulates the instantaneous power value of the permanent-magnet moving-coil linear compressor and the instantaneous power value of the energy of the permanent-magnet moving-coil linear compressor's external input, and divides the two; then, the efficiency of the permanent-magnet moving coil linear compressor is obtained. With an increase in simulation time, the efficiency value tends to a stable value, which is the efficiency value of the linear motion of the permanent-magnet moving-coil linear compressor.

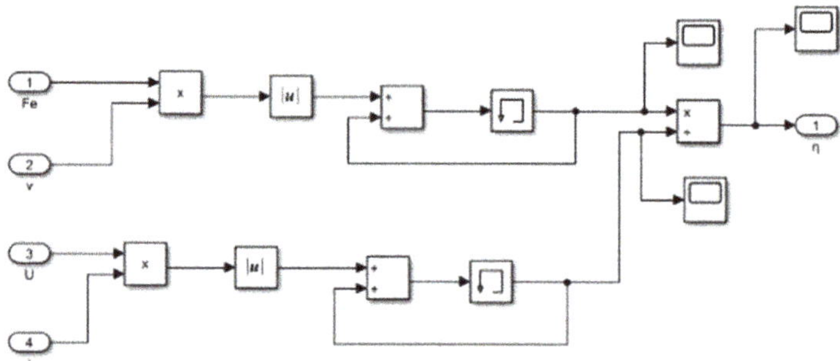

Figure 17. Efficiency calculation module's model diagram.

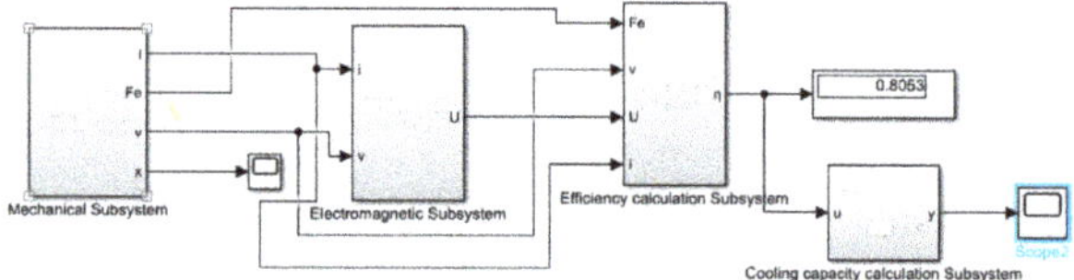

Figure 18. Simulink model diagram of the linear compressor.

The dynamic linear compressor in the Simulink model is composed of the mechanical subsystem, electromagnetic subsystem, and compressor cooling capacity calculation module. The refrigeration capacity is the product of the unit refrigeration capacity and the mass flow; the refrigeration capacity is linearly related to the mass flow, as the refrigeration capacity is proportional to the mass flow. Therefore, a linear function module is established, and the refrigeration capacity of the linear compressor [23] could be obtained. The Simulink model of the moving-coil linear compressor is shown in Figure 18. The efficiency of the linear motor currently reaches more than 80%, and there is still room for improvement in the motor efficiency after further control.

4.2.2. Dynamic Simulation Analysis

According to the above Simulink linear compressor model, the simulation results obtained using the trial method are as follows:

The push-to-push control strategy [24] is used to set the magnitude of the current in the coil. By feeding periodic DC power to the armature linear compressor, the DC generates periodic electromagnetic forces in different directions and is accompanied by damping forces. It has been calculated that the linear motor compressor requires at least a 450 N electromagnetic driving force. The working process of the direct-drive compressor with an electromagnetic actuator is shown in Figure 19.

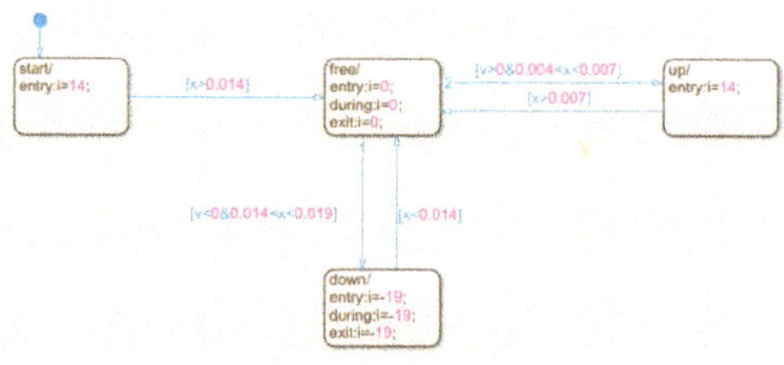

Figure 19. Signal input diagram of electromagnetic force current.

When the direct-drive compressor with an electromagnetic actuator is energized and starts working, the moving coil is first given a positive excitation current of 14 A. When the piston moves from the initial displacement of 11 mm to the displacement of 15 mm, the current in the moving coil becomes zero. Then, the piston moves under the action of inertia to a displacement of 20 mm. The piston then begins to move in a negative direction into the compression process. When the piston moves to a displacement of 19 mm, the current in the moving coil becomes −19 A until the piston moves to a displacement of 14 mm. When the current in the moving coil becomes zero, then the piston moves under the action of inertia to a displacement of 6.050 mm, and the compression process ends. The piston then enters the exhaust process under the action of inertia until it moves to a displacement of 1 mm, at which point the velocity is zero and the exhaust process ends. Under the action of the spring, the piston begins to move in a positive direction into the expansion process. When the piston moves to a displacement of 4 mm, the current becomes 14 A, and when the piston moves to 3.471 mm, the expansion process ends. Then, the piston still moves positively under the action of the electromagnetic force and enters the suction process; when the piston moves to a displacement of 7 mm, the current becomes zero. Then, the piston moves to 21 mm under the action of inertia. At this point, the speed is zero and the intake process ends. Then, the piston moves in a negative direction to enter the compression process, and the cycle begins again.

Using the trial method, each group of current is simulated in the whole system to obtain the appropriate current parameters, that is, this set of current data allows the linear compressor to work continuously to prove the feasibility of the compressor's control strategy accordingly. The data obtained by the trial method can be input into the system, and the simulation results of the direct-drive compressor with an electromagnetic actuator can be obtained, as shown in Figures 20–22.

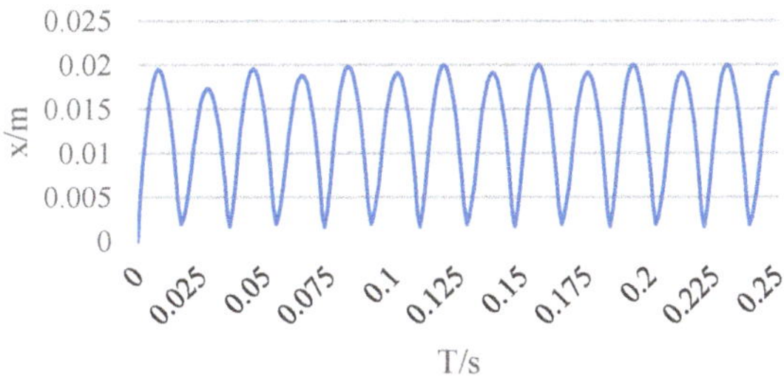

Figure 20. Dynamic linear compressor displacement simulation waveform.

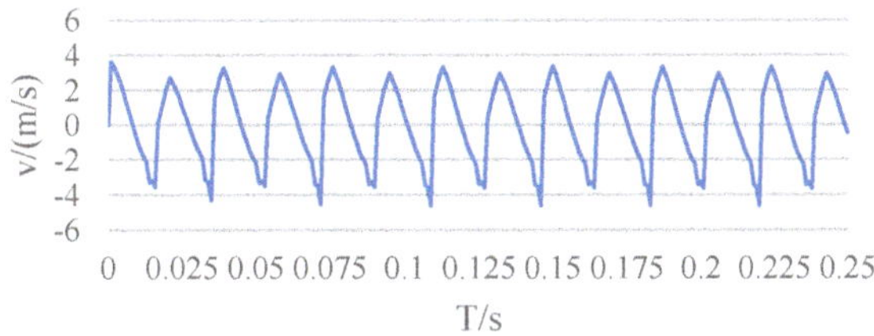

Figure 21. Dynamic coil linear compressor speed waveform diagram.

The piston carries out periodic movement in the cylinder. The amplitude of the piston is kept at about 20 mm, and there is a slight fluctuation, with a maximum amplitude between [−0.01, 0.020]. This error is within the allowable range. The result is fully in line with the working requirements of the moving-coil linear compressor.

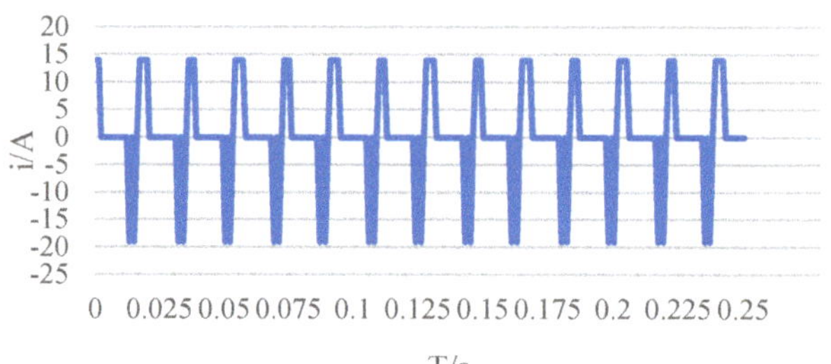

Figure 22. Current waveform diagram.

From the waveform diagram of the motion speed of the linear compressor piston, it can be seen that the simulation results of speed are kept within the $[-6, 6]$ range, the movement speed of the piston is kept within the allowable maximum speed of the compressor, and the movement speed is reasonable and meets the working requirements of the compressor.

The coil current is the input signal, and it can be seen from Figure 16 that the change law of the input coil current setting is specified when setting the model. Due to the unstable working state of the compressor at start-up, the input of the coil current fluctuates slightly, but as the compressor operates, the working state gradually stabilizes, and the coil current remains periodically changing between -19 A and 14 A. The electromagnetic force drives the compressor to work normally, and the current flow meets the basic requirements of the compressor.

According to the efficiency calculation principle of the linear compressor motor and Equation (10), from the output waveform diagram, it can be seen that the efficiency value increases rapidly, until it reaches 80.93%, and slowly becomes stable, which is in line with the basic requirements of compressor work, as shown in Figure 23.

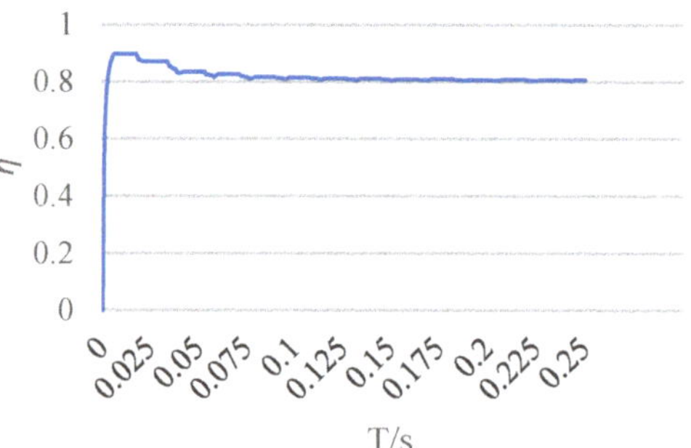

Figure 23. Linear motor efficiency waveform diagram.

In the figures below, Figure 24a–e shows the simulation results of the compressor motor efficiency under different strokes, different compression ratios, different coil resistance and damping, different mover masses, and different piston diameters, respectively.

By changing the piston stroke, for example, we set the piston stroke to 15 mm and 25 mm, we can see that the stroke has almost no effect on the efficiency of the motor. In subsequent work, we will control the model more accurately through a control algorithm to observe the factors influencing efficiency more clearly. Three sets of displacements are listed, which are 15 mm, 20 mm, and 25 mm, and the efficiency comparison chart is shown in Figure 24a.

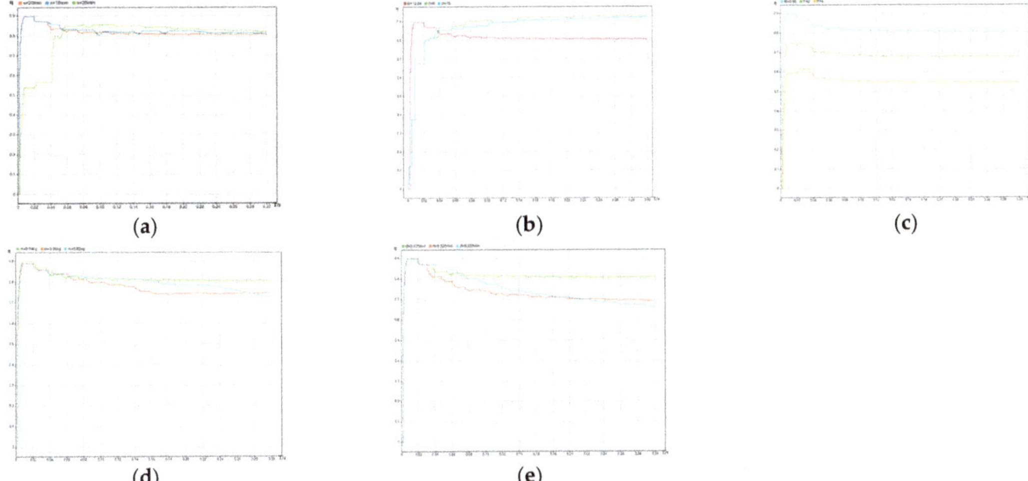

Figure 24. Efficiency under different conditions. (**a**) The efficiency of the compressor at different strokes; (**b**) The efficiency of the compressor at different compression ratios; (**c**) The efficiency of the compressor under different resistance and damping; (**d**) The efficiency of the compressor under different mover masses; (**e**) The efficiency of the compressor at different piston diameters.

It can be seen from Figure 24b that with an increase in the compression ratio, the compressor efficiency gradually increases, and the compression ratio has a greater impact on the compressor efficiency. However, it is not the case that larger compression ratio is better; the choice of compression ratio should be considered comprehensively in order to reduce energy consumption as much as possible, and the compression ratio should not be too large.

As shown in Figure 24c, the simulation curve of compressor efficiency under different coil resistances is comparable, and the influences of resistance and damping on efficiency are comparable. It can be seen from the figure that as the coil resistance increases, the compressor efficiency decreases. This is because the coil resistance directly affects the size of the driving current, which in turn affects the supply voltage, resulting in a decrease in compressor efficiency. The magnitude of the damping directly affects the acceleration of the system, thereby affecting the efficiency of the compressor. Therefore, in the design and manufacturing process of dynamic linear compressors, the damping friction between the piston and the cylinder and the coil resistance should be reduced as much as possible.

The mover mass of the linear compressor is an important factor affecting the efficiency of the compressor, as shown in the simulation curve of the compressor efficiency under different mover masses (Figure 24d). As the quality of the piston increases, the compressor becomes less and less efficient. This is because the greater the mass of the piston, the more energy the system loses and, thus, the lower the efficiency.

The piston diameter is also an important factor affecting the efficiency of the motor and the compressor, as shown in the simulation curve of the compressor efficiency under different piston diameters (Figure 24e). This is because as the piston diameter increases, the gas load force also increases, and the greater the force that the motor needs to overcome, the lower the compressor efficiency is.

5. Discussion and Conclusions

In this paper, a direct-drive compressor with an electromagnetic linear actuator is designed, and the moving-coil linear motor and linear compressor are simulated by the Maxwell software and Simulink software, respectively. This paper establishes a mathematical model for the direct-drive compressor with an electromagnetic actuator. However, there are still areas for improvement, such as the stepless adjustment of refrigeration capacity through control strategies, so that the efficiency of the compressor is further improved, so as to improve the refrigeration efficiency and reduce energy consumption. The main conclusions are as follows:

1. Based on the analysis of the working principle of moving-coil linear compressors, a direct-drive compressor with an electromagnetic linear actuator was designed. Compared to a traditional compressor, this paper adopts the direct-drive mode of high-performance electromagnetic linear actuator, which simplifies the structure, has no friction, reduces energy consumption, realizes the miniaturization and intelligent needs of the compressor, and improves the compressor efficiency.
2. The Maxwell and Simulink software were used to simulate the kinematics of the moving-coil linear compressor. The "push-to-push" control strategy was used to control the compressor current. The piston velocity displacement and electromagnetic force current exhibit periodic changes and meet the design requirements. Although the established design requirements are met, further experiments and improvements are needed in practical applications. This verifies the feasibility of a direct-drive compressor with a linear electromagnetic actuator.
3. Through the method of multiple trials, a set of data could be obtained to make the compressor work continuously, and the efficiency of the linear motor could reach 80%. In future work, through employing a control algorithm, the efficiency of the linear motor and the cooling efficiency can be improved.
4. Through simulation, the efficiency of the compressor under different strokes, different mover masses, different resistance and damping, different compression ratios and different piston diameters is obtained, which further verifies the feasibility of using electromagnetic actuators to directly drive compressors.

Author Contributions: J.Z. was responsible for modeling and simulating the compressor using the Maxwell and Matlab/Simulink software, as well as reviewing and revising the article to complete the final draft. M.X. wrote the article and completed the first draft. J.D. was responsible for analyzing and managing the simulation data and verifying the results. Z.Y. and J.Y. were responsible for the collection and collation of preliminary materials and preparations for later work. All authors have read and agreed to the published version of the manuscript.

Funding: This research was funded by the National Natural Science and Technology Foundation of China (Grant No. 51605183, 51975239); the Industry-University-Research Cooperation Project of Jiangsu Province (Grant No. BY2020695); and the Natural science research project of Jiangsu Province colleges and universities (Grant No. 19KJD460002).

Data Availability Statement: Not applicable.

Conflicts of Interest: We declare no conflict of interest. The funders had no role in the design of the study; in the collection, analyses, or interpretation of data; in the writing of the manuscript; or in the decision to publish the results.

References

1. Jin, T.; Zheng, S.; Xie, J.; Ma, Z. Research Status and Development of Linear Compressor. *China Mech. Eng.* **2004**, *15*, 89–93.
2. Hassan, A.; Bijanzad, A.; Lazoglu, I. Electromechanical modeling of a novel moving magnet linear oscillating actuator. *J. Mech. Sci. Technol.* **2018**, *32*, 4423–4431. [CrossRef]
3. Phadkule, S.; Inamdar, S.; Inamdar, A.; Jomde, A.; Bhojwani, V. Resonance analysis of opposed piston linear compressor for refrigerator application. *Int. J. Ambient. Energy* **2019**, *40*, 775–782. [CrossRef]
4. Chen, X.; Jiang, H.; Li, Z.; Liang, K. Modelling and Measurement of a Moving Magnet Linear Motor for Linear Compressor. *Energies* **2020**, *13*, 4030. [CrossRef]
5. Xie, J.; Jin, T.; Tong, S. Research Status and Development Trend of Linear Compressor. *Fluid Mach.* **2004**, *32*, 31–35.
6. Gao, Y.; Hong, Q.; Fan, X.; Zhang, Z.; Wang, L.; Gao, R. Design and Experimental Research of Lightweight High-Efficiency Linear Compressor. *Cryog. Supercond.* **2017**, *45*, 23–27.

7.   Fang, X.; Chen, X.; Bao, X. Thermodynamic Analysis and Experimental Verification of a New Oilless Linear Compressor. *Cryog. Supercond.* **2021**, *49*, 73–77.
8.   Zhang, K.; Chen, J.; Li, H.; Bi, X.; Zou, D.; Yang, Y.; Shi, H. Research on Support Performance of Flexible Spring Assembly for Linear Compressor. *Infrared Technol.* **2020**, *42*, 198–203.
9.   Chen, H.; Zhao, Z.; Wang, X.; Ning, H.; Chen, H. Electromagnetic Field Simulation and Magnetic Yoke Structure Parameter Optimization of Linear Compressor. *Mach. Des. Res.* **2020**, *36*, 169–174.
10.   Zhu, S. Movement of linear compressor and displacer in a displacer pulse tube refrigerator. *Cryogenics* **2018**, *97*, 70–76. [CrossRef]
11.   Yan, H.; Liu, Y.; Lu, L. Turbulence anisotropy analysis in a highly loaded linear compressor cascade. *Aerosp. Sci. Technol.* **2019**, *91*, 241–254. [CrossRef]
12.   Chen, L.; Li, L. Overview of the development of linear motors for compressors and their key technologies. *Proc. CSEE* **2013**, *33*, 52–68+15.
13.   Bijanzad, A.; Hassan, A.; Lazoglu, I.; Kerpicci, H. Development of a new moving magnet linear compressor. Part A: Design and modeling. *Int. J. Refrig.* **2020**, *113*, 70–79. [CrossRef]
14.   Luo, L.; Zhang, G.; Liang, J.; Wang, W.; Xu, Z.; Gu, X.; Guo, Y.; Liang, S. Study on the high response characteristics of moving coil linear motors. *Mach. Tool Hydraul.* **2016**, *44*, 54–58+120.
15.   Abdalla, I.I.; Ibrahim, T.; Nor, N.B.M. Development and optimization of a moving-magnet tubular linear permanent magnet motor for use in a reciprocating compressor of household refrigerators. *Electr. Power Energy Syst.* **2016**, *77*, 263–270. [CrossRef]
16.   Mao, J.; Wang, N.; Chen, H. Dynamic characteristics of kinetic linear compressors. *Mach. Des. Res.* **2022**, *38*, 205–209+214.
17.   Zou, H.; Li, C.; Tang, M.; Wang, M.; Tian, C. Online measuring method and dynamic characteristics of gas kinetic parameters of linear compressor. *Measurement* **2018**, *125*, 545–553. [CrossRef]
18.   Liu, Y. Design of Control System of Dynamic Magnetic Linear Motor for Compressor. Master's Thesis, Shenzhen University, Shenzhen, China, 2017.
19.   Wang, L. Dynamic Gas Analysis and Valve Research of Linear Compressors for Refrigeration. Master's Thesis, Hefei University of Technology, Hefei, China, 2013.
20.   Tan, J.; Dang, H. Effects of the driving voltage waveform on the performance of the Stirling-type pulse tube cryocooler driven by the moving-coil linear compressor. *Int. J. Refrig.* **2017**, *75*, 239–249. [CrossRef]
21.   Yang, X.; Lu, D.; Ma, C.; Zhang, J.; Zhao, W. Analysis on the multi-dimensional spectrum of the thrust force for the linear motor feed drive system in machine tools. *Mech. Syst. Signal Process.* **2017**, *82*, 68–79. [CrossRef]
22.   Zhu, J.; Liu, W.; Tang, W.; Li, J.; Li, H.; Xu, X.; Dai, J.; Wang, C. A Moving-Coil Linear Compressor. Patent CN212028006U, 27 November 2020.
23.   Kan, X.; Wu, W.; Yang, L.; Zhong, J. Effects of End-Bend and Curved Blades on the Flow Field and Loss of a Compressor Linear Cascade in the Design Condition. *J. Therm. Sci.* **2019**, *28*, 801–810. [CrossRef]
24.   Sun, Z.; Shi, Y.; Zhang, H. Research on linear compressor control strategy based on Push-to-Push principle. *Mech. Eng. Autom.* **2016**, *45*, 167–170.

 *actuators*

Article

Improvements on the Dynamical Behavior of a HiL-Steering System Test Bench

Alexander Haas [1], Benedikt Schrage [1], Gregor Menze [1], Philipp Maximilian Sieberg [2,*] and Dieter Schramm [2]

[1] Dr. Ing. h.c. F. Porsche AG, Porschestraße 911, 71287 Weissach, Germany; alexander.haas3@porsche.de (A.H.)
[2] Department Mechanical Engineering, University of Duisburg-Essen, Forsthausweg 2, 47057 Duisburg, Germany
* Correspondence: philipp.sieberg@uni-due.de; Tel.: +49-203-379-1862

Abstract: Shorter available development times and fewer available vehicle prototypes have increased the subsystem-based investigation on test rigs within the automotive development process. Steering systems exhibit a direct interface to the driver, therefore, posing high requirements to the control performance of a test bench, especially for the perception of steering feel. This work proposes three approaches to improve the force control performance of permanent magnet linear motors incorporated on a steering test bench. The first method improves control accuracy when a harmonic force signal is introduced into the steering system by adjusting the reference force signal based on the identified peak values of the measured and reference forces. The second method allows the inclusion of the actuator's inertia and the occurring ratios between steering wheel angle and rack displacement into the control scheme to reduce performance deterioration due to inertia. The third approach considers delay time in the actuator control and estimates its future position for delay compensation. A validation of the proposed methods is conducted, displaying an improvement for all three applications. The proposed methods extend the applicability of a steering test bench within the automotive development process by enabling more accurate and reproducible control performance.

Keywords: steering system test bench; permanent magnet linear synchronous motor (PMLSM); electric power steering (EPS); steering feedback; steering guidance behavior

Citation: Haas, A.; Schrage, B.; Menze, G.; Sieberg, P.M.; Schramm, D. Improvements on the Dynamical Behavior of a HiL-Steering System Test Bench. *Actuators* **2023**, *12*, 186. https://doi.org/10.3390/act12050186

Academic Editors: Qinfen Lu, Xinyu Fan, Cao Tan and Jiayu Lu

Received: 29 March 2023
Revised: 18 April 2023
Accepted: 23 April 2023
Published: 25 April 2023

1. Introduction

Modern steering systems for passenger vehicle cars possess an electric motor, which provides the necessary servo force to support the driver during steering maneuvers. During the automotive development process, the steering feel is improved, often in full vehicle tests. An alternative to road tests is the utilization of a steering system test bench, which allows the objective investigation of the steering system, without the requirement of a vehicle prototype.

Typical tests on steering test benches range from measurements of mechanical characteristics, such as the friction force of the steering gear, to approximations of full vehicle tests. Since electric power steering (EPS) systems exhibit large servo forces during parking, steering test benches for the investigation of such systems are required to provide sufficiently large actuator forces to imitate the wheel–road contact for real word applications. Currently, this goal is achieved by the usage of one or two permanent magnet linear synchronous motors (PMLSM).

Due to the high-power requirements, the utilized PMLSMs exhibit non-neglectable moving masses, deteriorating the dynamic behavior of the system. This is especially relevant when rapid steering wheel angle inputs are present. Furthermore, EPS systems represent nonlinear characteristics with friction, backlash, and natural frequencies within the investigated frequency range, which negatively influence the control performance, exemplarily when harmonic rack forces are introduced into the steering system. These shortcomings limit the applicability of steering test benches along the development process and require optimization.

Common approaches utilized to improve the dynamic control of regulated systems incorporate the inverse plant model into the control scheme [1,2]. In [3,4], a combined offline and online estimated transfer function is utilized to adapt the reference signal to the dynamics of the system under test, which is represented as a shaking table. For the online implementation, a recursive extended least square algorithm is used. A similar approach is described in [5], wherein an online recursive extended least square algorithm adaptively alters the transfer function for acceleration, force and position control of a shaking table. Another force control approach is introduced in [6], where the inverse model approach was successfully implemented within a flight simulator.

The estimation of a control transfer function on a steering test bench is connected to high efforts. Since the steering system represents an active component with friction, backlash and a variable driver input, steering systems are nonlinear systems. Consequently, the transfer function is only valid for one system state. Due to a great variety of variations in the system, transfer function-based approaches are not applicable for this purpose.

Similar, but alternative, approaches to control harmonic reference signals on a shaking table were introduced in [7,8]. Both methods utilize a least mean square filtering approach to weigh or adapt the input signal to compensate phase and amplitude deviation in the system response. Here, the adaptation occurs due to a comparison of the resulting and measured signals of the shaking table, without the explicit knowledge of the transfer function. Therefore, a similar approach can also be used on a steering test bench, where harmonic excitations are inserted into the steering rack.

Alternative approaches for the improvement of the dynamic characteristics of control systems are the usage of feedforward control signals based on velocity or acceleration information. In the discipline of hard disc drive control, there exist high requirements for the dynamic control performance, since the distance between a storage and the write/read pin needs to be small, but physical contact must not occur, even in the presence of external disturbances. For this purpose, [9] introduced an acceleration feedforward control (AFC), which detects external disturbances to improve the control performance. In [10], a similar approach was utilized to reduce the impact of external shocks to the drive system by an AFC in combination with a double disturbance observer, which successfully enhanced the control quality.

Within machining control applications, feedforward control approaches are exposed to changing boundary conditions, introduced by the workpiece characteristics, tool degradation and travel speed. Therefore, [11] implemented an adaptive feedforward control based on the reference position signal and the occurring error together with friction compensation to reduce the resulting position error. Approaches for inertia compensation by a feedforward control are even further restricted since the workpiece and direction of excitation vary over time. Consequently, [12] developed a velocity and AFC in combination with a jerk disturbance observer to compensate for varying inertia during the machining process. Similar approaches were conducted in [13–15], where observers and artificial intelligence are implemented to estimate the current inertia of the system.

Previously described approaches for enhanced dynamics are implemented for position or velocity control and for predefined reference signals. A torque control application with feedforward control for friction and inertia compensation was proposed in [16] on a rehabilitation device. The compensation is based on the currently measured angular velocity and the motor inertia, which successfully improves the control performance. A similar approach is also applicable for the implementation on a steering test bench, where the PMSLM is normally operated in force control mode.

Since steering system test benches are required to function within real-time applications, exemplarily together with a vehicle model, there are not only dynamic requirements in terms of magnitude accuracy but also concerning the time delay between the reference and measured signal. Such effects originate from dead time and the dynamic properties of the PMLSM control. A common method to deal with and improve the control performance of systems with dead time is the implementation of a Smith predictor [17]. Here, a model of the process is utilized to derive a mathematical formulation without a delay in the system, enabling the use of standard PID control tuning processes for systems without dead time. The work in [18] proposed an adaptation of a Smith predictor in combination with a fuzzy-PI controller on a brushless DC motor to account for modelling uncertainties, while an approach with an automatic dead time calculation is introduced in [19].

Another method to improve the dynamic behavior of a brushless DC motor incorporates time-shifting the voltage signal according to the electrical characteristics of the motor in terms of inductance and resistance [20]. The identification of the shifting time for the voltage signal is performed for different goals, such as performance [20] or acoustics [21]. A similar approach is proposed in [22], wherein a phase delay compensation is utilized to improve the efficiency of the electric motor and its generated torque. Here, the optimum shifting time is estimated based on a Fourier series. These approaches utilize information of the current system state to alter the control strategy to improve performance. The basic principle is also applicable for steering test bench applications in real-time operation.

For PMLSMs implemented in steering system test benches, the prevailing operation mode is force control, which can be subdivided into two categories, based on the origin of the reference force signal. First, a reference force signal is provided in advance of the test to the PMLSM for the excitation of the steering system. This setup is exemplarily utilized for the harmonic force excitation of the steering rack. Second, the force is calculated based on a vehicle model, where the rack displacement measured within the PMLSM functions as the input for the reference rack force calculation. For the most basic case, the PMLSM operates as a virtual spring and, therefore, provides a reference force proportional to the measured rack displacement.

In this paper, drawbacks of the control performance occurring for both control categories are identified, and associated improvement approaches are proposed. For a previously known harmonic reference signal, a similar methodology as in [7] is implemented to calculate a weighing factor to adapt the reference signal. Two additional methods, which are applicable for real-time testing, exemplarily combined with a vehicle model, use the steering wheel angle and steering rack-velocity information to provide the PMLSM control with a compensation signal for inertia and time delay.

The remaining article is structured as follows: Section 2 gives an overview of the steering system and the relevant steering test bench, followed by a problem formulation and the proposal of compensation approaches in Section 3. Results of the implemented control improvements are displayed and discussed in Section 4. Section 5 summarizes the conducted investigations.

2. Steering System and Test Bench

This section introduces both the structure and function of electric power steering systems as well as the steering system test bench used for the investigations.

2.1. Electric Power Steering System

Modern steering systems in vehicles are equipped with an electrical motor to assist the driver during driving maneuvers. Depending on the positioning of the electric motor, different variations of EPS exist [23]. In this work, the motor of the steering system is located parallel to the rack of the system, representing an axle parallel (apa) EPS. A simplified sketch of the system is displayed in Figure 1.

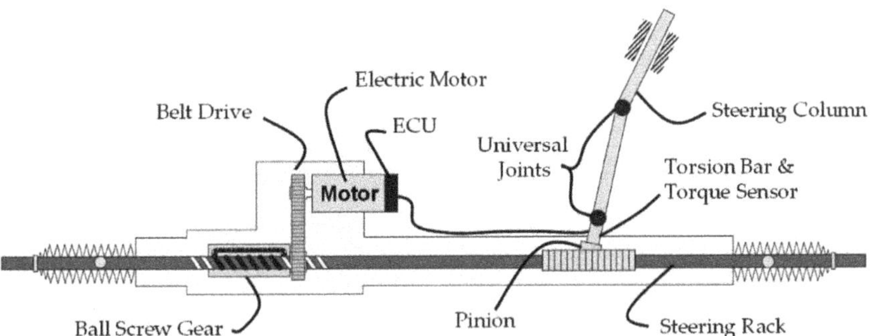

Figure 1. Axle-parallel electric power steering system. Adapted from [24].

The rotational driver input is introduced through the steering wheel into the steering column. Within the adjacent torsion bar, a sensor detects the applied steering torque. Universal joints in this path allow adjustment of the steering wheel position. Within the contact between the pinion and the steering rack, the rotational movement of the pinion is transformed into a translatory displacement of the steering rack. The ratio of the steering gear can be variable over the whole stroke, where the center range is generally less direct compared to larger steering wheel angles. This allows the reduction of steering angle demand during parking, while maintaining high stability under high-speed driving [23]. The translatory movement of the steering rack is then transferred by the tie rods to the wheel carrier, introducing a wheel rotation. Based on the measured steering torque at the steering systems torsion bar sensor and by different steering functions, a servo torque is calculated within the ECU and supplied by the electric motor. A belt drive and ball screw gear convert the assistance torque into an assistance force on the steering rack.

Due to the occurring inertia of the electric motor and the ball screw gear in combination with its elastic connection, the steering system exhibits a low-pass character for external excitations, originating from the tie rods [25,26]. In addition to the reduction of relevant road feedback to the driver [25], investigations also showed that this combination leads to a dominant natural frequency [26,27], complicating the force control during tests on a steering test bench.

2.2. Steering System Test Bench

Within this investigation, an electromechanical steering system test bench by dSPACE is utilized, which is described in [28], and consists of two PMLSMs for the introduction of road excitations to the steering rack and one steering wheel actuator (SWA) representing the driver input. Both PMLSMs are named according to their side of implementation, namely rod left actuator (RLA) and rod right actuator (RRA). The technical data are displayed in Table 1.

Table 1. Characteristics of the steering system test bench.

PMLSM			SWA		
Characteristic	**Unit**	**Value**	**Characteristic**	**Unit**	**Value**
max. Force	kN	20	max. Torque	Nm	50
max. Velocity	m/s	1	max. Velocity	°/s	3000
max. Frequency	Hz	30			

The PMLSM can be operated in position, velocity and force control. Here, only force control is considered. The SWA provides similar operation modes as position, velocity and torque control. These control tasks in addition to the relevant signal measurements are performed by a dSPACE Hardware-in-the-Loop (HiL) simulator [28]. A matlab/Simulink model allows the operator to adapt and extend the implemented control structures to match the requirements. Within this work, the Simulink model interface is utilized to implement the dynamic optimization methods.

Measurements on the steering system test bench range from the evaluation of the occurring friction in the passive system to HiL tests on full vehicle level, including a real-time vehicle dynamic model [26,27]. The setup of the test rig for HiL testing is shown in Figure 2.

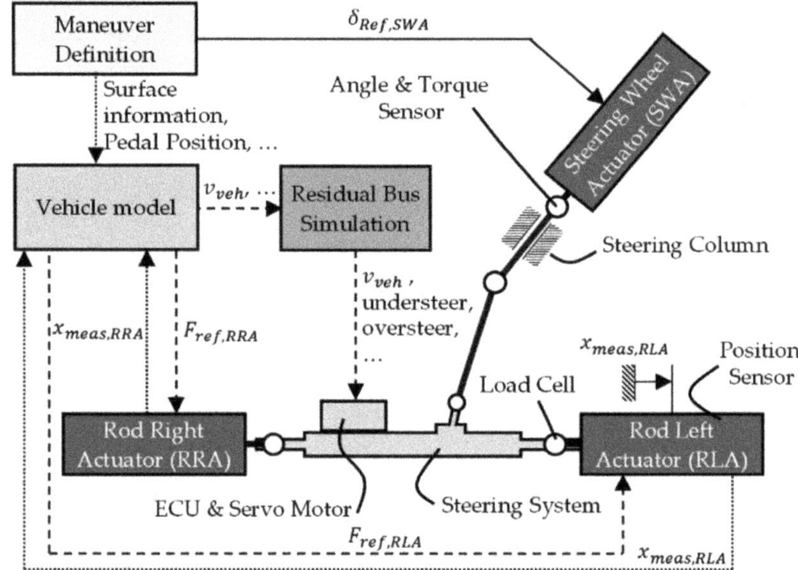

Figure 2. Setup of the steering system test bench for operation in combination with a vehicle model.

Based on the selected driving maneuver, the reference steering wheel angle $\delta_{Ref,SWA}$ is provided to the SWA. Further maneuver information, such as the surface geometry and pedal position, are sent to the vehicle model. The residual bus simulation provides the steering system with the necessary bus communication signals, such as vehicle speed $v_{veh'}$ for a proper function, such as within the actual vehicle.

As a response to the reference steering wheel angle, the SWA excites the steering column and induces a steering maneuver. Angle, velocity and torque sensors adjacent to the SWA detect the relevant quantities. Sensors incorporated within the PMLSMs measure the occurring rack displacement of the steering system, $x_{meas,RLA}$ as well as $x_{Meas,RRA'}$ and send this information to the vehicle model. Based on the rack displacement, the vehicle dynamic model calculates the wheel carrier rotation and the associated tire and rack forces. The latter are then utilized as the reference force signals $F_{ref,RRA}$ and $F_{ref,RLA}$ for the RRA and RLA, respectively. A load cell within the force path between the steering rack and the PMLSM detects the contact force for the closed-loop control. As a result, the required steering torque for the SWA to reach the reference angle, measured by an additional torque and angle sensor of the test rig close to the SWA, coincides with the steering torque perceived by a driver in a vehicle test.

Since the required rack forces for the previously described setup may originate from a vehicle model, the reference force signal is not known in advance to the test because it is derived from the occurring rack displacement, vehicle speed and surface condition. Alternative investigations, exemplarily the characterization of the feedback behavior of the steering system as explained in [26,27], utilize reference force signals with specific characteristics, which are defined as a time sequence before the test. Therefore, the reference force is not dependent on the steering system behavior and other surrounding conditions. Therefore, in this case, the displacement of the steering rack does not influence the desired reference force. The setup from Figure 2 remains similar, but the vehicle model is replaced by a time dependent signal generator, which provides a predefined reference force signal without incorporating the steering system behavior.

3. Problem Formulation and Compensation Approaches

In this section, two applications for the investigation of steering systems on a test bench are introduced, where dynamic insufficiencies deteriorate the measurement result. The first application describes the identification of the feedback behavior of the steering system, wherein a previously defined harmonic reference force signal is introduced into the steering rack. Due to the steering system's characteristics, the measured force exhibits deviation from the reference force. A second application investigates the guidance behavior of the steering system, where both the magnitude as well as the phase delay between the reference signal and the measured force do not adhere to the requirements. Solution approaches for both applications are motivated and derived.

3.1. Steering System Feedback

This subsection describes the feedback investigation test, motivates the occurring deviations and introduces a compensation approach.

3.1.1. Test Description

For the investigation of the feedback behavior of the steering system, only one PMLSM (the RLA) is connected to the steering rack, while the steering wheel is either swinging freely or locked at a constant angle, controlled by the SWA. Due to the high control performance of the SWA, the steering wheel is approximately fixed at its position and not rotating. Oscillations do occur for the case of a freely swinging steering wheel. A sine sweep signal with constant amplitude and increasing frequency is then used as the reference force signal for the connected PMLSM [26]. Based on the measured input force $F_{meas,RLA}$ and the steering torque sensed by the torsion bar M_{tb}, a transfer function $G_{FB,Steer}$ is calculated by a Fast Fourier Transformation (FFT) [29] of the input and output signal:

$$G_{FB,Steer} = \frac{FFT(M_{tb})}{FFT(F_{meas,RLA})}. \tag{1}$$

An exemplary bode plot of the identification result for a passive steering system is displayed in Figure 3, where a resonance frequency at 6 Hz is detected.

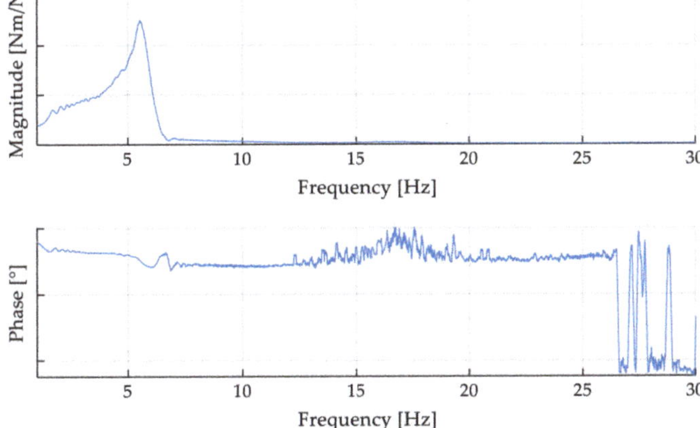

Figure 3. Bode diagram of the feedback behavior of a passive apa-EPS.

3.1.2. Motivation and Problem Formulation

Due to the occurring friction and backlash in the system in combination with an active servo motor, steering systems are nonlinear, which means that the identified transfer function is only valid for the exact condition of the test. To compare the results of different applications of whole steering systems, the boundary conditions of the measurement, such as the force amplitude, need to be held constant, requiring high dynamic performance control. Extrapolating the results outside the reference state is only valid for linear, time-invariant systems [29]. This requirement becomes clear when considering the support force of the servo motor, which does not correlate linearly to the rack force. Therefore, reduced or increased rack forces lead to different torsion bar torques and, therefore, to varying assistance forces, which change the share of the required steering torque of the driver. Consequently, the magnitude of the transfer functions increases or decreases, complicating a comparison for different rack force amplitudes.

As a result of the steering system characteristics, the resulting rack force amplitude exhibits deviations from the reference amplitude along the frequency range. An exemplarily resulting normed rack force course in the time domain is presented in Figure 4. Here, the reference amplitude indicates the desired amplitude of the input force sine sweep reference signal from 1 Hz to 30 Hz and ideally coincides with the amplitudes of the measured rack forces. At 13.8 s, the rack force displays an overshoot of 43.6% compared to the reference amplitude. For increasing time, which is equivalent to increasing frequency, the amplitude drops to 81% of the reference signal. These deviations hinder reliable comparison.

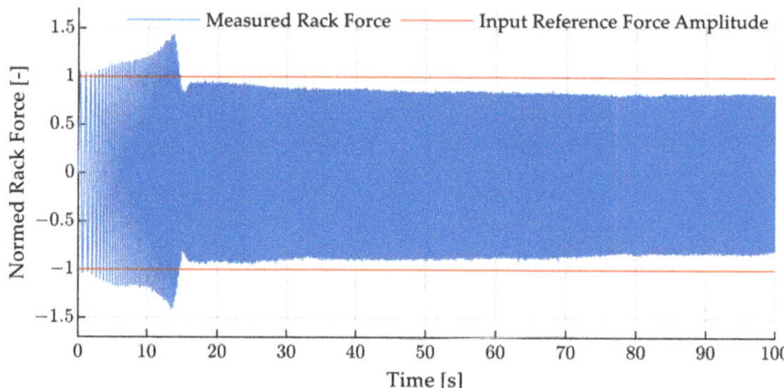

Figure 4. Time course of the measured rack force for a feedback behavior investigation on a steering system test bench.

3.1.3. Solution Approach

As the force control performance is strongly influenced by the characteristics of the steering system, a transfer-function-based approach, as introduced in [1–4], represents a valid compensation approach. Since the steering system is a nonlinear component, a transfer function for the test bench control must be derived for each operating point of the steering system, which is associated with a large effort. Therefore, an online approach which solely requires the reference and the measured force signal is proposed. To allow for wide applicability of the solution approach, an improvement is required for the following conditions:

- Harmonic signal with constant and/or increasing frequency;
- Harmonic signal with constant and/or varying amplitude;
- Harmonic signal with or without a varying and previously unknown offset value and their combination.

The latter case is especially important when the steering test bench is used in combination with a vehicle model and external disturbances are superposed to the rack forces from the vehicle model.

The basic principle of the proposed solution approach is, similar to [7], the adaption of the reference signal to consider the current dynamic properties of the complete control system. This means that the amplitude of the reference signal is increased for decreased measured rack force amplitudes and vice versa. The magnitude of the correction factor depends on the detected difference between the reference and measured rack force amplitudes. Therefore, a real-time capable algorithm is introduced which calculates a correction factor for the reference signal based on the currently measured force amplitudes, estimated by a peak value identification (PVID). A visualization of the PVID algorithm to detect a high point is displayed in Figure 5.

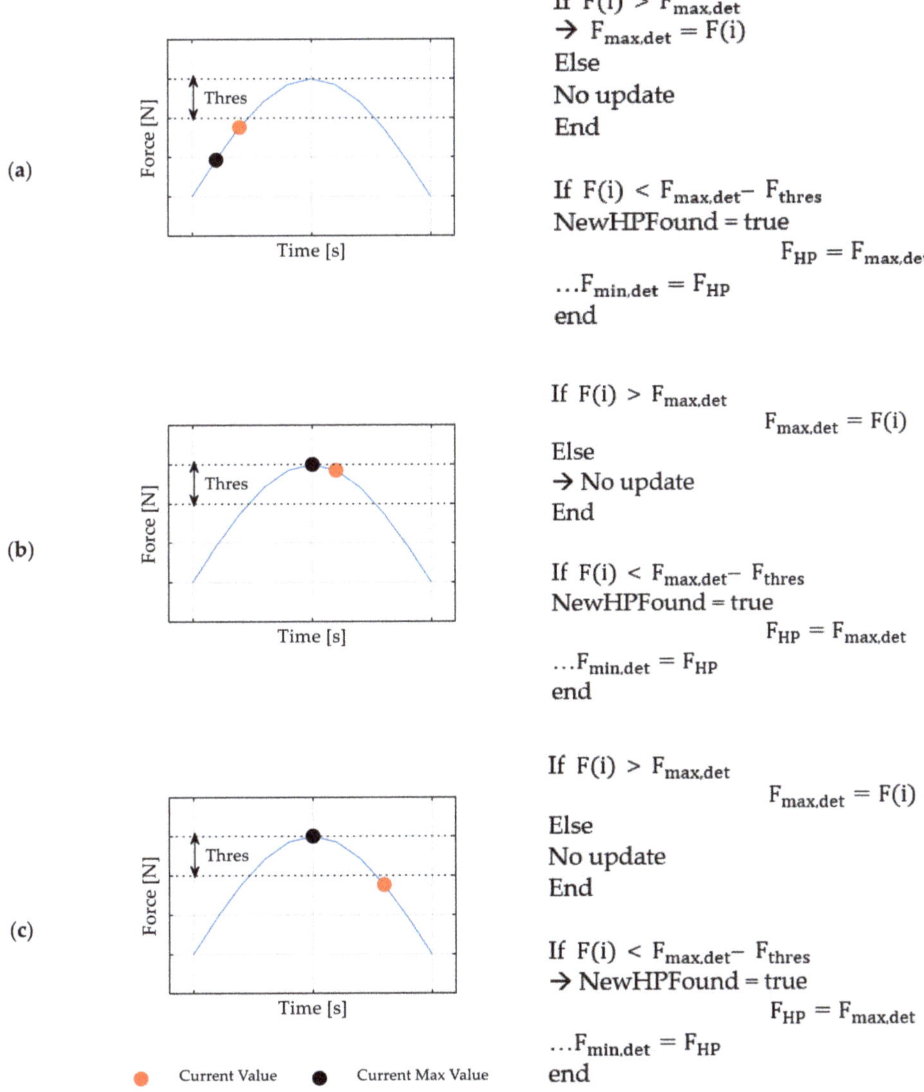

Figure 5. Visualization of the proposed peak value identification algorithm and pseudo code explanation for high point detection. (**a**) Update of the temporary maximum value; (**b**) search for further maximum values; and (**c**) trigger that a new high point is found for further calculations.

If the search for a new high point is triggered, a comparison of the current value $F(i)$ and the stored maximum value $F_{max,det}$ is performed for each timestep (see Figure 5a).

An update happens if the current value is larger than the previously stored value, otherwise the update is suspended (see Figure 5b). Since the force measurement is accompanied with measurement noise from the load cell, a threshold value is defined, which allows a distinction between the discovery of a new high point and noise-induced force reductions. As soon as the measured force value drops below a threshold, represented as the value which is F_{thres} lower than the currently stored maximum, a high point detection is triggered to update the currently reached force amplitude F_{HP}, as displayed in Figure 5c. Furthermore, the currently detected minimum value $F_{min,det}$ is initialized with the high point value. Then, the search for a low point is performed identically.

The presented algorithm can be applied to the measured force signal as well as the reference signal to obtain both the desired LP_{ref}, HP_{ref} and achieved LP_{meas}, HP_{meas} force amplitudes. Here, HP_i and LP_i stand for the force high point and low point, respectively. Subsequently, the correction ratio $\lambda(i+1)$ for both amplitudes is updated according to

$$\lambda(i+1) = \frac{\lambda(i)\cdot HP_{ref}}{HP_{meas}} \tag{2}$$

or

$$\lambda(i+1) = \frac{\lambda(i)\cdot LP_{ref}}{LP_{meas}} \tag{3}$$

and used to adapt the reference signal to the current situation (see Equation (4)). Since this approach adapts the reference signal based on the occurring maximum and minimum values, it is called peak value control (PVC).

$$F_{harm,ref,new} = F_{harm,ref}\cdot\lambda(i+1) \tag{4}$$

Here, $F_{harm,ref}$ and $F_{harm,ref,new}$ stand for the harmonic reference signal and the adapted harmonic reference signal, respectively.

Note that the harmonic signal without offset $F_{harm,ref}$ is separated from the reference offset force $F_{off,ref}$. This distinction allows the previously introduced superposition of rack forces from a vehicle model and the harmonic signal. A block model of the whole algorithm together with the force control structure of the steering test bench is displayed in Figure 6.

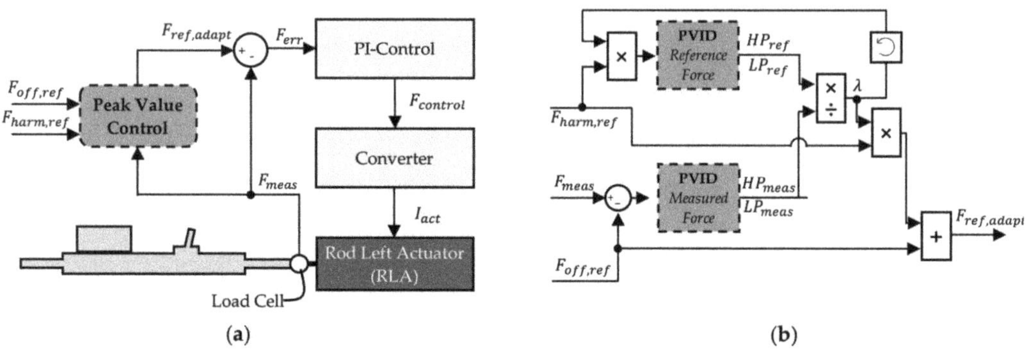

Figure 6. Implementation of the peak value control. (**a**) Location within the complete control structure of the steering system test bench. (**b**) Detailed representation of the algorithm.

The adaption of the reference signal is conducted before the PI control structure of the test bench and includes information from the measured force F_{meas} and the reference forces $F_{off,ref}$ and $F_{harm,ref}$ (see Figure 6a). Figure 6b displays a block diagram of the peak value control implementation). The harmonic reference signal is a multiplication with the currently active correction ratio, directly fed to the PVID algorithm, whereas the measured force is offset adjusted by subtracting the reference offset value before being analyzed by the PVID. The correction ratio λ is calculated according to Equation (2) or Equation (3) and multiplied with the harmonic reference signal. Finally, the adapted harmonic reference signal is added to the offset reference value, representing the adapted reference signal, which is fed to the control structure. The separated analysis is conducted to only adapt the high-frequency inputs, since the low-frequency signals possess high control accuracy.

The advantages of this approach are that no further information other than the measured and reference force value are required. Furthermore, the calculation can be performed for any steering system with different applications and boundary conditions (e.g., vehicle speed). A disadvantage is that the calculation of a correction factor is only possible when a new peak value is updated. Firstly, this leads to a time delay since the threshold value must first be undershot. Secondly, it results in a lower limit up to which amplitudes can be resolved. This value is determined by the measurement noise and the resulting threshold parameter.

Here, phase correction is not implemented, since only the transfer from the measured rack force to the measured steering wheel torque is considered. A phase delay between the measured and reference force signal is, therefore, irrelevant for the feedback characterization of the steering system.

3.2. Steering System Guidance Dynamics

In this subsection, a description of the test setup and the derived control performance limitation are explained. The description of two solution approaches, one for magnitude improvement and another for delay compensation, conclude the subsection.

3.2.1. Test Description

For the investigation of the dynamic guidance behavior of steering systems, one or two PMLSMs are connected to the steering rack. In contrast to the previously described feedback investigation, the reference force signal is calculated based on the rack displacement of the steering system and is not previously defined. For simplicity in the evaluation of the results, a virtual spring is implemented in the control scheme instead of a vehicle model. Consequently, the reference force generated is proportional to the rack displacement.

The identification of the guidance behavior of the steering system is adapted from a standard full vehicle test, the frequency response test as described in ISO 7401 [30]. For this maneuver, a sine steer input with slowly increasing frequency is introduced into the steering wheel. For the investigation of the steering system, exemplary relevant objective parameters are the transfer function from steering wheel angle to lateral acceleration and from the steering wheel angle to the vehicle yaw velocity [27]. Consequently, the frequency response test represents a relevant investigation for a steering test bench.

3.2.2. Motivation and Problem Formulation

To demonstrate the occurring control performance deterioration during the test, measurements with a steering wheel angle of $20°$ and an implemented virtual spring stiffness k_{virt} of 400 N/mm are conducted. Within 50 s, the frequency is exponentially increased from 0.1 Hz up to 5 Hz, which lays above the reported frequency input ranges of 2.0 Hz in [30], 2.8 Hz in [27], 3 Hz in [23,31] and 4 Hz in [32] and is, therefore, considered sufficiently dynamic. Figure 7 depicts the bode plot of the transfer function from measured rack displacement $x_{meas,RLA}$ to the measured rack force $F_{meas,RLA}$, introduced by the RLA, calculated as the feedforward control performance $G_{FF,Perf}$

$$G_{FF,Perf} = \frac{FFT(F_{meas,RLA})}{FFT(x_{meas,RLA})}. \tag{5}$$

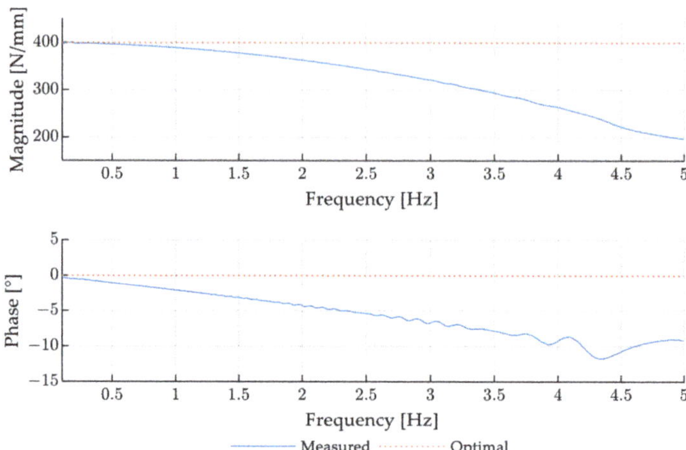

Figure 7. Bode plot of the control performance for the dynamic guidance evaluation with a steering wheel angle input of $20°$ and a virtual spring stiffness of $k_{virt} = 400$ N/mm.

Ideally, the magnitude course equals the defined spring stiffness, and the phase delay is zero.

Two deviations from the desired test bench behavior can be observed. First, the effective stiffness decreases for increasing input frequency. At 3 Hz, a stiffness reduction of 20% is measured which increases up to 50% at 5 Hz. That means that the desired rack forces are not reached for the setup. Second, the phase delay increases in magnitude over the frequency range. Consequently, the force applied to the steering system no longer matches the steering wheel angle and the rack position, which negatively influences the steering torque buildup and, therefore, the overall impression.

3.2.3. Solution Approach

Since the proposed approach is required for applications where the reference force signal is not known in advance, adjustments to the reference signal cannot be conducted beforehand. To overcome the previously introduced deficits, a two-part approach is presented. The first method allows an improvement of the magnitude course over the frequency range, whereas the second solution reduces the occurring time or phase delay.

Inertia Compensation

The basic assumption to improve the magnitude control performance of the PMLSM is that its moving mass is not incorporated in the control scheme and, therefore, deteriorates the performance during high-dynamic maneuvers. To improve these dynamic properties, inertia compensation (IC) is implemented in the control scheme by considering the acceleration of the steering system, similar to [16]. For this purpose, two locations for acceleration measurement are available. The first is located directly at the PMLSM and detects the acceleration of the RLA or RRA. A second possibility measures the steering wheel angle and derives the currently occurring steering wheel angle acceleration. For the investigated setup, the acceleration of the steering rack represents the result of the steering wheel input. Due to the elasticity of the torsion bar and the inertia of the steering rack, a time delay is expected to occur within the transfer path. As the acceleration information should be available as soon as possible, the steering wheel angle φ_{SW} and, more concrete, the steering wheel angle speed $\dot{\varphi}_{SW}$ is selected as the relevant signal.

Since the angular velocity is transformed into a translatory displacement of the steering rack, the steering ratio as well as the gimbal error due to the universal joints within the steering column need to be incorporated in the control structure. For this purpose, the steering system is steered from the maximum to the minimum steering wheel angle after the initial setup, and the resulting overall steering ratio is measured and mapped to a position dependent lookup table.

The measured steering wheel velocity $\dot{\varphi}_{SW}$ is then multiplied with the relevant steering ratio, resulting in the estimated rack speed $\dot{\hat{x}}_{RLA}$ or $\dot{\hat{x}}_{RRA}$. To receive an estimated rack acceleration signal $\ddot{\hat{x}}_{RLA}$ and $\ddot{\hat{x}}_{RRA}$, a derivative transfer function, according to the recommendation in [33], is utilized as

$$\frac{sK_D}{1 - sT_D}.$$

(6)

K_D and T_D stand for the gain and time constant of the filter, respectively. For the implementation, a discrete transfer function for a sample time of 125 µs is derived, setting $K_D = 1$ and $T_D = 1.25$ ms.

To reduce the influence of the PMLSMs inertia, the estimated acceleration is then multiplied with the mass of the respective motor m_{RLA} or m_{RRA}, resulting in the inertia compensation force signal $F_{comp,in}$, which is added to the reference force signal from the spring model F_{spring}. The overall implementation of the approach is displayed in Figure 8a and the more detailed representation of the IC in Figure 8b.

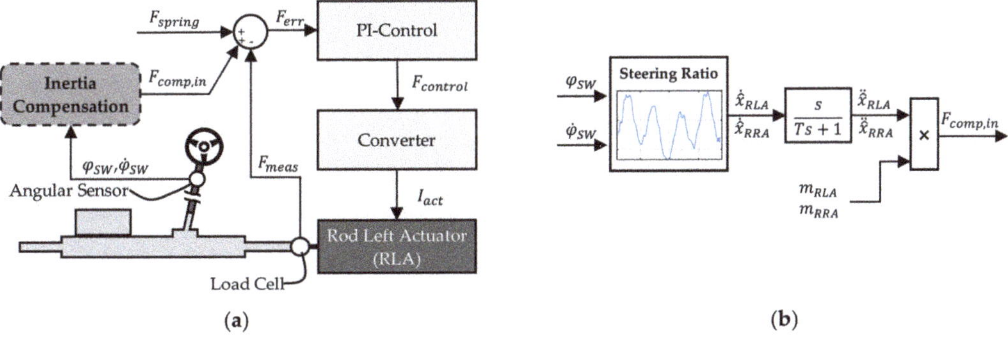

Figure 8. Implementation of inertia compensation. (**a**) Location within the complete control structure of the steering system test bench. (**b**) Detailed representation of the algorithm.

In cases of active steering with steer-by-wire systems, as in [34], where rack displacement occurs without steering wheel movement, the rack acceleration signal can be utilized instead of the steering wheel angle displacement. The rack acceleration is then directly available without the incorporation of a steering ratio in the control structure. Since EPS systems with a mechanical link between the steering wheel and steering gear represent the state of the art, this work focuses on a fixed and predefined ratio between steering wheel angle and rack displacement.

Delay Compensation

The second approach aims to reduce the phase delay between the rack displacement and the resulting contact force and is referred to as delay compensation (DC). Therefore, a similar basic principle, as in [22], is incorporated. To compensate for the resulting delay, the currently available reference signal is adapted so that the dynamic limitations of the PMLSMs and the overlying control structure are considered. Therefore, the following considerations are made exemplarily for a connected RLA.

Assuming the occurring displacement for the investigation of the dynamic guidance behavior of the steering system can be expressed as a harmonic sine in the form of

$$x_{meas,RLA}(t) = s_{\max} \cdot \sin(2\pi f t),$$

(7)

where s_{max} is the amplitude of the excitation, f the current frequency and t is the time. The phase delay in the resulting measured rack force $F_{meas}(t)$ is then

$$F_{meas}(t) = k_{virt} \cdot s_{max} \sin\left(2\pi f\left(t - t_{delay}\right)\right) = k_{virt} x_{meas,RLA}\left(t - t_{delay}\right). \tag{8}$$

The phase delay is considered here by incorporating t_{delay} into the formulation. The goal is to introduce an additional term $\Delta x(t)$ to the measured position $x_{meas,RLA}(t)$ to compensate the delay, so that

$$F_{meas}(t) = k_{virt} \cdot \left(x_{meas,RLA}\left(t - t_{delay}\right) + \Delta x\left(t - t_{delay}\right)\right) \overset{!}{=} k_{virt} x_{meas,RLA}(t). \tag{9}$$

Together with the phase delay φ_{delay}

$$\varphi_{delay} = 2\pi f t_{delay} \tag{10}$$

and the previously introduced definitions, the two right hand terms in Equation (9) can be written as

$$s_{max} \sin\left(2\pi f t - \varphi_{delay}\right) + \Delta x\left(t - t_{delay}\right) \overset{!}{=} s_{max} \cdot \sin(2\pi f t). \tag{11}$$

Adding a zero in the sine term on the right side, it can be reformulated as

$$s_{max} \sin\left(2\pi f t - \varphi_{delay} + \varphi_{delay}\right) = s_{max}\left(\sin\left(2\pi f t - \varphi_{delay}\right)\cos\left(\varphi_{delay}\right) + \cos\left(2\pi f t - \varphi_{delay}\right)\sin\left(\varphi_{delay}\right)\right) \tag{12}$$

For small delay angles φ_{delay}, the approximations $\sin\left(\varphi_{delay}\right) \approx \varphi_{delay}$ and $\cos\left(\varphi_{delay}\right) = 1$ can be implemented. For the investigated delay here, a maximum deviation of 1.15% for the sine term and 3.5% for the cosine term at $15°$ phase delay is observed, which is sufficiently accurate. Therefore, Equation (11) can be simplified to

$$s_{max} \sin\left(2\pi f t - \varphi_{delay}\right) + \Delta x\left(t - t_{delay}\right) \overset{!}{=} s_{max}\left(\sin\left(2\pi f t - \varphi_{delay}\right) \cdot 1 + \cos\left(2\pi f t - \varphi_{delay}\right) \cdot \varphi_{delay}\right), \tag{13}$$

and further to

$$\Delta x\left(t - t_{delay}\right) \overset{!}{=} \cos\left(2\pi f t - \varphi_{delay}\right) \cdot \varphi_{delay}. \tag{14}$$

Replacing the phase delay by its definition from Equation (10)

$$\Delta x\left(t - t_{delay}\right) \overset{!}{=} s_{max} \cos\left(2\pi f\left(t - t_{delay}\right)\right) \cdot 2\pi f \cdot t_{delay} \tag{15}$$

with the derivative of the measured rack position

$$\frac{d}{dt}\dot{x}_{meas,RLA}\left(t - t_{delay}\right) = \frac{d}{dt}\left(s_{max} \sin\left(2\pi f\left(t - t_{delay}\right)\right)\right) = s_{max} \cos\left(2\pi f\left(t - t_{delay}\right)\right) \cdot 2\pi f \tag{16}$$

Equation (15) can be rewritten as

$$\Delta x\left(t - t_{delay}\right) \overset{!}{=} \dot{x}_{meas,RLA}\left(t - t_{delay}\right) \cdot t_{delay}. \tag{17}$$

Consequently, the additional term from Equation (9) can be introduced by considering the current velocity of the RLA multiplied with the occurring delay time.

$$\Delta x(t) \overset{!}{=} \dot{x}_{meas,RLA}(t) \cdot t_{delay}. \tag{18}$$

Therefore, the derived term represents an estimation of the future position of the steering rack. To incorporate a frequency-dependent delay time, a simple frequency estimation model, which is valid for sinusoidal excitation, is introduced. Since both the acceleration $\ddot{x}_{meas,RLA}$ and position of the PMLSM $x_{meas,RLA}$ are measured, their ratio is utilized for frequency estimation

$$\frac{\ddot{x}_{meas,RLA}}{x_{meas,RLA}} = \frac{s_{\max}(2\pi f)^2 \sin(2\pi f t)}{s_{\max}\sin(2\pi f t)} \tag{19}$$

with a zero-value exception, the frequency is calculated as

$$f_{estim} = \frac{\sqrt{\left|\frac{\ddot{x}_{meas,RLA}}{x_{meas,RLA}}\right|}}{2\pi} \tag{20}$$

The approach with a variable delay time is called variable delay compensation and is abbreviated as DCvar for distinction from the constant delay time solution.

For the DC and DCvar approach, the measured position of the PMLSM is added to the look ahead term Δx and utilized for the evaluation of the spring force F_{spring}, as displayed in Figure 9a. The calculation of the look ahead term is illustrated in detail in Figure 9b for a frequency-dependent estimated delay time $\hat{t}_{delay}$. In case of a constant delay time, the lookup table is replaced by the constant value.

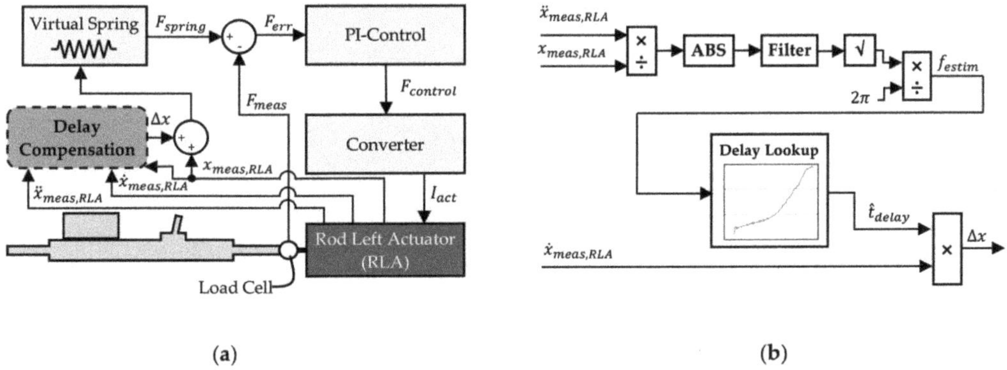

(a)   (b)

Figure 9. Implementation of delay compensation. (**a**) Location within the complete control structure of the steering system test bench. (**b**) Detailed representation of the algorithm.

Both DC and DCvar can be utilized for steer-by-wire systems, since only rack information and no steering wheel information are required. That means that there is no need for a fixed link between steering wheel angle and rack displacement.

4. Results and Discussion

In this section, the comparison of the proposed approaches and the reference state is conducted.

4.1. Feedback Control Performance

For the evaluation of the control performance during feedback characterization, the transfer function from the reference force signal to the measured force is utilized as the objective criteria. Identical to previously introduced transfer functions, the estimation is performed with an FFT to receive the feedback control performance $G_{FB,Perf}$

$$G_{FB,Perf} = \frac{FFT(F_{meas,RLA})}{FFT\left(F_{ref,RLA}\right)}. \tag{21}$$

Here, the magnitude of $G_{FC,Perf}$ represents the relevant parameter. For a setup where the steering wheel is fixed in its position, the vehicle velocity is set to 70 km/h and a force excitation amplitude of 600 N is defined; the magnitude of the feedback control performance is displayed in Figure 10. The PVC successfully reduces the deviation from the desired amplitude, which is indicated by the red horizontal line with a magnitude of one. To display the improvement for multiple setups, the maximum and minimum magnitude are identified for all available setups. The relevant values for the displayed setup are marked with a triangle in Figure 10.

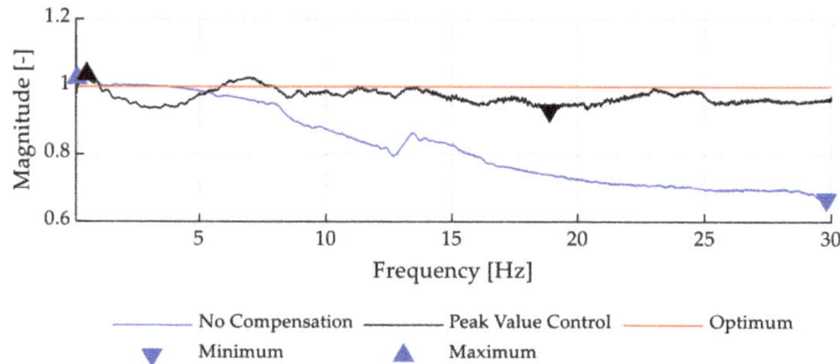

Figure 10. Control performance displayed as the magnitude of the transfer function from the reference to the measured rack force during steering feedback evaluation.

An overview on the boundary conditions for four investigated setups is displayed in Table 2. The results in Figure 10 represent S1.

Table 2. Overview of the investigated setups for feedback evaluation.

	Setup Abbreviation			
Parameter	**S1**	**S2**	**S3**	**S4**
Velocity [km/h]	70	70	120	Passive EPS
Amplitude [N]	600	800	800	600
Offset Value [N]	0	0	1000	0
Steering Lock	Blocked	Free Steering	Angle Control	Angle Control

Table 3 summarizes the minimum and maximum occurring magnitudes for the initial control structure without compensation (noComp) and the PVC for four setups. Additionally, the last row displays the maximum deviation from the reference course. The average and median for the investigated setups are calculated in the last column. On average, the PVC is able to maintain the amplitude within a range of $\pm10.4\%$ deviation compared to the reference signal. For the reference setup, an average deviation of 26.7% is detected. This means that the control performance is improved by more than 50%.

Table 3. Resulting maximum and minimum control magnitudes for the investigated setups with the initial control structure and the PVC.

	S1		S2		S3		S4		Average/Median			
	noComp	PVC	noComp	PVC	noComp	PVC	noComp	PVC	noComp		PVC	
Min	0.668	0.930	0.687	0.913	0.810	0.828	0.766	0.915	0.733	0.727	0.896	0.914
Max	1.024	1.035	1.028	1.098	1.028	1.051	1.294	1.052	1.094	1.028	1.059	1.052
Total	0.332	0.070	0.313	0.098	0.190	0.172	0.294	0.085	0.267	0.273	0.104	0.090

Despite the improvement in the average control performance, the maximum magnitudes when PVC is applied are larger compared to the initial control approach. This means that the maximum occurring force amplitudes exhibit increased deviation from the reference signal in comparison to the reference setup. Since the PVC is able to limit the maximum positive deviations for all applications below 10%, especially for setup S3, where the reference exhibits overshoot of 30%, the performance of the control scheme is improved by the application of PVC.

Figures 10 and 11 display the control performance for the feedback investigation for an alternative steering system with different inertia, friction and application. An improvement in the control performance for the whole frequency range and the peak values is achieved by the implementation of PVC. The qualitative deviation of the magnitude course over the frequency range between Figures 10 and 11 is a result of the altered dynamic behavior of the new steering system.

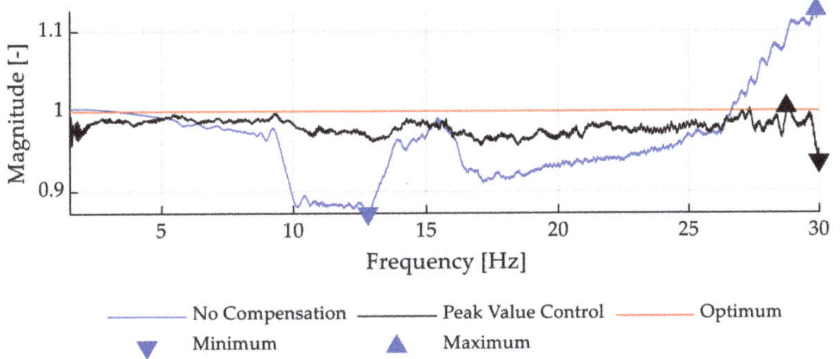

Figure 11. Control performance displayed as the magnitude of the transfer function from the reference to the measured force for an alternative steering system.

To demonstrate the performance of the PVC for time-varying offset values, a low-frequency sine for the offset force is superposed with a high-frequency sine, representing the harmonic excitation. The offset force reference signal is a sine function with an amplitude of 1000 N and 0.5 Hz frequency, while the harmonic signal possesses an amplitude of 600 N and an increasing frequency from 3 Hz to 30 Hz. The time course of the reference signal displayed in Figure 12a. Figure 12b shows the objective control performance. An improvement by the PVC control for both maximum and average deviation is achieved. Consequently, the PVC is also applicable for these investigations.

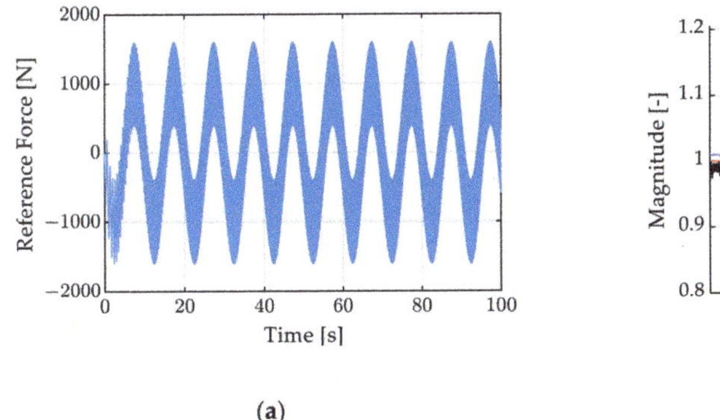

(a)

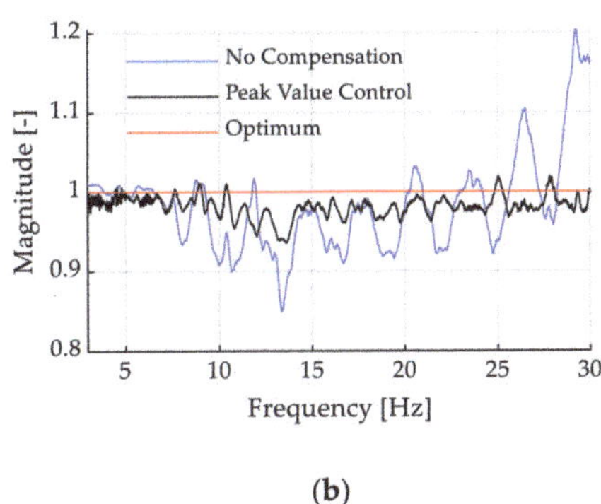

(b)

Figure 12. Control performance for a harmonic signal superposed to a variable offset signal. (a) Time course of the reference force signal. (b) Resulting magnitude of the transfer functions from reference to measured rack force.

After the verification of improved control performance by the PVC, a final experiment is conducted to demonstrate the importance of the feedback characterization of steering systems. Figure 13a depicts the identified feedback behavior of a steering system for three different excitations, while Figure 13b represents the associated control performance.

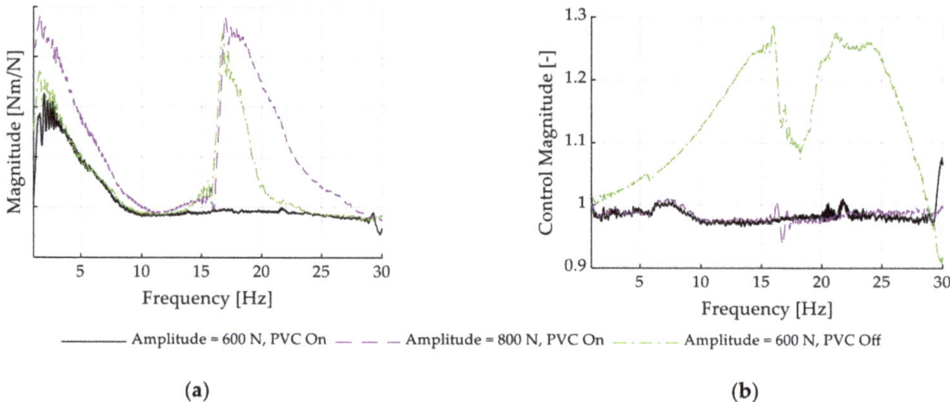

Figure 13. Comparison of the identified feedback behavior of a steering system with and without PVC. (**a**) Derived feedback behavior. (**b**) Magnitude of the transfer function from reference to measured rack force.

In Figure 13a, two different feedback magnitudes for the same excitation force amplitude of 600 N are calculated, which mainly differ at 16.5 Hz, where one result exhibits a resonance phenomenon, while the other solution remains constant. The difference between the results is the implementation of PVC. An explanation for the deviation can be found when considering the occurring control performance. At 16.5 Hz, where the peak in the feedback behavior is located, the measured rack force without PVC is 25% larger than the reference value. Therefore, instead of 600 N, a force of 750 N is present at the steering rack. For PVC, the deviation is less than 5%. A second experiment with 800 N amplitude and PVC is introduced. It displays a similar resonance at 16.5 Hz like the course for 600 N without PVC, while the control performance is close to one due to PVC implementation. Consequently, the measured peak without the PVC with an amplitude of 600 N is not a characteristic of the steering system but a result of the test bench control performance, because the higher prevailing force amplitude is responsible for the behavior. If we now consider benchmark investigations with different steering systems where the amplitude setpoint is derived based on vehicle applications, the importance of the implementation of an approach, such as PVC, is obvious to avoid unwanted influences.

4.2. Feedforward Control Performance

Since the derived IC, DC and DCvar approaches exhibit various parameters, a parameterization process is conducted.

For the selection of an adequate inertia compensation mass, a variation study of the PMLSMs mass is performed. The experimental setup for the investigation of the dynamic guidance behavior is utilized. Based on information from the test bench manufacturer, a physical mass of 200 kg is assumed. Therefore, a range from 150 kg to 250 kg is considered for identification. Figure 14a displays the sum of the root square error of the resulting spring stiffness according to Equation (5), from the reference value of 400 N/mm, normed to the solution without inertia compensation.

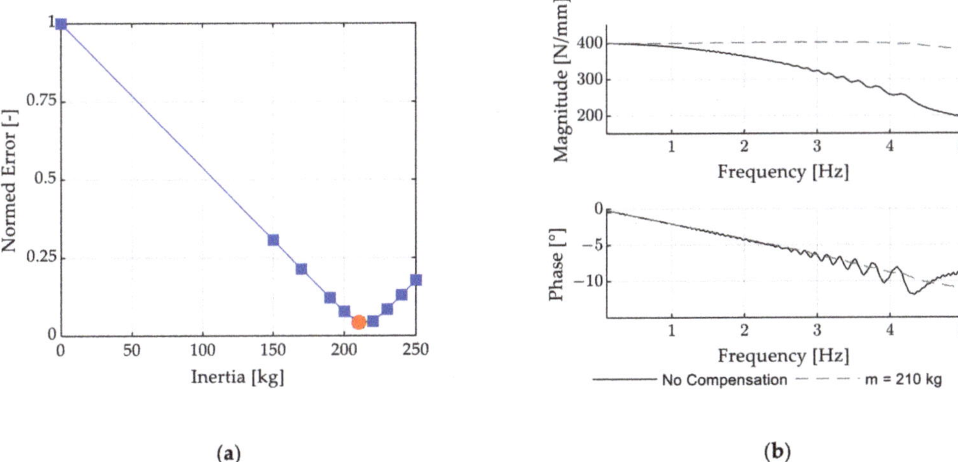

(a)

(b)

Figure 14. Derivation of the PMLSM inertia value for inertia compensation (IC) approach. (**a**) Relative stiffness error for different compensation masses (blue rectangles) and optimal solution (red circle). (**b**) Comparison of the resulting control bode plot.

The occurring minimum error from the reference stiffness is obtained for a mass of 210 kg, where the error is reduced by more than 95%. This improvement is also visible for the resulting stiffness in Figure 14b. Note that the phase delay is still identical to the initial result, where no compensation is introduced. Consequently, a second identification is performed to define the relevant delay time Δt_{delay} for the DC and DCvar approach. The results of the parameter optimization for a delay time variation between 5.7 ms and 6.2 ms are displayed in Figure 15.

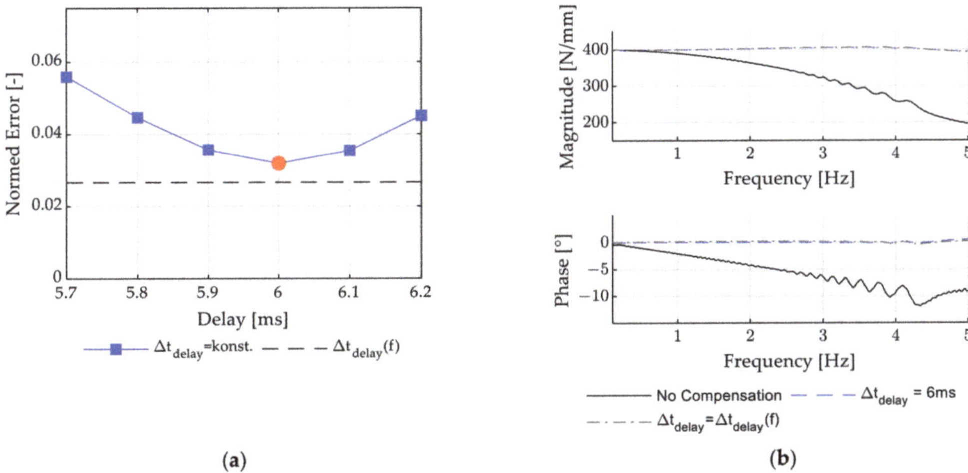

(a)

(b)

Figure 15. Derivation of the time delay value for the delay compensation (DC) approach. (**a**) Relative phase error for different delay time (blue rectangles) and optimal solution (red circle). (**b**) Comparison of the resulting control bode plot.

Based on the results from Figure 15a, the constant delay time Δt_{delay} is set to 6 ms. The frequency-dependent lookup table is identified as the inverse phase delay from Figure 14b and exhibits a reduced error in comparison to the constant time delay compensation. The resulting bode plot in Figure 15b now displays good agreement for the desired spring stiffness due to IC, as well as phase delay for the DC and DCvar approaches. The values for the phase delay compensation for the frequency-dependent lookup table are displayed in Table 4.

Table 4. Frequency breakpoints and lookup values for the frequency-dependent delay time.

Breakpoint	1	2	3	4	5	6	7	8	9
Frequency [Hz]	0.52	1.02	1.52	2.02	2.52	3.02	3.52	4.02	4.52
Δt_{delay} [ms]	5.6	5.7	5.7	5.8	5.8	5.9	6	6.1	6.2

To obtain an objective comparison of the different control performances, Tables 5 and 6 summarize the maximum and average deviations of the control performance magnitude and phase delay from the desired behavior, respectively. In total, four different setups for the steering guidance test are investigated where the vehicle velocity v_{veh}, the spring stiffness k_{virt} and the steering wheel angle amplitude φ_{SW} are varied. The setups are abbreviated as

- V1: $v_{veh} = 70\,\text{km/h}$ $k_{virt} = 400\,\text{N/mm}$ $\varphi_{SW} = 20°$
- V2: $v_{veh} = 70\,\text{km/h}$ $k_{virt} = 800\,\text{N/mm}$ $\varphi_{SW} = 20°$
- V3: $v_{veh} = 180\,\text{km/h}$ $k_{virt} = 800\,\text{N/mm}$ $\varphi_{SW} = 20°$
- V4: $v_{veh} = 180\,\text{km/h}$ $k_{virt} = 800\,\text{N/mm}$ $\varphi_{SW} = 10°$

Table 5. Summary of the measured magnitudes of the transfer function from rack displacement to measured rack force and the difference from the spring stiffness for the compensation methods.

Compensation	None		IC		IC + DC		IC + DCvar	
Test	Max	Avg.	Max	Avg.	Max	Avg.	Max	Avg.
V1	207.46	74.75	23.47	3.15	10.12	3.43	11.68	3.44
V2	220.9	78.39	25.06	4.24	22.56	8.62	21.74	8.69
V3	229.78	78.82	25.40	8.50	38.23	14.42	36.99	14.5
V4	234.81	84.97	56.72	26.79	79.43	33.12	77.56	32.62
Average	223.24	79.23	32.66	10.67	37.59	14.90	36.99	14.81

Table 6. Summary of the phase delay of the transfer function from rack displacement to measured rack force for the compensation methods.

Compensation	None		IC		IC + DC		IC + DCvar	
Test	Max	Avg.	Max	Avg.	Max	Avg.	Max	Avg.
V1	12.32	5.62	11.35	5.53	1.11	0.21	0.74	0.17
V2	10.84	5.72	11.94	5.85	0.97	0.34	1.28	0.36
V3	10.97	5.74	11.45	5.50	1.00	0.15	0.89	0.17
V4	10.83	5.47	10.79	4.89	2.93	1.23	2.97	1.14
Average	11.24	5.64	11.38	5.44	1.50	0.48	1.47	0.46

Table 5 demonstrates that the IC approach successfully reduces the resulting deviations between the measured magnitude and the reference spring stiffness. The best solution concerning the magnitude accuracy is derived for IC without phase compensation. Incorporating the DC and DCvar approach leads to an increase of 15% and 13.3% for the maximum deviation, respectively. On average, the phase compensation exhibits a 50% increase compared to IC alone. Still, all three approaches reduce the maximum occurring stiffness error by at least 83% and on average by 81.2%, demonstrating the improved control performance for the investigated application.

For the phase deviation in Table 6, the same observation as in Figure 14b is valid. Although the resulting measured spring stiffness is thoroughly improved by the implementation of IC alone, the phase delay is maintained at a constant level compared to the reference measurement without compensation. For the integration of DC and DCvar, the phase delay is reduced by 86.7% for the maximum and up to 91.5% for the average phase delays of the measured force signal to the displacement signal. Here, DCvar exhibits slightly improved performance if compared to DC, but the added value is neglectable. Therefore, the following comparisons exclude the IC + DCvar approach and concentrate on the IC and IC methods. Still, both DC and DCvar improve the phase accuracy of the test bench.

A second experiment is conducted to validate the applicability of the control improvements for applications, wherein their parameters are not optimized. Here, k_{virt} = 400 N/mm and v_{veh} = 70 km/h are chosen for the maneuver conditions. Instead of a sine sweep input, which is utilized for guidance behavior identification, a steering wheel reference signal is implemented, which is derived from a real driving maneuver from a driving simulator. The complete reference force signal for the maneuver is depicted in Figure 16a, wherein the relevant regions are highlighted with the indices "A" and "B". In "A", a steering wheel excitation similar to an impulse is introduced, while region "B" represents a high-frequency sine excitation. The resulting measured forces for the relevant regions A and B are displayed in detail in Figure 16b,c to demonstrate the performance of the different compensation approaches.

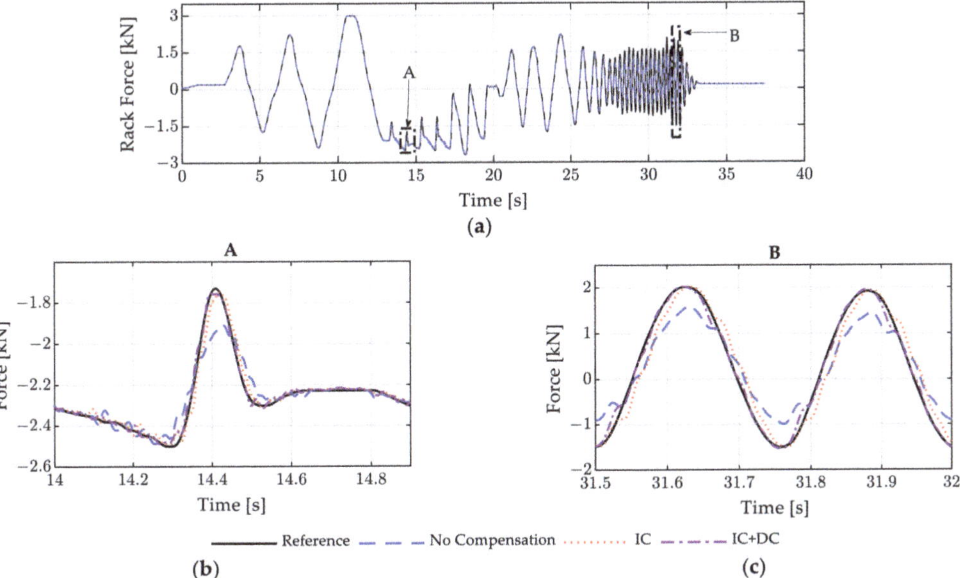

Figure 16. Comparison of the compensation approaches for the steering angle signal recorded on a steering driving simulator. (**a**) Overview for the complete maneuver. (**b**) Zoom to an impulse-like steering input. (**c**) Zoom to a harmonic steering input.

The contribution of the individual compensation approaches can be understood by the measured courses. In the case of no implemented compensation approach, the PMLSM inertia causes deviations at 14.3 s in Figure 16b, where an increase of the contact force is observed in advance to the rise of the reference force. Here, negative contact forces represent pressure on the load cell. When the steering maneuver is initiated at 14.3 s, the steering system moves to the center region, therefore, reducing the rack force. Since the PMLSM is not able to follow the movement immediately, an overshoot occurs where the decrease in the rack force is primarily caused by the missing movement due to inertia of the PMLSM. At the maximum reference force of the time course at 14.4 s, both the amplitude and phase deviation of the measured force without compensation are visible. When incorporating the IC approach, the measured force deviation is reduced for both previously described points, resulting in a better approximation of the reference signal. The phase delay, as evaluated earlier, is still present. When IC and DC are implemented simultaneously, the reference force is represented accurately by the measured force for both amplitude accuracy and the phase delay. A similar tendency can be observed in Figure 16c. Here, due to the implemented control improvements, the reference signal and the measured force with IC + DC are almost indistinguishable. These observations also underline the improved dynamic performance of the force control for an application where the tuning parameters were not derived.

For the third comparison, a basic setup is investigated, which is utilized for the evaluation of the steering power on a test bench (see [27]). For these tests, the steering wheel angle is defined to obtain a constant steering wheel velocity over a large portion of the complete steering stroke. The rack force is held at a constant value. For each new test, the rack force is increased until the torsion bar torque exceeds a predefined limit. At this point, the steering power limit is reached. For the herein investigated test setup, a constant force of 0 N is predefined, while the steering wheel maneuvers at $800°/s$. Since the rack force reference signal is constant over time, no DC is implemented since it has no effect on the control performance. The time courses of the measured rack force with and without IC is displayed in Figure 17.

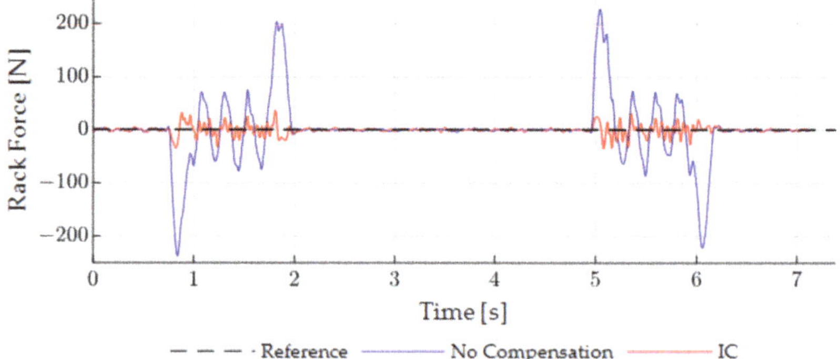

Figure 17. Comparison of the force control performance for steering system power measurement at $800°/s$ steering wheel angle velocity and 0 N reference force.

The fluctuations of the measured forces without compensation originate from the gimbal fault and the steering wheel angle acceleration and deceleration at the beginning and at the end of the steering wheel movement. Since the IC incorporates both the steering ratio as well as the gimbal fault and provides an inertia compensation signal, these influences are successfully reduced.

5. Conclusions

Within this paper, an automotive steering system testbench, consisting of two PMLSMs and one steering wheel actuator, is examined concerning its dynamic performance. The study presents three methods, each of which improves one defined use case. The first approach deals with the situation when a harmonic reference signal is previously known but the steering system characteristics hinder the PMLSM from reaching the desired force course. An online real-time capable algorithm uses the peak value information of the harmonic reference signal and the measured peak forces to calculate a ratio and adapt the input signal accordingly to receive a good amplitude accuracy. A second approach is applicable when the steering test bench operates in closed loop operation, meaning that the rack force is calculated based on the rack displacement. Due to high inertia of the PMLSM, the magnitude and phase delay of the PMLSM are deteriorated. Therefore, the steering wheel angle signal is used to provide an inertia compensation signal, which incorporates not only the moving mass of the PMLSM but also the occurring steering ratio and gimbal error in the column. It is demonstrated that the approach successfully improves the dynamic control performance by reaching the desired force amplitudes. To compensate for the phase delay in the same setup, a phase delay compensation approach is motivated, which ideally avoids the occurrence of phase delay between the reference and measured force signal. The latter approach can be combined with the inertia compensation to significantly improve the test bench control. Final validation tests demonstrate the applicability of the proposed methods, also for new applications where the parameterization was not performed. The introduced methods represent an improvement of the steering test bench control to allow subsystem-based development, even for the investigation of steering feel, which poses strict requirements due to the close contact to the human driver. Future improvements in the dynamic control performance may include model predictive control approaches for the closed loop operation of the steering system test bench. Additionally, automated, online transfer function estimation represents an approach to include the system behavior into the control of previously defined excitation signals. In this method, the representative transfer function can be selected based on information of the surrounding conditions of the experiment and used to alter the reference signal accordingly.

Author Contributions: Conceptualization, A.H., B.S., G.M., P.M.S. and D.S.; methodology, A.H. and B.S.; validation, A.H. and B.S.; writing—original draft preparation, A.H. and P.M.S.; writing—review and editing, D.S.; supervision, D.S. and P.M.S.; project administration, G.M.; resources, G.M. All authors have read and agreed to the published version of the manuscript.

Funding: This research received no external funding.

Data Availability Statement: Restrictions apply to the availability of these data. Data was obtained from Dr.-Ing. h.c. F. Porsche AG and are available from the authors with the permission of Dr.-Ing. h.c. F. Porsche AG.

Acknowledgments: We acknowledge support by the Open Access Publication Fund of the University of Duisburg-Essen.

Conflicts of Interest: The authors declare no conflict of interest. The funders had no role in the design of the study; in the collection, analyses, or interpretation of data; in the writing of the manuscript; or in the decision to publish the results.

Abbreviations

AFC	Acceleration Feedforward Control
Apa	Axle Parallel
DC	Delay Compensation
DCvar	Frequency Variable Delay Compensation
EPS	Electric Power Steering
FFT	Fast-Fourier-Transformation
HiL	Hardware-in-the-Loop
IC	Inertia Compensation
PMLSM	Permanent Magnet Linear Synchronous Motor
PVC	Peak Value Control
PVID	Peak Value Identification
RLA/RRA	Rod Left Actuator/Rod Right Actuator
S1–S4	Setups for Steering Feedback Investigation
V1–V4	Setups for Feedforward Investigations
SWA	Steering Wheel Actuator

Nomenclature

$F_{ref,RLA/RRA}$	Reference Force at RRA/RLA
$F_{meas,RLA}$	Measured Rack Force at RLA
φ_{SW}	Steering Wheel Angle
$x_{RLA/RRA}$	Rack Displacement
λ	Correction Ratio
G_{FF}	Transfer Function of Feedforward Control Performance
G_{FB}	Transfer Function of Feedback Control Performance
$m_{RLA/RRA}$	Mass of the RLA/RRA
f	Frequency
LP_{ref}/HP_ref	Reference Low- and Highpoint Amplitude Value
LP_{meas}/HP_{meas}	Measured Low- and Highpoint Amplitude Value

References

1. Shen, G.; Zhu, Z.; Li, X.; Li, G.; Tang, Y.; Liu, S. Experimental evaluation of acceleration waveform replication on electrohydraulic shaking tables: A review. *Int. J. Adv. Robot. Syst.* **2016**, *13*, 1–25. [CrossRef]
2. Smolders, K.; Volckaert, M.; Swevers, J. Tracking control of nonlinear lumped mechanical continuous-time systems: A model-based iterative learning approach. *Mech. Syst. Signal Process.* **2008**, *22*, 1896–1916. [CrossRef]
3. Shen, G.; Lv, G.-M.; Ye, Z.-M.; Cong, D.-C.; Han, J.-W. Implementation of electrohydraulic shaking table controllers with a combined adaptive inverse control and minimal control synthesis algorithm. *IET Control. Theory Appl.* **2011**, *5*, 1471–1483. [CrossRef]
4. Shen, G.; Lv, G.-M.; Ye, Z.-M.; Cong, D.-C.; Han, J.-W. Feed-forward inverse control for transient waveform replication on electro-hydraulic shaking table. *J. Vib. Control* **2011**, *18*, 1474–1493. [CrossRef]
5. Liu, G.-D.; Li, G.; Shen, G. Experimental evaluation of the parameter-based closed-loop transfer function identification for electro-hydraulic servo systems. *Adv. Mech. Eng.* **2017**, *9*, 1–12. [CrossRef]
6. Zhao, J.; Shen, G.; Zhu, W.; Yang, C.; Agrawal, S.K. Force tracking control of an electro-hydraulic control loading system on a flight simulator using inverse model control and a damping compensator. *Trans. Inst. Meas. Control* **2016**, *40*, 135–147. [CrossRef]
7. Yao, J.-J.; Fu, W.; Hu, S.-H.; Han, J.-W. Amplitude phase control for electro-hydraulic servo system based on normalized least-mean-square adaptive filtering algorithm. *J. Cent. South Univ.* **2011**, *18*, 755–759. [CrossRef]
8. Yao, J.; Di, D.; Jiang, G.; Gao, S. Acceleration amplitude-phase regulation for electro-hydraulic servo shaking table based on LMS adaptive filtering algorithm. *Int. J. Control* **2012**, *85*, 1581–1592. [CrossRef]
9. Jinzenji, A.; Sasamoto, T.; Aikawa, K.; Yoshida, S.; Aruga, K. Acceleration feedforward control against rotational disturbance in hard disk drives. *IEEE Trans. Magn.* **2001**, *37*, 888–893. [CrossRef]
10. Kim, J.-G.; Hwang, H.-W.; Park, K.-S.; Park, N.-C.; Yang, H.; Park, Y.-P.; Jeong, J. Improved Air Gap Control With Acceleration Feedforward Controller Using Time Delay for Solid Immersion Lens-Based Near-Field Storage System. *IEEE Trans. Magn.* **2011**, *47*, 556–559. [CrossRef]
11. Cao, J.; Zhang, J. Trajectory tracking control method for high-speed and high-acceleration machine tool. In Proceedings of the 27th Chinese Control and Decision Conference, Qingdao, China, 23–25 May 2015. [CrossRef]
12. Kim, J.-H.; Choi, J.-W.; Sul, S.-K. High precision position control of linear permanent magnet synchronous motor for surface mount device placement system. In Proceedings of the Power Conversion Conference, Osaka, Japan, 2–5 April 2002. [CrossRef]
13. Jee, S.; Lee, J. Real-time inertia compensation for multi-axis CNC machine tools. *Int. J. Precis. Eng. Manuf.* **2012**, *13*, 1655–1659. [CrossRef]
14. Li, S.; Liu, Z. Adaptive Speed Control for Permanent-Magnet Synchronous Motor System With Variations of Load Inertia. *IEEE Trans. Ind. Electron.* **2009**, *56*, 3050–3059. [CrossRef]
15. Zhang, Y.; Kim, D.; Zhao, Y.; Lee, J. PD Control of a Manipulator with Gravity and Inertia Compensation Using an RBF Neural Network. *Int. J. Control Autom. Syst.* **2020**, *18*, 3083–3092. [CrossRef]
16. Weiss, P.; Zenker, P.; Maehle, E. Feed-forward friction and inertia compensation for improving backdrivability of motors. In Proceedings of the 12th International Conference on Control Automation Robotics & Vision (ICARCV), Guangzhou, China, 5–7 December 2012. [CrossRef]
17. Smith, J.M. Closer Control of Loops with Dead Time. *Chem. Eng. Prog.* **1957**, *53*, 217–219.
18. Xia, C.; Gao, G. Brushless DC Motors Control Based on Smith Predictor Modified by Fuzzy-PI Controller. In Proceedings of the Fifth International Conference on Fuzzy Systems and Knowledge Discovery, Jinan, China, 18–20 October 2008. [CrossRef]
19. Veronesi, M. Performance Improvement of Smith Predictor through Automatic Computation of Dead Time. Available online: https://web-material3.yokogawa.com/rd-tr-r00035-007.pdf (accessed on 16 March 2023).
20. Yahaya, N.Z.; Abu-Bakar, M.N.; Mohd, M.S. Study on Phase Advance Angle Control (PAAC) Technique for Brushless DC (BLDC) Motor. In Proceedings of the IEEE Conference on Control, Systems and Industrial Informatics (ICCSII), Bandung, Indonesia, 23–26 June 2013.

21. Lee, S.-J.; Hong, J.-P.; Jang, W.-K. Characteristics comparison of BLDC motor according to the lead angles. In Proceedings of the 2012 IEEE Vehicle Power and Propulsion Conference, Seoul, Republic of Korea, 9–12 October 2012. [CrossRef]

22. Lee, M.; Kong, K. Fourier-Series-Based Phase Delay Compensation of Brushless DC Motor Systems. *IEEE Trans. Power Electron.* **2018**, *33*, 525–534. [CrossRef]

23. Harrer, M.; Pfeffer, P.E. *Steering Handbook*; Springer International Publishing: Cham, Switzerland, 2017; ISBN 978-3-319-05449-0.

24. Lunkeit, D. Ein Beitrag zur Optimierung des Rückmelde-und Rückstellverhaltens Elektromechanischer Servolenkungen. Ph.D. Thesis, Universität Duisburg-Essen, Duisburg, Germany, 2014.

25. Grau, J.; Nippold, C.; Bossdorf-Zimmer, B.; Küçükay, F.; Henze, R. Objective Evaluation of Steering Rack Force Behaviour and Identification of Feedback Information. *SAE Int. J. Passeng. Cars-Mech. Syst.* **2016**, *9*, 1279–1304. [CrossRef]

26. Düsterloh, D. Funktionsoptimierung und Komplexitätsbeherrschung im Entwicklungsprozess Mechatronischer Fahrwerksysteme am Beispiel Elektromechanischer Lenksysteme. Ph.D. Thesis, Universität Duisburg-Essen, Duisburg, Germany, 2018.

27. Schimpf, R. *Charakterisierung von Lenksystemen mit Hilfe Eines Lenksystemprüfstands. Ph.D. Thesis*; TU Wien: Wien, Austria, 2016.

28. Uselmann, A.; Preising, E.; Schrage, B.; Düsterloh, D. A Test of Character for Steering Systems. *dSPACE Magazine*, 2 November 2016; 24–31.

29. Isermann, R.; Münchhof, M. *Identification of Dynamic Systems: An Introduction with Applications*; Springer-Verlag: Berlin/Heidelberg, Germany, 2011; ISBN 978-3-540-78879-9.

30. *ISO 7401:2011*; Road Vehicles—Lateral Transient Response Test Methods—Open-Loop Test Methods. International Organization for Standardization: Geneva, Switzerland, 2011.

31. Zschocke, A.K.; Albers, A. Links between subjective and objective evaluations regarding the steering character of automobiles. *Int. J. Automot. Technol.* **2008**, *9*, 473–481. [CrossRef]

32. Huneke, M. Fahrverhaltensbewertung Mit Anwendungsspezifischen Fahrdynamikmodellen. Ph.D. Thesis, Technische Universität Braunschweig, Braunschweig, Germany, 2012.

33. *IEEE Std 421.5-2016*; IEEE Recommended Practice for Excitation System Models for Power System Stability Studies. IEEE: New York, NY, USA, 2016. [CrossRef]

34. Shao, K.; Zheng, J.; Huang, K. Robust active steering control for vehicle rollover prevention. *Int. J. Model. Identif. Control.* **2019**, *32*, 70–84. [CrossRef]

Article

Multi-Objective Optimal Design of μ–Controller for Active Magnetic Bearing in High-Speed Motor

Yuanwen Li and Changsheng Zhu *

Electrical Engineering Department, University of Zhejiang, Hangzhou 310027, China; 11610052@zju.edu.cn
* Correspondence: zhu_zhang@zju.edu.cn; Tel./Fax: +86-0571-8795-1625

Abstract: In this paper, a control strategy based on the inverse system decoupling method and μ-synthesis is proposed to control vibration in a rigid rotor system with active magnetic bearings that are built into high-speed motors. First, the decoupling method is used to decouple the four-degrees-of-freedom state equation of the electromagnetic bearing rigid rotor system; the strongly coupled and nonlinear rotor system is thus decoupled into four independent subsystems, and the eigenvalues of the subsystems are then configured. The uncertain parametric perturbation method is used to model the subsystem, and the multi-objective ant colony algorithm is then used to optimize the sensitivity function and the pole positions to obtain the optimal μ-controller. The closed-loop system thus has the fastest possible response, the strongest internal stability, and the best disturbance rejection capability. Then, the unbalanced force compensation algorithm is used to compensate for the high-frequency eccentric vibration; this algorithm can attenuate the unbalanced eccentric vibration of the rotor to the greatest extent and improve the robust stability of the rotor system. Finally, simulations and experiments show that the proposed control strategy can allow the rotor to be suspended stably and suppress its low-frequency and high-frequency vibrations effectively, providing excellent internal and external stability.

Keywords: active magnetic bearing (AMB); gyroscopic effect; μ-synthesis; D-K iterations; multi-objective optimization; unbalance compensation

Citation: Li, Y.; Zhu, C. Multi-Objective Optimal Design of μ–Controller for Active Magnetic Bearing in High-Speed Motor. *Actuators* **2023**, *12*, 206. https://doi.org/10.3390/act12050206

Academic Editor: Zongli Lin

Received: 6 April 2023
Revised: 11 May 2023
Accepted: 16 May 2023
Published: 17 May 2023

1. Introduction

High-speed motors have advantages that include high efficiency, low volume, and high power density, and they can be connected directly to high-speed mechanical equipment without an additional gearbox, which improves efficiency and reduces the overall size of the machine. These motors are thus widely used in turbo boosters, air compressors, grinders, and flywheel energy storage devices [1].

An active magnetic bearing (AMB) has notable features such as an absence of mechanical friction, zero lubrication, a long service life, and controllable dynamic behavior, compared with traditional mechanical bearings. Therefore, an AMB has an irreplaceable position in the application of high-speed rotor systems and has become the most important rotor support unit for high-speed rotating machinery [2].

Because the air gap between the AMB stator and the rotor is very small, the controller of the AMB rotor system must be stable enough to withstand vibrations and maintain system operation. Therefore, when designing this controller, it is necessary not only to meet the basic requirement of providing a stable rotor system, but also to optimize the controller as much as possible to achieve high dynamic performance for the rotor system. The design of this controller thus becomes a key step in AMB engineering design practice.

However, the high-speed motor rotor system is still a strongly coupled system. As the system's rotational speed increases, the gyroscopic torque that is coupled between the system's four degrees of freedom (4-DOF) will not only destroy the applied control action, but may even affect the overall stability of the AMB rotor system [3]. Gong [4] used a polarity switch tracking filter and a disturbance observer to achieve the active control of a magnetically levitated rotor system over its full rotational speed range; however, the core of this method was still based on use of a proportional–integral–derivative (PID) controller, which has low robustness and stability, and coupling still occurred between the degrees of freedom of the rotor system, meaning that the closed-loop system could not achieve accurate decoupling control in the higher speed range. Therefore, to realize high-precision control at higher speeds, more sophisticated control strategies must be developed. Zhao [5] used the feedforward decoupling method and a time-optimal tracking differentiator to realize decoupling of the rotor system at the cost of a minimum calculation amount, but this method will cause the decoupling subsystem to be not entirely symmetrical, and this is not conducive to controller design for high-precision control. Chen [6] used a combined strategy involving the feedback decoupling method and a 2-DOF PID controller to realize radial displacement control in the higher speed range. This method can also realize complete decoupling of the rotor system, but the 2-DOF PID controller shows a weakly robust performance, and it cannot overcome relatively strong external disturbance forces.

At present, PID controllers are widely used in AMB rotor systems to provide a mature control algorithm. Usually, a PID controller can allow the rotor to run stably at certain speeds, but when the plant parameters change or when the disturbance factors are uncertain, it is difficult to obtain good control performance, and the system may even become unstable. To solve the problems described above, Zhang [7] designed an H_∞ robust controller that showed a strong ability to suppress external disturbances and effectively attenuated a fluid surge disturbance force acting on the impeller of a centrifugal compressor. However, this method cannot measure and stabilize uncertain parametric perturbations of the control system itself. Kuseyri [8] modeled the rotor's eccentric unbalanced disturbance force and the uncertain parametric perturbation structure of the AMB system and then synthesized an H_∞ controller using a linear matrix inequality (LMI) method, which achieved exciting experimental results. This method suppressed more than 95% of the unbalanced disturbance vibrations, but it is essentially an empirical selection method without the multi-objective optimization process. As a rising star of the robust control family, the μ-controller provides an effective way to solve these problems. The μ-controller can model an uncertain parametric perturbation structure when it is synthesized, and the μ-analysis can even measure the stability margin for the uncertain parametric perturbation system to ensure that the entire closed-loop system remains stable with respect to the bounded parametric perturbation. At the same time, because of the fast response time of the μ-controller, the adjustment time required after multi-objective optimization can be less than 10 ms, and the controller also has a strong ability to suppress disturbances. When the rotor is disturbed by an external force, the controller will then generate a compensation force to offset the disturbance, and the bounded disturbance will be attenuated completely to zero within 10 ms. Therefore, the μ-controller performs particularly rigidly during the experiments, and regardless of the disturbance or even after a collision with the rotor from the exterior, the rotor remains perfectly centered. In contrast, the suspension force of the PID-controlled AMB has a "soft" characteristic, and the anti-interference performance of the rotor system is weak. Because the "P" parameter that represents rigidity in PID control cannot be increased without limit, too large a value of "P" will lead to system instability.

In this paper, the major contributions are as follows:

(1) An inverse system decoupling method is used to decompose the radial 4-DOF state equation of the nonlinear AMB-rigid rotor system into four double-integrator subsystems to eliminate the gyroscopic effects that are coupled at high speeds. Pole reconfiguration of the subsystem is performed to overcome the limitations of the μ-synthesis DK iteration method. The subsystem is modeled using an uncertain parametric perturbation method, and the μ-controller is synthesized via a 6-DK iterative method. Then, a closed-loop robust system is designed that is not overly conservative at the expense of its performance.

(2) To avoid the frequent use of trial-and-error to find a suitable weighting function, this paper introduces a multi-objective ant colony algorithm to search automatically for the optimal sensitivity function and pole position. The controller obtained using this method thus has the fastest response speed possible, the highest stability margin, and the strongest external disturbance attenuation.

(3) To eliminate the unbalanced eccentric disturbance force that occurs during high-speed operation, this paper also proposes an unbalanced compensation module that is added to the current channels of the AMBs. This module searches for the eccentric position in four directions with high efficiency and can completely compensate for the unbalanced eccentric vibration.

2. Dynamic Model of Rotor System

As shown in Figure 1, the magnetically levitated high-speed motor studied in this paper is a horizontal structure. The rotor is supported by two AMBs (AMB-A and AMB-B). The rotor's radial positions are measured using a total of four eddy current displacement sensors on the left and right.

Figure 1. Sketch of the magnetically levitated high-speed motor.

The dynamic model of the high-speed motor rotor system is depicted in Figure 2. In order to analyze it rigorously, some assumptions are set: due to the rotor's first-order bending speed being much higher than its rated speed, the AMB rotor can be regarded as a rigid rotor; the left and right AMBs are installed at the same axial positions as the corresponding sensors; the magnetic coupling interaction between the radial and axial coordinates is ignored. The origin O is the geometric center of the rotor, and the right-handed spiral stator coordinate system O-XYZ is established. The Z axis is on the line connecting the two radial AMB geometric centers O_A and O_B. In addition, the radial plane AMB stator coordinate systems $O_A X_A Y_A$ and $O_B X_B Y_B$ are established to describe the operating state of the rotor at the positions where AMBs are located. The distances from the left and right radial AMB planes to the origin point are l_A and l_B, respectively.

During operation, the rotor's spatial position is described using the translational displacements x and y of the mass center O along the X- and Y-axes and the rotation angles θ_x and θ_y around the X- and Y-axes. The positive directions for θ_x and θ_y are shown in Figure 2.

Based on the rotor dynamics, the equation for the radial 4-DOF horizontal AMB-rigid rotor system can be given as:

$$M\ddot{Z} + G\dot{Z} = LF + f_u \tag{1}$$

where $Z = [\theta_y\ x\ \theta_x\ y]^T$, and M and G are the mass matrix and the gyroscopic effect matrix of the rotor, respectively. L and F are the arm coefficient matrix of the rotor and the electromagnetic force vector, respectively, and f_u is the unbalanced eccentric force vector; these parameters are:

$$M = \begin{bmatrix} J & 0 & 0 & 0 \\ 0 & m & 0 & 0 \\ 0 & 0 & J & 0 \\ 0 & 0 & 0 & m \end{bmatrix}, G = \begin{bmatrix} 0 & 0 & -J_z\omega & 0 \\ 0 & 0 & 0 & 0 \\ J_z\omega & 0 & 0 & 0 \\ 0 & 0 & 0 & 0 \end{bmatrix}, F = \begin{bmatrix} f_{xA} \\ f_{xB} \\ f_{yA} \\ f_{yB} \end{bmatrix}, L = \begin{bmatrix} l_A & -l_B & 0 & 0 \\ 1 & 1 & 0 & 0 \\ 0 & 0 & -l_A & l_B \\ 0 & 0 & 1 & 1 \end{bmatrix}, f_u = \begin{bmatrix} mu_z\varepsilon(\omega^2\cos\theta + \dot\omega\sin\theta) \\ m\varepsilon(\omega^2\cos\theta + \dot\omega\sin\theta) - \frac{\sqrt{2}}{2}mg \\ mu_z\varepsilon(\dot\omega\cos\theta - \omega^2\sin\theta) \\ m\varepsilon(\omega^2\sin\theta - \dot\omega\cos\theta) - \frac{\sqrt{2}}{2}mg \end{bmatrix}$$

where m is the rotor mass; J is the transverse inertia moment of the rotor; J_z is the polar inertia moment of the rotor; $f_{xA}, f_{xB}, f_{yA},$ and f_{yB} are the electromagnetic forces generated by the radial AMB-A and AMB-B in the x and y directions, respectively; θ is the rotation angle; $\theta = \omega t + \varphi$, where ω is the angular velocity; φ is the initial angle; g is the acceleration due to gravity; ε is the projection of the distance between the mass center and the geometric center on the OXY surface; and u_z is the projection of the distance between the mass center and the geometric center on the OZ axis.

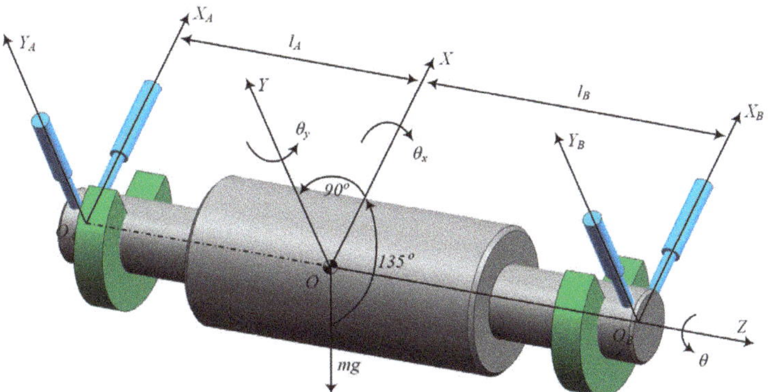

Figure 2. Model of the AMB-rigid rotor system.

As shown in Figure 3, the AMB stator uses a novel 12-pole coil structure. When compared with the traditional eight-pole coil structure, its magnetic force distribution is more uniform, thus producing less electromagnetic noise and vibration, and the maximum magnetic levitation capacity is improved significantly.

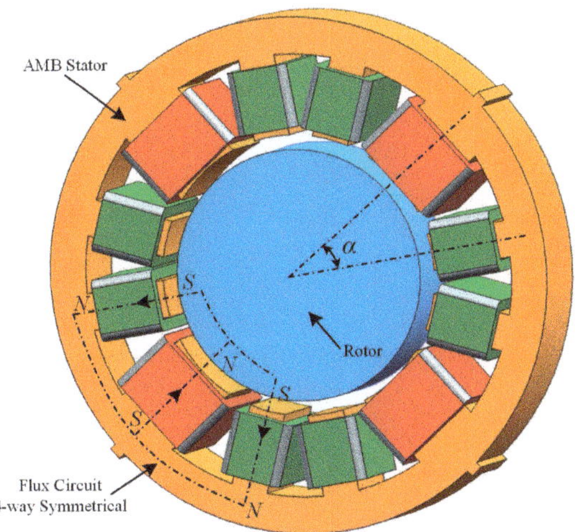

Figure 3. Sketch of the AMB stator coils.

There are a total of four magnetic poles built into the 12-pole AMB stator, where each magnetic pole includes a main pole yoke and two secondary pole yokes, and the opposing pairs of magnetic poles constitute a radial channel. In theory, the four radial channels are independent of each other; the electromagnetic force for each channel is generated by a differential current, and the related differential equations are

$$
\begin{cases}
f_{xA} = \mu_0 A N^2 \left(\dfrac{(I_0+i_{xA})^2}{(\delta-x_a\cos\alpha)^2} - \dfrac{(I_0-i_{xA})^2}{(\delta+x_a\cos\alpha)^2} \right)\cos\alpha + \mu_0 A N^2 \left(\dfrac{(I_0+i_{xA})^2}{(\delta-x_a)^2} - \dfrac{(I_0-i_{xA})^2}{(\delta+x_a)^2} \right) \\[3mm]
f_{xB} = \mu_0 A N^2 \left(\dfrac{(I_0+i_{xB})^2}{(\delta-x_b\cos\alpha)^2} - \dfrac{(I_0-i_{xB})^2}{(\delta+x_b\cos\alpha)^2} \right)\cos\alpha + \mu_0 A N^2 \left(\dfrac{(I_0+i_{xB})^2}{(\delta-x_b)^2} - \dfrac{(I_0-i_{xB})^2}{(\delta+x_b)^2} \right) \\[3mm]
f_{yA} = \mu_0 A N^2 \left(\dfrac{(I_0+i_{yA})^2}{(\delta-y_a\cos\alpha)^2} - \dfrac{(I_0-i_{yA})^2}{(\delta+y_a\cos\alpha)^2} \right)\cos\alpha + \mu_0 A N^2 \left(\dfrac{(I_0+i_{yA})^2}{(\delta-y_a)^2} - \dfrac{(I_0-i_{yA})^2}{(\delta+y_a)^2} \right) \\[3mm]
f_{yB} = \mu_0 A N^2 \left(\dfrac{(I_0+i_{yB})^2}{(\delta-y_b\cos\alpha)^2} - \dfrac{(I_0-i_{yB})^2}{(\delta+y_b\cos\alpha)^2} \right)\cos\alpha + \mu_0 A N^2 \left(\dfrac{(I_0+i_{yB})^2}{(\delta-y_b)^2} - \dfrac{(I_0-i_{yB})^2}{(\delta+y_b)^2} \right)
\end{cases}
\tag{2}
$$

where μ_0 is the free-space permeability; A is the cross-sectional area of the secondary magnetic pole; N is the number of turns of the coil; I_0 is the bias current; δ is the AMB air gap length; and i_{xA}, i_{xB}, i_{yA}, and i_{yB} are the control currents of AMB-A and AMB-B in the x and y channels, respectively.

Let Y and U be the radial displacement vector and the control current vector of the AMBs, respectively:

$$
Y = \begin{bmatrix} x_a & x_b & y_a & y_b \end{bmatrix}^T, U = \begin{bmatrix} i_{xA} & i_{xB} & i_{yA} & i_{yB} \end{bmatrix}^T
$$

The relationship between Y and Z is

$$
Y = L^T Z
\tag{3}
$$

The electromagnetic force is linearized around the operating point (I_0, i_{x0}, i_{y0}, $x_a = x_b = y_a = y_b = 0$) to give

$$
\begin{cases}
k_x = 4\mu_0 A N^2 \dfrac{I_0^2+i_{x0}^2}{\delta^3}(1+\cos^2\alpha) \\[2mm]
k_y = 4\mu_0 A N^2 \dfrac{I_0^2+i_{y0}^2}{\delta^3}(1+\cos^2\alpha) \\[2mm]
k_i = 4\mu_0 A N^2 \dfrac{I_0}{\delta^2}(1+\cos\alpha)
\end{cases}
\tag{4}
$$

where k_x and k_y are the displacement stiffness coefficients of the x and y channels, respectively; i_{x0} and i_{y0} are the compensation currents of the x and y channels to cancel the gravity, respectively; and k_i is the current stiffness coefficient of the x and y channels.

The linearized electromagnetic force can then be expressed as

$$
F = K_s Y + K_i U
\tag{5}
$$

where K_s and K_i are the force–displacement matrix and force–current matrix of the AMB, respectively, and are

$$
K_s = \begin{bmatrix} k_{sxA} & 0 & 0 & 0 \\ 0 & k_{sxB} & 0 & 0 \\ 0 & 0 & k_{syA} & 0 \\ 0 & 0 & 0 & k_{syB} \end{bmatrix}, K_i = \begin{bmatrix} k_{ixA} & 0 & 0 & 0 \\ 0 & k_{ixB} & 0 & 0 \\ 0 & 0 & k_{iyA} & 0 \\ 0 & 0 & 0 & k_{iyB} \end{bmatrix}
$$

where $k_{sxA} = k_{sxB} = k_x$; $k_{syA} = k_{syB} = k_y$; and $k_{ixA} = k_{ixB} = k_{iyA} = k_{iyB} = k_i$.

3. Inverse Decoupling Method and Eigenvalue Assignment

By combining (1), (3) and (5), the following can be obtained:

$$
M(L^T)^{-1}\ddot{Y} + G(L^T)^{-1}\dot{Y} = LF + f_u
\tag{6}
$$

The following can be obtained by converting from (6):

$$\ddot{Y} = -L^T M^{-1} G\dot{Z} + L^T M^{-1} LF + L^T M^{-1} f_u \tag{7}$$

When only considering the coupling relationship inside the system, the external disturbance f_u can be removed first; then, after the expansion of (7), it can be found that:

$$\ddot{Y} = \begin{bmatrix} \ddot{Y}_1 \\ \ddot{Y}_2 \\ \ddot{Y}_3 \\ \ddot{Y}_4 \end{bmatrix} = \begin{bmatrix} \frac{l_A J_z \omega \dot{\theta}_x}{J} + (\frac{1}{m} + \frac{l_A^2}{J})f_{xA} + (\frac{1}{m} - \frac{l_A l_B}{J})f_{xB} \\ -\frac{l_B J_z \omega \dot{\theta}_x}{J} + (\frac{1}{m} + \frac{l_B^2}{J})f_{xB} + (\frac{1}{m} - \frac{l_A l_B}{J})f_{xA} \\ \frac{l_A J_z \omega \dot{\theta}_y}{J} + (\frac{1}{m} + \frac{l_A^2}{J})f_{yA} + (\frac{1}{m} - \frac{l_A l_B}{J})f_{yB} \\ -\frac{l_B J_z \omega \dot{\theta}_y}{J} + (\frac{1}{m} + \frac{l_B^2}{J})f_{yB} + (\frac{1}{m} - \frac{l_A l_B}{J})f_{yA} \end{bmatrix} \tag{8}$$

Because F is a function of U, $\ddot{Y}$ can be written as a function of U. Then, the question is whether U can be written as a function of $\ddot{Y}$ inversely, i.e., $\ddot{Y}$ also exists if U exists. The reversibility of (8) is now derived as follows:

Let

$$D = \frac{\partial \ddot{Y}}{\partial U} = \begin{bmatrix} \frac{\partial \ddot{Y}_1}{\partial U_1} & \frac{\partial \ddot{Y}_1}{\partial U_2} & \frac{\partial \ddot{Y}_1}{\partial U_3} & \frac{\partial \ddot{Y}_1}{\partial U_4} \\ \frac{\partial \ddot{Y}_2}{\partial U_1} & \frac{\partial \ddot{Y}_2}{\partial U_2} & \frac{\partial \ddot{Y}_2}{\partial U_3} & \frac{\partial \ddot{Y}_2}{\partial U_4} \\ \frac{\partial \ddot{Y}_3}{\partial U_1} & \frac{\partial \ddot{Y}_3}{\partial U_2} & \frac{\partial \ddot{Y}_3}{\partial U_3} & \frac{\partial \ddot{Y}_3}{\partial U_4} \\ \frac{\partial \ddot{Y}_4}{\partial U_1} & \frac{\partial \ddot{Y}_4}{\partial U_2} & \frac{\partial \ddot{Y}_4}{\partial U_3} & \frac{\partial \ddot{Y}_4}{\partial U_4} \end{bmatrix} = \begin{bmatrix} k_{ixA}(\frac{1}{m} + \frac{l_A^2}{J}) & k_{ixB}(\frac{1}{m} - \frac{l_A l_B}{J}) & 0 & 0 \\ k_{ixA}(\frac{1}{m} - \frac{l_A l_B}{J}) & k_{ixB}(\frac{1}{m} + \frac{l_B^2}{J}) & 0 & 0 \\ 0 & 0 & k_{iyA}(\frac{1}{m} + \frac{l_A^2}{J}) & k_{iyB}(\frac{1}{m} - \frac{l_A l_B}{J}) \\ 0 & 0 & k_{iyA}(\frac{1}{m} - \frac{l_A l_B}{J}) & k_{iyB}(\frac{1}{m} + \frac{l_B^2}{J}) \end{bmatrix} \tag{9}$$

It can be concluded from the above that

$$\det(D) = \frac{k_{ixA} k_{ixB} k_{iyA} k_{iyB} (l_A + l_B)^4}{m^2 J^2} \neq 0 \tag{10}$$

The relative order of the system is $\alpha = (\alpha 1, \alpha 2, \alpha 3, \alpha 4) = (2, 2, 2, 2)$, which satisfies $\alpha 1 + \alpha 2 + \alpha 3 + \alpha 4 = 8 \leq n$, where n is the number of state variables. According to inverse system theory [9], the system is invertible.

Introducing a new variable $V = [V_1 \; V_2 \; V_3 \; V_4]^T$, the new variable and the second derivative of the output variable are made to satisfy the following relationship:

$$V_1 = \ddot{Y}_1, V_2 = \ddot{Y}_2, V_3 = \ddot{Y}_3, V_4 = \ddot{Y}_4 \tag{11}$$

Then, the transfer function from V to Y is given by

$$\frac{Y_i}{V_i} = \frac{1}{s^2}, i = 1, 2, 3, 4 \tag{12}$$

Equation (12) shows that the nonlinear and strongly coupled AMB-rigid rotor system is decoupled via the inverse decoupling method, the radial 4-DOF rotor system is decomposed into four translational radial degrees of freedom, and on each degree of freedom is a pseudo-linear subsystem with a transfer function of $1/s^2$.

When designing the controller for the subsystem, the *D-K* iterations method will indicate that the μ-controller cannot be synthesized because the plant contains two zero eigenvalues located on the imaginary axis. To overcome this limitation, it is necessary to configure the subsystem poles. As shown in Figure 4, the closed-loop negative state feedback is adopted. Let the configured poles be s_1 and s_2, and the characteristic equation be $D(s) = (s - s_1)(s - s_2)$; then, the coefficients of the state feedback are

$$c = -(s_1 + s_2), k = s_1 \cdot s_2 \tag{13}$$

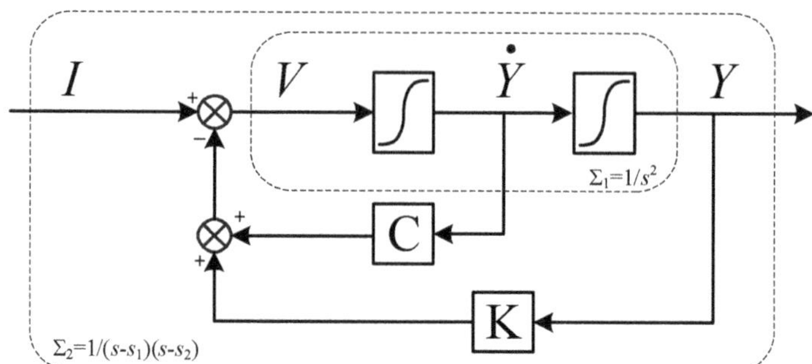

Figure 4. Block diagram of second-order subsystem eigenstructure assignment.

4. Uncertainties Model and μ-Synthesis Method

Because the motion equations of the four radial channels of the AMB-rigid rotor system when decoupled using the inverse system method are symmetrical and are independent of each other, only the subsystem structure attributed to a single channel is studied here. The designed μ-controller can be distributed symmetrically for each channel. The motion equation for one single-channel subsystem after pole configuration can be given as

$$\ddot{Y} + c\dot{Y} + kY = I \tag{14}$$

During actual operation, and especially during high-speed operation, the AMB-rigid rotor system has a high-amplitude sinusoidal noise signal in the output from the displacement sensors, which is attributed to the unbalanced eccentric vibration. This signal will affect the stability of the decoupled subsystem; equivalently, it can also be regarded as the main source of the uncertain parametric perturbation structure of the subsystem.

Suppose that during high-speed operation, there is an unbalanced vibration in one radial channel given by $\bar{d}(t) = \bar{r}\sin(\omega t) = p_k\delta_k\overline{Y}$, where $\bar{r}$ is the vibration amplitude, which is generally less than 0.05 mm over the full rotational speed range, and the radial displacement amplitude $\overline{Y}$ is less than 0.5 mm of the bearing air gap; thus, we set $p_k = 10\%$ and $-1 \leq \delta_k \leq 1$. By introducing the vibration displacement $Y = \overline{Y} + \bar{d}$, we find that

$$(1 + p_m\delta_m)\ddot{\overline{Y}} + c(1 + p_c\delta_c)\dot{\overline{Y}} + k(1 + p_k\delta_k)\overline{Y} = I \tag{15}$$

where $\overline{Y}$ is the ideal displacement variable when the vibration signal $\bar{d}$ is not considered in the total radial displacement Y, $p_m = p_c = p_k = 10\%$ and $-1 \leq \delta_m, \delta_c, \delta_k \leq 1$.

Equation (15) can be represented by the upper linear fractional transform (LFT) with M_m, M_c, M_k, δ_m, δ_c, and δ_k, and its state equation is

$$\begin{cases} \begin{bmatrix} y_m \\ \ddot{\overline{Y}} \end{bmatrix} = M_m \begin{bmatrix} u_m \\ I - v_c - v_k \end{bmatrix}, u_m = \delta_m y_m \\[2mm] \begin{bmatrix} y_c \\ v_c \end{bmatrix} = M_c \begin{bmatrix} u_c \\ \dot{\overline{Y}} \end{bmatrix}, u_c = \delta_c y_c \\[2mm] \begin{bmatrix} y_k \\ v_k \end{bmatrix} = M_k \begin{bmatrix} u_k \\ \overline{Y} \end{bmatrix}, u_k = \delta_k y_k \end{cases} \tag{16}$$

where $M_m = \begin{bmatrix} -p_m & 1 \\ -p_m & 1 \end{bmatrix}$, $M_c = \begin{bmatrix} 0 & c \\ p_c & c \end{bmatrix}$, $M_k = \begin{bmatrix} 0 & k \\ p_k & k \end{bmatrix}$

The subsystem structure with uncertainties can be described as shown in Figure 5.

From Figure 5, the state equation for the subsystem with structured uncertainty can be derived, and by eliminating variables v_c and v_k, the system matrix S, which describes the dynamic characteristics of the subsystem, is

$$
\begin{bmatrix} \dot{y} \\ \ddot{y} \\ -- \\ y_m \\ y_c \\ y_k \\ y \end{bmatrix} = \left[\begin{array}{ccccc|c} 0 & 1 & 0 & 0 & 0 & 0 \\ -k & -c & -p_m & -p_c & -p_k & 1 \\ \hline -k & -c & -p_m & -p_c & -p_k & 1 \\ 0 & c & 0 & 0 & 0 & 0 \\ k & 0 & 0 & 0 & 0 & 0 \\ \hline 1 & 0 & 0 & 0 & 0 & 0 \end{array} \right] \begin{bmatrix} y \\ y \\ u_m \\ u_c \\ u_k \\ -- \\ I \end{bmatrix}, \quad \begin{bmatrix} u_m \\ u_c \\ u_k \end{bmatrix} = \begin{bmatrix} \delta_m & 0 & 0 \\ 0 & \delta_c & 0 \\ 0 & 0 & \delta_k \end{bmatrix} \begin{bmatrix} y_m \\ y_c \\ y_k \end{bmatrix}, \quad (17)
$$

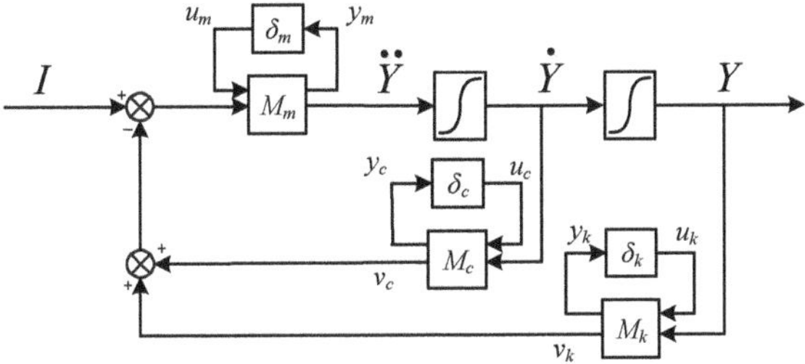

Figure 5. Block diagram of subsystem structure with uncertain parameters.

The block diagram of the closed-loop system that shows the feedback structure and includes elements reflecting the model uncertainty and performance requirements obtained at this stage is as shown in Figure 6, where P is the plant, K is the controller, and Δ is the uncertainty matrix. In addition, W_p is the performance weighting function, and W_u is the control weighting function.

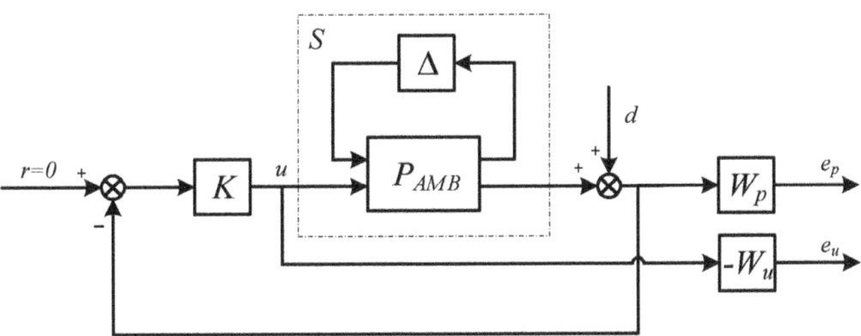

Figure 6. Block diagram of AMB closed-loop system.

The μ-synthesis method was proposed to assort systems with structured uncertainties. The uncertain behavior of the original system can be described by the LFT representation, and the controller synthesis is performed with an uncertain closed-loop system model using the structured singular value μ and the LFT framework. As a result, a μ-control closed-loop system can be designed that is not excessively conservative at the expense of performance [10]. The diagram of the μ-control framework for the AMB-rotor system is shown in Figure 7.

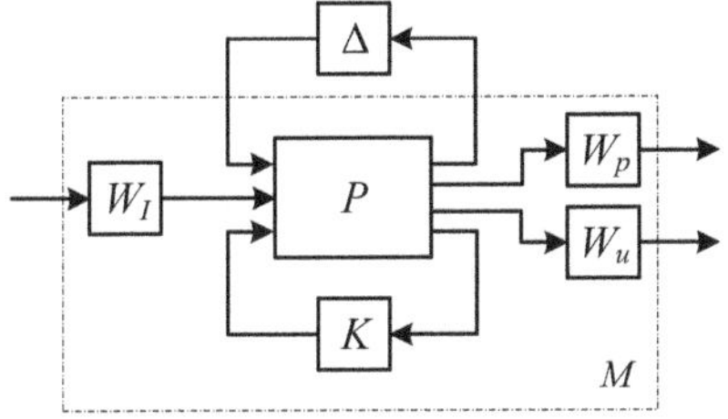

Figure 7. Block diagram of μ-synthesis method.

The structure of the matrix Δ is the result of the locations at which each parametric uncertainty occurs in the system model. The weighting functions W_p and W_u are selected to constrain the frequency-based performance requirements in closed-loop systems. The matrix M is the lower LFT of the plant P and the controller K.

$$M = F_l(P, K) \tag{18}$$

The μ-synthesis framework can be represented by the upper LFT of the weighted closed-loop system M and the uncertain perturbation matrix Δ, which maps disturbance input w to performance response z.

$$z = F_u(M, \Delta)w \tag{19}$$

The defined structure Δ destabilized the system M, and the stability of the entire system can be evaluated using the maximum singular value μ.

$$\mu_\Delta^{-1}(M) = \min_{\Delta \in \Delta}\{\overline{\sigma}(\Delta) : \det(I - M\Delta) = 0\} \tag{20}$$

If the structured singular value is less than unity, this indicates that a greater perturbation than the set uncertain perturbation is required to destabilize the system. Therefore, the closed-loop system with the synthesized controller is stable and robust with respect to the bounded uncertainties [11].

Finding appropriate weighting functions is a critical step in robust controller design and usually requires many trials. For complex systems, significant effort is thus required. Therefore, a multi-objective ant colony algorithm is introduced to search automatically for the optimal sensitivity function. The μ-controller is synthesized using the dksyn tool integrated in the MATLAB Robust Control Toolbox. In this tool, a 6-DK iterative algorithm [12] is used. The flow chart of the algorithm is shown in Figure 8.

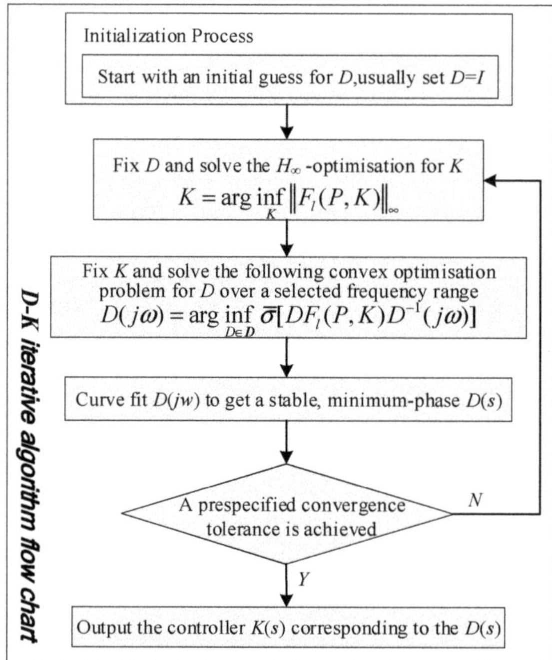

Figure 8. Flow chart of D-K iterations method.

5. Multi-Objective Optimization of μ-Controller

5.1. Multi-Objective Function

When optimizing the μ-controller, it requires the explicit definition of some targets to optimize the general performance of the controller. The response time, the tracking error, and the disturbance attenuation of the controller can be used as one objective function. The multi-objective optimization process frequently calls several objective functions simultaneously. The results of using these single objective functions are usually contradictory, hence they cannot be simply compared as being superior or inferior. Generally, the better result for one target will be the worse result for the other target, and enhancement to any objective value will deteriorate at least another objective value. One such result is called the Pareto solution [13].

The multi-objective function for μ-controller optimization used in this paper is defined as:

$$f = a \cdot (R_{\max} - R)/(R_{\max} - R_{\min}) + b \cdot (S - S_{\min})/(S_{\max} - S_{\min}) + c \cdot (D_{\max} - D)/(D_{\max} - D_{\min}), a + b + c = 1 \quad (21)$$

where a, b, and c are weight factors that represent the importance of the relevant objectives, and different ratios will have different effects on the controller performance. R, R_{min}, and R_{max} are the settling time of the closed-loop system to a step input and the minimum and maximum values of this settling time, respectively; S, S_{min}, and S_{max} are the stability margin of the closed-loop system and the minimum and maximum values of this stability margin, respectively. D, D_{min}, and D_{max} are the disturbance attenuation of the closed-loop system and the minimum and maximum values of this attenuation, respectively.

This function includes the step input reference response performance, the robust stability, and the robust performance of the μ-controller multiplied by their respective weighting factors. The essential feature of the ant colony algorithm is that it maximizes the evaluation value of the objective function. Equation (21) thus shows that the optimization objectives are the shortest response time, the largest stability margin, and the strongest disturbance attenuation. In addition, according to ISO standard 14839-3, the penalty constraint set for the optimization algorithm is that the maximum magnitude of the sensitivity function for the closed-loop control system should be less than 9.5 dB [14].

5.2. Variable Constraints

It can easily be determined from Figure 6 that:

$$\begin{bmatrix} e_p \\ e_u \end{bmatrix} = \begin{bmatrix} W_p(I + SK)^{-1} \\ W_u K(I + SK)^{-1} \end{bmatrix} d \quad (22)$$

Therefore, the design criterion for robust performance is that all transfer functions from d to e_p and e_u should be small in the sense of H_∞ for all possible uncertain transfer matrices Δ.

$$\left\| \begin{bmatrix} W_p(I + SK)^{-1} \\ W_u K(I + SK)^{-1} \end{bmatrix} \right\|_\infty < \gamma \quad (23)$$

The weighting functions W_p and W_u are used to reflect the relative significance of the performance requirements over the different frequency ranges. The performance weighting function W_p is selected to have a second-order form as follows:

$$W_p(s) = x_1 \cdot \frac{s^2 + x_2 s + x_3}{s^2 + x_4 s + x_5} \quad (24)$$

The value of x_1 shows a strong relationship with the response speed and the quality of the waveform of the closed-loop system relative to the reference input. Both x_2 and x_4 affect the turning frequency of the weighting function W_p and thus indirectly and slightly affect some of the closed-loop system performance characteristics. Equation (24) shows that in the low frequency range, the reciprocal of W_p tends toward x_5/x_3, and the designed controller sensitivity function will be below the $1/W_p$ curve; this means that the value of x_5/x_3 is related to the disturbance attenuation and the steady-state tracking error. Therefore, x_5/x_3 is usually set to be equal to 0.001 or less to meet the performance requirements.

The control weighting function W_u is selected simply as the scalar form:

$$W_u(s) = x_6 \tag{25}$$

Equation (13) shows that the subsystem poles affect the damping c and the stiffness k of the controlled plant. A low level of system damping will accelerate the controller's response. Greater system stiffness leads to a greater bearing control force and a better active control effect, but the required controller is not easy to synthesize. Therefore, it is necessary to use a multi-objective optimization algorithm to search for the best parametric combination. The multi-objective algorithm uses eight variables, as listed in Table 1, with upper and lower bounds that are selected carefully for the optimization process.

Table 1. Ant variables and bounds.

Variable	Description	Lower Bound	Upper Bound
x_1	proportional term of W_p	0.01	1.0
x_2	first-order term of the W_p numerator	0.5×10^3	2×10^3
x_3	constant term of the W_p numerator	1.5×10^4	3×10^4
x_4	first-order term of the W_p denominator	10	20
x_5	constant term of the W_p denominator	0.1	0.5
x_6	proportional term of W_u	1×10^{-8}	12×10^{-8}
x_7	positive real pole	450	550
x_8	negative real pole	-550	-450

5.3. Ant Colony Algorithm and Optimization

The artificial ant colony algorithm, which simulates an ant colony's intelligence, has features that include distributed computation, positive feedback, and heuristic searching. The ant colony algorithm has shown many good performance aspects through use of its inherent pheromone search mechanism. Its positive feedback and synergy make it suitable for use in distributed systems, and its implicit parallelism offers strong development potential. The problems that it can solve have gradually expanded to include some constrained problems and multi-objective problems [15–17]. When the ant colony algorithm was initially proposed, it was used for discrete domain optimization problems. Therefore, the μ-synthesis problem, which is a continuous domain optimization problem, requires the original ant colony algorithm to be modified.

In this paper, a grid scaling method is used for the continuous domain ant colony algorithm. Each grid point corresponds to a variable space state, and each ant crawls between the grid points and leaves certain amounts of pheromone information to influence the action of future generations of ants. When all ants in one generation have finished crawling, for grid points that satisfy the constraints, the objective function values are compared, the optimal individual is recorded, and the pheromone matrix is updated. Then, at the beginning of the next generation, the variable range near the grid point is scaled down by a ratio r (0.5–0.9), and the new generation of ants is placed to start crawling. By repeating the above process, until the grid spacing is below the given precision.

When each variable range is divided into N parts, there are $N + 1$ nodes, and n variables have $(N + 1)^n$ grid points. The calculations of these grid points can become an

n-level decision process. In the decision process, the calculation of ant transition probability can use the following formula:

$$p_{ij}^k(t) = \frac{[\tau_{ij}(t)]^\alpha [\eta_{ij}(t)]^\beta}{\sum\limits_{k \subset A_k} [\tau_{is}(t)]^\alpha [\eta_{is}(t)]^\beta}, if j \in A_k \tag{26}$$

where $p^k_{ij}(t)$ is ant k's transition probability from the ith to jth node, A_k is the node collection that ant k can reach, $\tau_{ij}(t)$ is determined using the pheromone matrix, and $\eta_{ij}(t)$ is determined using the heuristic matrix, i.e., $\eta_{ij}(t)$ is set as a constant (1.5).

The equation to update the pheromone matrix component is:

$$\tau_{ij}(t+1) = (1 - \rho) \cdot \tau_{ij}(t) + \Delta\tau_{ij}(t) \tag{27}$$

where ρ is the evaporation rate, and $\Delta\tau_{ij}(t)$ is determined by the pheromone increment matrix.

The equation to update the pheromone increments is:

$$\Delta\tau_{ij}(t) = \sum_{k=1}^{m} \Delta\tau_{ij}^k(t) + Q \cdot f \tag{28}$$

where Q is set as a constant (0.05), and f is the multiple objective function as in (21).

5.4. Optimization Solution

Figure 9 shows that the multi-objective ant colony algorithm converges fully after 16 iterations, which confirms the high efficiency of the algorithm sufficiently.

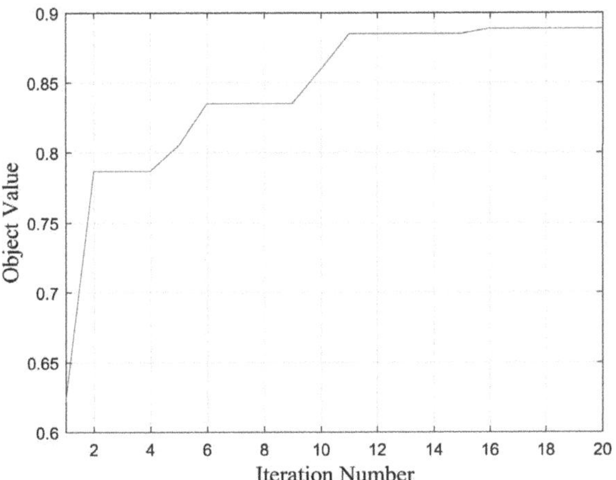

Figure 9. Evolution of the multi-objective value.

After the closed-loop system is modeled and the optimization objectives are defined, it is interesting to observe the results of optimization algorithm execution. Figure 10 depicts the individual sampling process and the iterative evolution of the three targets. It is known that the optimization objectives are the settling time of the step response of the closed-loop system, the stability margin of the uncertain perturbation structure, and the controller design index γ. The settling time can be measured directly using the stepinfo function. The stability margin is obtained via the frequency sweep method, where the starp function is used first to connect the controller to form a closed-loop system; then, the frequency sweep function, frsp, is used to obtain the frequency response of the closed-loop system; finally, the pkvnorm function is used to obtain the upper bound of the μ function of the perturbation structure, and the reciprocal of this bound is the lower bound of the stability margin. This means that at least the stability margin times of the perturbation can be tolerated to ensure that the closed-loop system is robust and stable. The frequency sweep method is conservative to an extent but can be regarded as a fast and stable evaluation method.

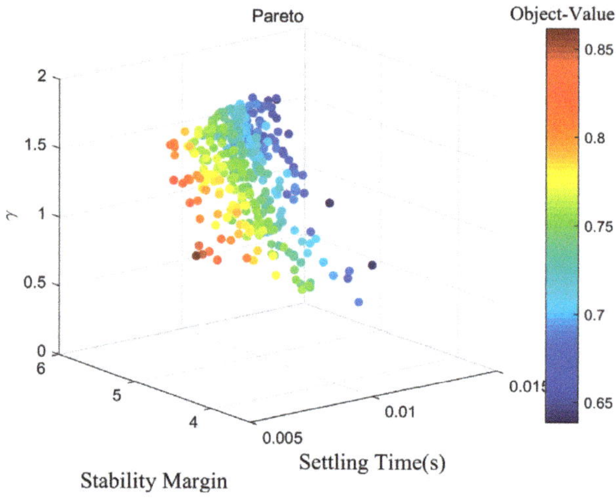

Figure 10. Optimal designs with color plot for the object values.

Smaller settling times give faster control system responses, and a larger stability margin indicates stronger system robustness to the structured uncertainty. The γ value is an important index for controller design. A larger γ value means that the controller is more difficult to synthesize. When the γ value exceeds 1, it means that a stable closed-loop controller cannot be synthesized. As the value of γ decreases, it becomes easier to design the controller, and a smaller sensitivity function amplitude for the closed-loop system indicates stronger system attenuation of external disturbances. The optimization results show clear consistency between the response speed and the stability margin, and there is also a clear contradiction between the response speed and γ or between the stability margin and γ. It is easy to understand that when the controller is faster, it is then more stable, but it is also more difficult to synthesize.

Finally, the compromise solution that was calculated after the comparison of the object values is shown in the lower left corner of Figure 10. This solution should provide a comprehensive optimal performance, and its variables and performance parameters are listed in Table 2.

Table 2. Optimized ant solutions.

x_1	x_2	x_3	x_4	x_5	x_6	x_7	x_8	Settling Time	Stability Margin	γ	Object Value
0.5791	1923.5	2.607×10^4	14.076	0.3603	1.3977×10^{-8}	452.151	-471.994	8.45 ms	5.1108	0.8821	0.8887

6. Unbalanced Eccentric Vibration Compensation Method

Because of machining technology limitations and assembly tolerances, the mass center of a rigid rotor is not consistent with its geometric center. During high-speed operation, unbalanced vibration will be generated by the centrifugal force. The frequency of this unbalanced vibration is the same as the rotational speed, and the excitation force is proportional to the square of the rotational speed. When the rotational speed increases, the frequency of this excitation force also increases. Figure 14 (see Section 7.1) shows that when the frequency exceeds 100 Hz (6000 rpm), the disturbance suppression force of the robust controller gradually decreases. This results in saturation of the controller's control force, causing the rotor vibration force to increase sharply, and this seriously affects the operational stability of the high-speed motor rotor system. Therefore, it is necessary to compensate for this unbalanced force.

6.1. Principle of Unbalanced Vibration Compensation Method

The basic principle of the compensation method proposed in this paper is to extract the same frequency vibration signal from the rotor, from which a compensation signal can be generated and injected into the output channel of the controller; then, a compensation force that is equal and opposite to the rotor's centrifugal force is generated by the AMB to compensate for the centrifugal force, and the rotor is forced to rotate around its geometric axis. The aim of this method is to minimize the rotor vibration amplitude, and it begins by identifying the position of the rotor mass center. Here, a correlation method in the signal processing technique is applied that can extract the amplitude of the fundamental frequency of the vibration signal accurately from the radial channels of the AMBs.

The displacement vibration signal of the radial AMB channel can be given as

$$f(t) = a_1 \cos \omega t + b_1 \sin \omega t + a_2 \cos 2\omega t + b_2 \sin 2\omega t + \cdots \tag{29}$$

when only the fundamental frequency vibration is considered, and the following can be obtained by performing a Fourier series expansion:

$$\begin{cases} a_1 = \frac{\omega}{\pi} \int_0^{\frac{2\pi}{\omega}} f(t) \cos \omega t \, dt \\ b_1 = \frac{\omega}{\pi} \int_0^{\frac{2\pi}{\omega}} f(t) \sin \omega t \, dt \end{cases} \tag{30}$$

Then, according to digital signal processing theory, the sampling period is set as T_s, and the Fourier coefficient of the same frequency vibration signal at this time is:

$$\begin{cases} a_x = \frac{\omega}{\pi} \sum_{k=1}^{\frac{2\pi}{\omega T_s}} f(kT_s) \cos(k\omega T_s) T_s \\ b_x = \frac{\omega}{\pi} \sum_{k=1}^{\frac{2\pi}{\omega T_s}} f(kT_s) \sin(k\omega T_s) T_s \end{cases} \tag{31}$$

The vibration signal amplitude at the rotor A end can be defined as $E_a = \sqrt{a_x^2 + b_x^2}$, which represents the intensity of the vibration at the rotor A end at the current sampling time.

As illustrated in Figure 11a, the detailed steps for the quadrangle search method for the unbalanced mass center position are as follows:

(1) Randomly define a search starting point, usually by selecting the origin point $(\alpha,\beta)_0 = (0,0)$, $r_0 = R\angle 0$, where $\alpha = \varepsilon \cos\varphi$ and $\beta = \varepsilon \sin\varphi$;

(2) The kth step, $r_{k-1} = R\angle\psi_{k-1}$, $(\alpha,\beta)_k = (\alpha,\beta)_{k-1} + r_{k-1}$;

(3) $\begin{cases} \psi_k = \psi_{k-1}, r_k = R\angle\psi_k, if(E_a(k) \leq E_a(k-1)) \\ \psi_k = \psi_{k-1} + \frac{\pi}{2}, r_k = R\angle\psi_k, if(E_a(k) > E_a(k-1)) \end{cases}$;

(4) $\begin{cases} \text{Output the search result}(\alpha,\beta)_k, if(E_a(k) < E_t), \\ \text{Return to step 2, continue search}, k = k+1, if(E_a(k) \geq E_t) \end{cases}$.

Iterative searching with a fixed step size can be used to identify unbalanced parameters, but the identification precision of the algorithm is only R (i.e., one step size). To improve the algorithm's convergence precision, the size of R must be reduced, but a reduction in R will increase the search time. Therefore, the precision and speed of the convergence of the unbalanced search method with the fixed step size are contradictory. To solve this problem, a variable step size algorithm is used in this work. In the initial stage of the search, a large step size is used, and the step size is then reduced gradually. Finally, the target position is approached with an infinitely small error to ensure high convergence precision.

The core variable step size algorithm Is

$$R(k) = R(0)E_a(k) \tag{32}$$

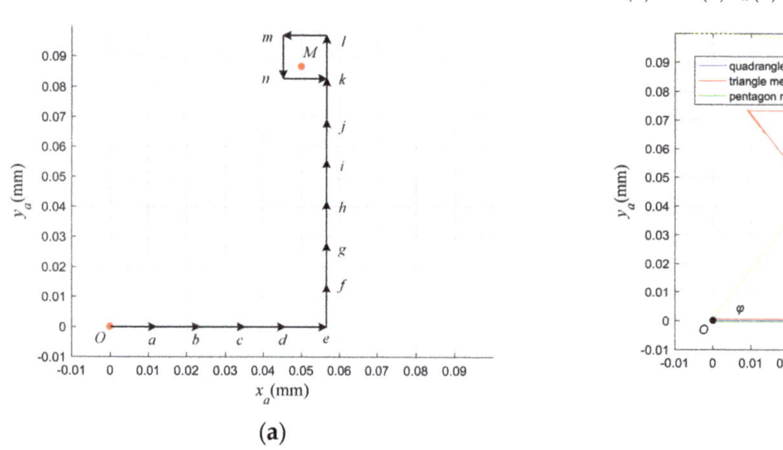

(a)

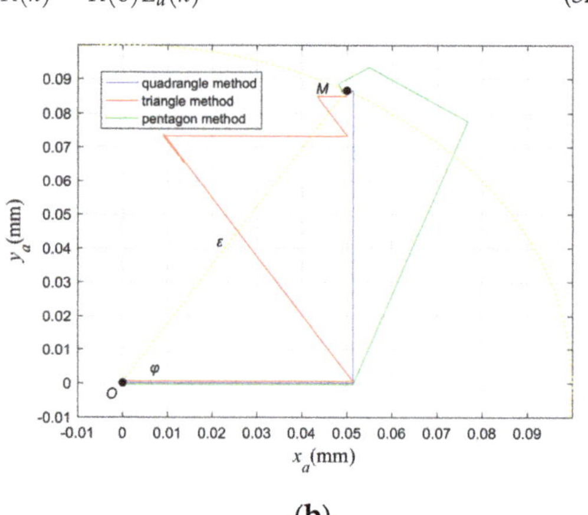

(b)

Figure 11. Diagrams of fixed and variable step size search processes. (**a**) Fixed step size search. (**b**) Variable step size search.

The actual effect of this algorithm is illustrated in Figure 11b. The simulation shows that the search path of the quadrangle method is more direct and more efficient than those of other search methods such as the triangle and pentagon methods. If the starting threshold is set to zero, it converges fully to obtain the exact unbalanced parameters.

6.2. Realization of Unbalanced Vibration Compensation

From (7) and the inverse system decoupling method, it can be found that

$$\ddot{Y} + c\dot{Y} + kY = I + K_i^{-1}L^{-1}f_u \tag{33}$$

It is only necessary to inject the compensation current $I_u = -K_i^{-1}L^{-1}f_u$ into the current channels, and then, precise compensation of the unbalanced vibration can be achieved.

7. Simulation and Experiments

7.1. Simulation with Simulink

7.1.1. Response Performance Testing of Closed-Loop System

When the AMB-rigid rotor system uses closed-loop control, each radial independent channel adopts a μ-controller that is tuned using the multi-objective optimal method. When simulating a rotor that is statically suspended (i.e., $n = 0$ rpm) at the working point, square wave signals with amplitudes of 0.1 mm are added to the input of each channel, and the transient time response at the output of each channel is shown in Figure 12. The simulation results show that the transient time response to the reference input is relatively fast: the rise time is less than 1 ms, the settling time is less than 10 ms, and the tracking error is almost zero. Hence, the target of high-precision control is realized.

7.1.2. Robust Performance Testing of Closed-Loop System

When the rotor is suspended stably in the center position and rotates at its rated speed ($n = 3000$ rpm), a sinusoidal signal with a frequency of 50 Hz and eccentricity ε of 1×10^{-4} m is injected into the control system at t = 0.1 s to test the anti-disturbance performance of the rotor system. The corresponding output signal waveform is shown in

Figure 13. The green line represents the equivalent vibration caused by the eccentric force at the system's output end, which has an amplitude of 0.1 mm; the red line is the output waveform of the decentralized PID controller. The design and adjustment of this controller were achieved by defining the equivalent stiffness and equivalent damping of the AMB and by analyzing the influences of bias currents on them and on the critical speed of the rotor system. Then, numerical simulations and experiments were carefully carried out to tune the PID parameters for this test rig (P = 6500, I = 1000, D = 4.5) [18,19]; the blue line is the output waveform from the μ-controller. These results show that the μ-controller has a strong disturbance suppression capability, with a disturbance attenuation rate of 0.21, and its displacement vibration peak value is 65.8% smaller than that of the PID controller.

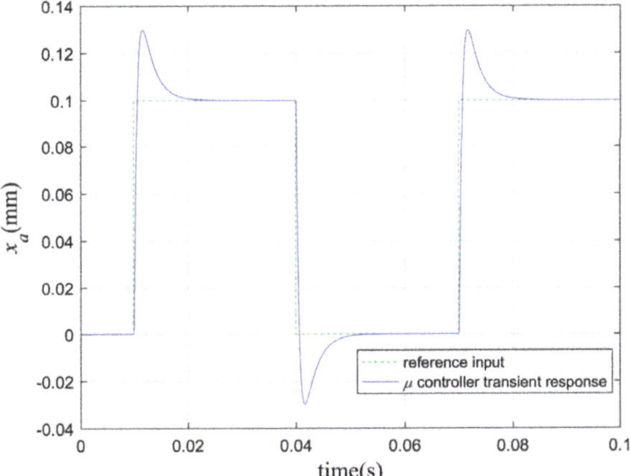

Figure 12. Static rotor simulation.

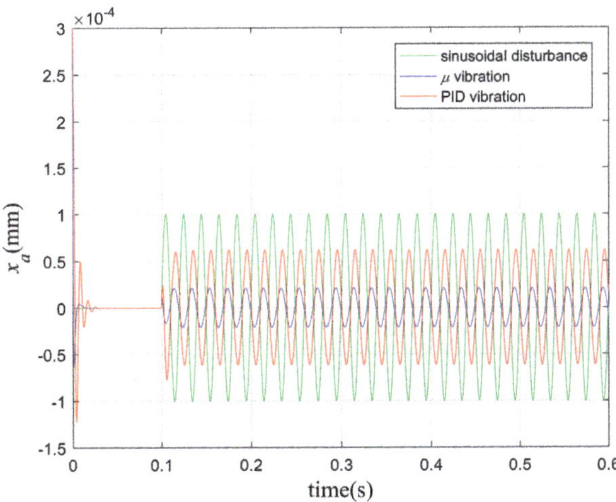

Figure 13. Anti-disturbance simulation of the μ-controller.

In Figure 14, the singular value curve of the optimal μ-controller shows that the μ-controller has a strong suppression force relative to the external disturbance below a frequency of 50 Hz, and stronger suppression occurs as the frequency decreases.

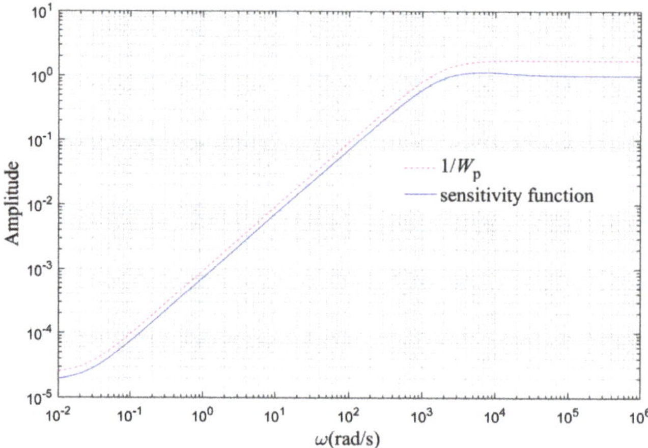

Figure 14. Sensitivity function of the closed loop system.

7.2. Experimental Results

The magnetically levitated high-speed motor rig is shown in Figure 15. The motor is a 75 kW permanent magnet synchronous motor, where the rotor is supported by two AMBs in the radial direction and by a pair of permanent magnet bearings in the axial direction. The rig control system consists of a dSPACE controller, a two-level switching power amplifier, four non-contact eddy current sensors, and a host computer. The parameters used in the experiments are given in Table 3.

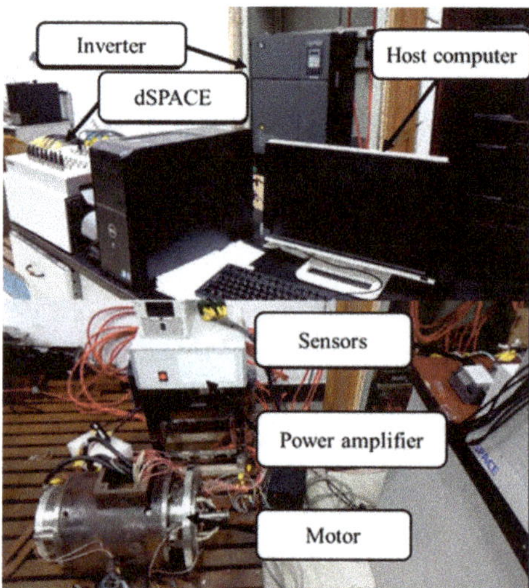

Figure 15. AMB-rigid rotor platform.

7.2.1. Static Suspension Test

When the rotor is actually suspended in the central position on the rig ($n = 0$ rpm), the transient response of x_a is as shown in Figure 16. In the figure, the red line represents the response waveform of the decentralized PID controller, and the blue line represents the response waveform of the optimal μ-controller. The figure shows that the overshoot of the μ-controller is smaller, its settling time is shorter, there is no floating transition time, and its overall performance is thus better.

Table 3. Parameters of the experimental platform.

Symbol	Description	Value
V_{DC}	DC-link voltage	100 V
I_0	maximum bias current	2.5 A
P_t	total available power	500 VA
m	mass of rotor	18.09 kg
δ	nominal air gap length	0.5 mm
J	transverse moment of inertia	0.2 kg·m^2
J_z	polar moment of inertia	0.0223 kg·m^2
l_A	distance between the AMB-A and geometric center	140 mm
l_B	distance between the AMB-B and geometric center	120 mm
k_x, k_y	displacement stiffness	542,464 N/m
k_i	current stiffness	256.872 N/A

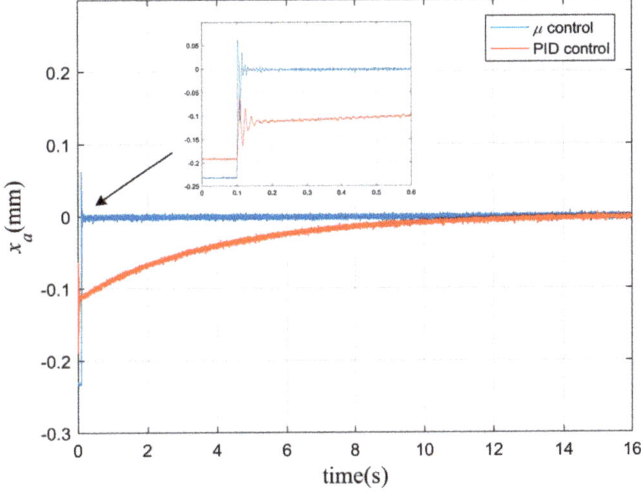

Figure 16. Transient response of the static rotor.

7.2.2. Decoupling Performance Test

At 6000 rpm, the reference input of x_a stepped up from 0 to 0.1 mm. The radial displacements of the AMB-rigid rotor system in this case are as shown in Figure 17. These results demonstrate that the step change in the radial displacement x_a does not lead to variations in the other three radial displacements, i.e., the four radial displacements of the control system are decoupled completely. These experimental results are consistent with the theoretical analysis, and thus, the expected goal of the control system design has been realized.

7.2.3. Active Vibration Control of the AMB-Rigid Rotor System

After decoupling via the inverse system method, the gyro coupling effect in the radial channels of the closed-loop rotor system is eliminated completely, and there is only the eccentric disturbance vibration with the same frequency as the speed. As Figures 18 and 19 show, the rotor vibration amplitudes before compensation were slightly less than 0.05 mm and slightly greater than 0.05 mm at 6000 rpm and 10,000 rpm, respectively.

Figure 18 shows that as a result of the anti-disturbance effect of the μ-controller and the effect of the compensation algorithm, the unbalanced vibration is effectively suppressed at 6000 rpm, and thus, the rotor motion trajectory after compensation is very small. Figure 19 also shows that even at a higher rotational speed, the unbalanced vibration can largely be controlled, and active displacement vibration control is achieved.

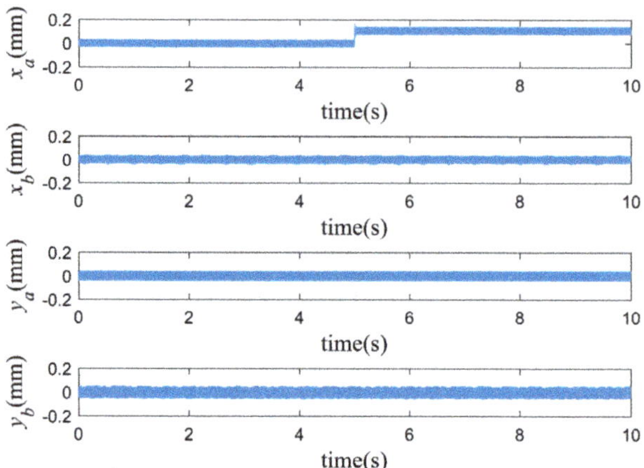

Figure 17. Radial displacements at a speed of 6000 rpm.

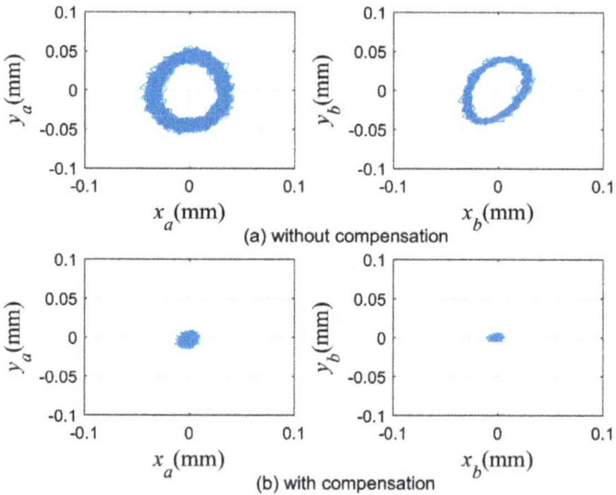

Figure 18. Rotor trajectories without/with unbalanced vibration compensation at 6000 rpm.

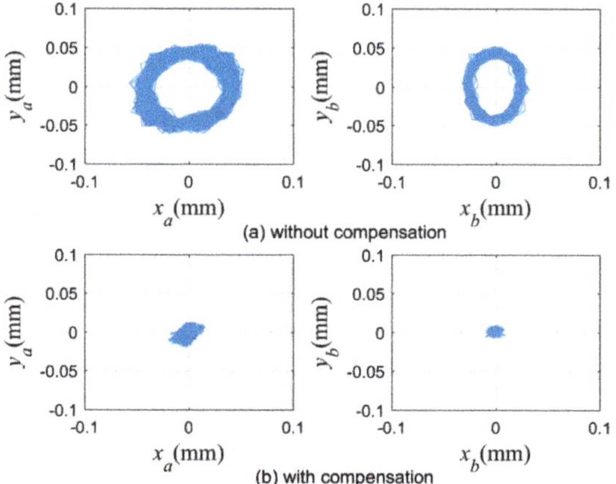

Figure 19. Rotor trajectories without/with unbalanced vibration compensation at 10,000 rpm.

To test the effectiveness of the proposed compensation algorithm over the full rotational speed range, uniform acceleration running testing of the AMB-rigid rotor system was performed. During the test, the motor accelerated from 0 rpm to 12,000 rpm with a constant acceleration of $2\pi \ \mathrm{rad/s^2}$, and the results are shown in Figure 20.

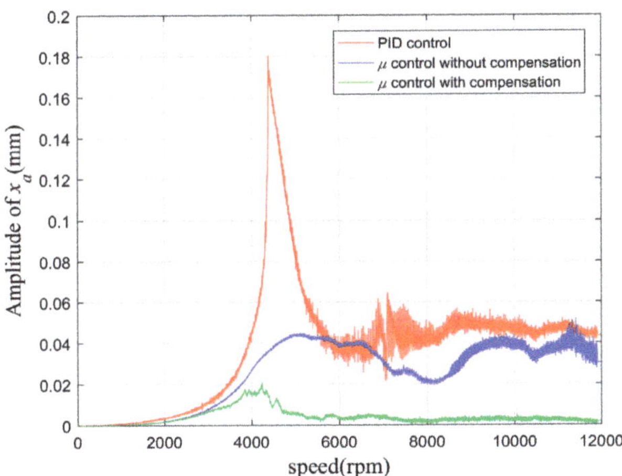

Figure 20. Unbalanced response curve of the rotor system over the full rotational speed range.

As Figure 20 shows, μ-control does not have a first-order vibration peak like that of PID control, and the acceleration curve is both continuous and smooth. With the application of the compensation algorithm, the rotor vibration is greatly suppressed to within 0.02 mm and is later close to zero. As the speed continues to increase beyond 12,000 rpm, the vibration displacement of the rotor tends to show a constant amplitude; this is called the "self-centering" effect. The rotor rotates around its inertial axis and operates in a true force-free state. The high-speed motor rotor system has thus achieved active vibration control over its full rotational speed range.

8. Conclusions

In this paper, a μ-synthesis strategy for AMBs in a high-speed motor is proposed. The determination of the appropriate weight functions W_p and W_u represents a key step in the μ-control scheme. By defining a second-order weighting function W_p with higher degrees of freedom, a multi-objective ant colony algorithm based on this function can be used to search for the optimal sensitivity function to achieve the fastest possible response speed, the highest stability margin, and the strongest external disturbance attenuation for the closed-loop system under study. By using the perturbation method for uncertain parameters, it is theoretically guaranteed that the control system will be robust and stable up to its rated speed ($n = 12{,}000$ rpm), and the stability margin is 5.11 times (i.e., it is stable within a vibration amplitude of 0.25 mm). The simulation results also show that the optimal μ-controller has an excellent disturbance suppression ability, which weakens the unbalanced disturbance vibration of the rotor system at low speeds by as much as 65.8% when compared with the classical PID controller. At higher rotational speeds, a compensation algorithm based on real-time variable step size iterative searching for eccentric positions is applied to enhance the disturbance rejection of the controller. Further experiments show that the algorithm can realize unbalanced displacement vibration compensation over the full rotational speed range. The entire control strategy can not only ensure stable rotor suspension but also can suppress the disturbance vibration strongly. The μ-controller's response performance is also greatly improved when compared with the PID. The μ-control strategy performs excellently in both static and dynamic conditions and is the preferred choice to replace PID. The realization of an optimal μ-controller also provides effective information and a reference for AMB engineering design practice.

Author Contributions: Y.L. and C.Z. conceived the proposed method and designed the experiment. Then, Y.L. conducted the experiment and analyzed the data. All authors have read and agreed to the published version of the manuscript.

Funding: This work was supported in part by Fundamental Research Project under Grant 2019110C026, Key Fundamental Research under Grant 2020-ZD-232-00 and National Natural Science Foundation of China under Grant 11632015.

Data Availability Statement: Not applicable.

Conflicts of Interest: The authors declare no conflict of interest.

References

1. Koo, B.; Kim, J.; Nam, K. Halbach array PM machine design for high speed dynamo motor. *IEEE Trans. Magn.* **2021**, *57*, 8202105. [CrossRef]
2. Le, Y.; Wang, D.; Zheng, S. Design and optimization of a radial magnetic bearing considering unbalanced magnetic pull effects for magnetically suspended compressor. *IEEE/ASME Trans. Mechatron.* **2022**, *27*, 5760–5770. [CrossRef]
3. Schweitzer, G. Dynamics of the Rigid Rotor. In *Magnetic Bearings: Theory, Design, and Application to Rotating Machinery*, 1st ed.; Schweitzer, G., Maslen, E.H., Eds.; Springer: Berlin, Germany, 2009; pp. 167–174.
4. Gong, L.; Zhu, C. Vibration suppression for magnetically levitated high-speed motors based on polarity switching tracking filter and disturbance observer. *IEEE Trans. Ind. Electron.* **2021**, *68*, 4667–4678. [CrossRef]
5. Zhao, H.; Zhu, C. Feedforward decoupling control for rigid rotor system of active magnetically suspended high-speed motors. *IET Electr. Power Appl.* **2019**, *13*, 1298–1309. [CrossRef]
6. Chen, L.; Zhu, C.; Zhong, Z.; Wang, C.; Li, Z. Radial position control for magnetically suspended high-speed flywheel energy storage system with inverse system method and extended 2-DOF PID controller. *IET Electr. Power Appl.* **2020**, *14*, 71–81. [CrossRef]
7. Zhang, S.Y.; Wei, C.B.; Li, J.; Wu, J.H. Robust H_∞ controller based on multi-objective genetic algorithms for active magnetic bearing applied to cryogenic centrifugal compressor. In Proceedings of the 2017 29th Chinese Control and Decision Conference, Chongqing, China, 28–30 May 2017.
8. Kuseyri, I.S. Robust control and unbalance compensation of rotor/active magnetic bearing systems. *J. Vib. Control* **2011**, *18*, 817–832. [CrossRef]
9. Dai, X.; He, D.; Zhang, T. MIMO system invisibility and decoupling control strategies based on ANN αth-order inversion. *IEE Proc.–Control Theory Appl.* **2001**, *148*, 125–136. [CrossRef]
10. Pesch, A.H.; Sawicki, J.T. Active magnetic bearing online levitation recovery through μ-Synthesis robust control. *Actuators* **2017**, *6*, 2. [CrossRef]
11. Zhou, K.; Doyle, J. *Essentials of Robust Control*, 1st ed.; Horton, M., Dworkin, A., Eds.; Prentice-Hall: Upper Saddle River, NJ, USA, 1998; pp. 187–192.
12. Gu, D.-W.; Petkov, P.; Konstantinov, M.M. *Robust Control Design with MATLAB*, 2nd ed.; Grimble, M.J., Johnson, M.A., Eds.; Springer: London, UK, 2013; pp. 205–209.
13. Mansouri, A.; Smairi, N.; Trabelsi, H. Multi-objective optimization of an in-wheel electric vehicle motor. *Int. J. Appl. Electromagn. Mech.* **2016**, *50*, 449–465. [CrossRef]
14. ISO 14839—3; Mechanical Vibration-Vibration of Rotating Machinery Equipped with Active Magnetic Bearings-Part 3: Evaluation of Stability Margin. ISO: Geneva, Switzerland, 2006.
15. Shi, Z.; Kumar, R.; Tomar, R. Multi-objective optimization of smart grid based on ant colony algorithm. *Electrica* **2022**, *22*, 395–402. [CrossRef]
16. Wang, W.; Zhao, J.; Li, Z.; Huang, J. Smooth path planning of mobile robot based on improved ant colony algorithm. *J. Rob.* **2021**, *2021*, 4109821. [CrossRef]
17. Li, Y.; Soleimani, H.; Zohal, M. An improved ant colony optimization algorithm for the multi-depot green vehicle routing problem with multiple objectives. *J. Cleaner Prod.* **2019**, *227*, 1161–1172. [CrossRef]
18. Wang, Z.; Mao, C.; Zhu, C. A design method of PID controller for active magnetic bearings-rigid rotor systems. *Proc. CSEE* **2018**, *38*, 6154–6163. (In Chinese)
19. Psonis, T.K.; Nikolakopoulos, P.G.; Mitronikas, E. Design of a PID controller for a linearized magnetic bearing. *Int. J. Rotating Mach.* **2015**, *2015*, 656749. [CrossRef]

 *actuators*

MDPI

Article

Currents Analysis of a Brushless Motor with Inverter Faults—Part I: Parameters of Entropy Functions and Open-Circuit Faults Detection

Cristina Morel [1,*], Sébastien Rivero [2], Baptiste Le Gueux [2], Julien Portal [2] and Saad Chahba [1]

[1] Ecole Supérieure des Techniques Aéronautiques et de Construction Automobile, ESTACA'Lab Paris-Saclay, 12 Avenue Paul Delouvrier—RD10, 78180 Montigny-le-Bretonneux, France; saad.chahba@estaca.fr
[2] ESTACA Campus Ouest, Rue Georges Charpak—BP 76121, 53 009 Laval, France; sebastien.rivero@estaca.eu (S.R.); baptiste.legueux@estaca.eu (B.L.G.); julien.portal@estaca.eu (J.P.)
* Correspondence: cristina.morel@estaca.fr

Abstract: In the field of signal processing, it is interesting to explore signal irregularities. Indeed, entropy approaches are efficient to quantify the complexity of a time series; their ability to analyze and provide information related to signal complexity justifies their growing interest. Unfortunately, many entropies exist, each requiring setting parameter values, such as the data length N, the embedding dimension m, the time lag τ, the tolerance r and the scale s for the entropy calculation. Our aim is to determine a methodology to choose the suitable entropy and the suitable parameter values. Therefore, this paper focuses on the effects of their variation. For illustration purposes, a brushless motor with a three-phase inverter is investigated to discover unique faults, and then multiple permanent open-circuit faults. Starting from the brushless inverter under healthy and faulty conditions, the various possible switching faults are discussed. The occurrence of faults in an inverter leads to atypical characteristics of phase currents, which can increase the complexity in the brushless response. Thus, the performance of many entropies and multiscale entropies is discussed to evaluate the complexity of the phase currents. Herein, we introduce a mathematical model to help select the appropriate entropy functions with proper parameter values, for detecting open-circuit faults . Moreover, this mathematical model enables to pick up many usual entropies and multiscale entropies (bubble, phase, slope and conditional entropy) that can best detect faults, for up to four switches. Simulations are then carried out to select the best entropy functions able to differentiate healthy from open-circuit faulty conditions of the inverter.

Keywords: open-circuit; inverter; brushless motor; entropy; multiscale; fault detection

Citation: Morel, C.; Rivero, S.; Le Gueux, B.; Portal, J.; Chahba, S. Currents Analysis of a Brushless Motor with Inverter Faults—Part I: Parameters of Entropy Functions and Open-Circuit Faults Detection. *Actuators* **2023**, *12*, 228. https://doi.org/10.3390/act12060228

Academic Editors: Qinfen Lu, Xinyu Fan, Cao Tan and Jiayu Lu

Received: 4 May 2023
Revised: 25 May 2023
Accepted: 28 May 2023
Published: 31 May 2023

1. Introduction

One of the most powerful tools to assess the dynamical characteristics of time series is entropy. Entropy used in several kinds of applications is able to account for vibrations of rotary machines [1] (electric machines), to detect battery faults [2] (short-circuit and open-circuit faults), to reveal important information about seismically actives zones [3] (electroseismic time series), to measure financial risks [4] (economic sciences), to categorize softwood species under uniform and gradual cross-sectional structures [5] (biology) and to categorize benign and malignant tissues of different subjects [5] (biomedical).

Various entropy measures have been established over the past two decades. Pincus [6] proposed the approximation entropy $ApEn$, which calculates the complexity of data and measures the frequency of similar patterns of data in a time series. However, $ApEn$ also has some disadvantages: due to self-matching, the bias of $ApEn$ is important for small time series and depends on the entropy parameters. To avoid self-matching, Richman [6] defined the sample entropy $SampEn$. Since the introduction of $ApEn$ [6], other entropies have been proposed, such as Kolmogorov entropy $K2En$, conditional entropy $CondEn$, dispersion entropy *DispEn*, cosine similarity entropy *CoSiEn*, bubble entropy *BubbEn*, fuzzy entropy *FuzzEn*, increment entropy *IncrEn*, phase entropy *PhasEn*, slope entropy *PhasEn*, entropy of entropy *EnofEn*, attention entropy *AttEn* and several other multiscale entropies.

Entropy is now widely applied to analyze signals in various fields having universal applications. A combination of wavelet transformation and entropy is proposed and applied in power grid fault detection [7]. Wavelet transform is commonly used to extract characteristic quantities, and to analyze transient signals, while entropy is ideal for the measurement of uncertainty. The approximate entropy features of a multiwavelet transform [8] combined with an artificial neural network recognizes transmission line faults. The multi-level wavelet Shannon entropy was proposed to locate single-sensor fault [9]. Guan [10] developed a precise diagnosis method of structural faults of rotating machinery based on a combination of empirical mode decomposition, *SampEn* and deep belief network. Entropy measures [11] are used in machine fault diagnosis. In [12], the variational mode decomposition energy entropy for each phase current cycle is calculated to accurately diagnose the arc fault: the noise component is removed according to the permutation entropy. [13] proposed to diagnose multi-circuit faults of three-phase motors. This method only needs to collect phase currents to diagnose multi-circuit points accurately: it improves the independence of diagnosis. Based on signal feature extraction, a combination of the empirical mode decomposition entropy and index energy methods is adopted in [14] to extract the Draft Tube's dynamic feature information for a water turbine. Open-circuit fault diagnosis of a multilevel inverter [15] uses the fast fault detection algorithm based on two sample techniques and the fault localization algorithm using the entropy of wavelet packets as a feature. The authors in [16] presented a fast feature extraction technique including wavelet packet decomposition, an entropy of wavelet packets for fault detection and classification of IGBT-based converters.

Open-circuit fault diagnosis methods can be divided into voltage-based methods and current-based methods, according to different fault characteristics. Voltage-based methods [17,18] can be implemented with external hardware or modeled. Recently, current-type methods based on current waveform analysis have attracted much attention [19–21].

An effective open-circuit fault diagnosis using the phase current performance of a brushless motor or inverters is shown in [22]. Other practical current-based diagnostic algorithms are addressed in [19,23,24]: they identify the reference current errors and the average absolute value of currents. Then, the average value of the current error and the average absolute value of the motor phase current are used to realize the diagnostic variable. A fast approach based on the amplitude of the d-q axis referential currents is proposed by [21]. The development of intelligent algorithms, such as fuzzy logic [25], sliding mode observer [26], neural networks [27], machine learning [28], an optimized support vector machine method [29] and wavelet transform [30], which allows to detect and identify faulty switches.

A mathematical model of healthy and faulty conditions is developed by [31]: it detects an open-circuit in interleaved boost converters with the Filippov method. The stable range of the load variation is extended using an original fault-tolerant strategy based on this model. In [32], one or a maximum of two open-circuit faults are detected by entropy functions. Seven entropies are investigated, but only sample and fuzzy entropies are able to differentiate healthy from open-circuit faulty conditions of the AC-DC-AC converter considered in [32].

We now propose a fault-detection method for a brushless motor with a three-phase inverter. The occurrence of faults in an inverter leads to atypical characteristics of phase currents, specific to the drive circuit. Usual and multiscale entropies are then used to detect multiple open-circuit faults. In this paper, we broaden the spectrum of investigation to 52 entropies, to evaluate their ability to differentiate healthy states from open-circuit faulty conditions. This is why we herein introduce a mathematical model to select the appropriate entropy functions with an appropriate parameter combination for open-circuit faults detection. The entropy calculation has several parameters, such as data length N, embedding dimension m, time lag τ, tolerance r and scale s. However, the dependence of the entropy effectiveness on the choice of parameters used for the phase currents analysis has not yet been investigated for a brushless motor. Moreover, using this mathematical model, we are able to pick up many usual entropies and multiscale entropies (bubble, phase, slope and conditional entropy) that can better detect faults for up to four switches. Our goal herein is to be able to select the appropriate entropy.

The paper is organized as follows. The usual entropies and multiscale entropies are introduced in Section 2. Sections 3 and 4 present the brushless motor and the dataset we used of output currents under healty state, with one, two, three and four open-circuit faults. Then, Section 5 illustrates the evaluation of the different entropies under variation of the data length, embedding dimension, time lag, tolerance and scale. We end with a Conclusion in Section 7.

2. Entropy Methods

- Sample entropy $SampEn$ and approximate entropy $ApEn$ are the most commonly used measures for analyzing time series. For a time series $\{x_i\}_{i=1}^{N}$ with a given embedding dimension m, tolerance r and time lag τ, the embedding vector $x_i^m = [x_i, x_{i+\tau}, \dots, x_{i+(m-1)\tau}]$ is constructed. The number of vectors x_i^m and x_j^m, close to each other, in Chebyshev distance:

$$ChebDist_{i,j}^m = max_{k=\overline{1,m}}\{|x_i^m[k] - x_j^m[k]|\} \le r \tag{1}$$

is expressed by the number $P_i^m(r)$. This number is used to calculate the local probability of occurrence of similar patterns:

$$B_i^m(r) = \frac{1}{N-m+1} P_i^m(r). \tag{2}$$

The global probability of the occurrence of similar patterns is:

$$B^m(r) = \frac{1}{N-m+1} \sum_{i=1}^{N-m+1} B_i^m(r) \tag{3}$$

with a tolerance r. For $m+1$:

$$B^{m+1}(r) = \frac{1}{N-m} \sum_{i=1}^{N-m} B_i^{m+1}(r). \tag{4}$$

The approximation entropy is:

$$ApEn(m,\tau,r,N) = ln\frac{B^m(r)}{B^{m+1}(r)}. \tag{5}$$

- Kolmogorov entropy [33]—$K2En$ is defined as the probability of a trajectory crossing a region of the phase space: suppose that there is an attractor in phase space and that the trajectory $\{x_i\}_{i=1}^{N}$ is in the basin of attraction. $K2En$ defines the probability distribution of each trajectory, calculated from the state space, and computes the limit of Shannon entropy. The state of the system is now measured at intervals of time. The time series $\{x_i\}_{i=1}^{N}$ is divided into a finite partition $\alpha = \{C_1, C_2, \dots, C_k\}$, according to $C_k = [x(i\tau), x((i+1)\tau), \dots, x((i+k-1)\tau)]$. The Shannon Entropy of such a partition is given by:

$$K(\tau,k) = -\sum_{C \in \alpha} p(C) \cdot \log p(C). \tag{6}$$

$K2En$ is then defined by:

$$K2En = - \sup_{\alpha \ finite \ partition} \lim_{N\to\infty} \frac{1}{N} \sum_{n=0}^{N-1} (K_{n+1}(\tau,k) - K_n(\tau,k)). \tag{7}$$

- Conditional entropy [34]—$CondEn$ quantifies the variation of information necessary to specify a new state in a one-dimensional incremented phase space. Small Shannon entropy values are obtained when a pattern appears several times. $CondEn$ uses the normalization:

$$x(i) = \frac{X(i) - av[X]}{std[X]}, \tag{8}$$

where $av[X]$ is the series' mean and $std[X]$ is the standard deviation of the series. From the normalized series, the vector $x_L(i) = [x(i), x(i-1), \ldots, x(i-L+1)]$ of L consecutive pattern is constructed in L dimensional phase space. With a variation in the Shannon entropy of $x_L(i)$, the $CondEn$ is obtained as:

$$CondEn(L) = - \sum_L p_L \cdot log \ p_L + \sum_{L-1} p_{L-1} \cdot log \ p_{L-1}. \tag{9}$$

- Dispersion entropy [35,36]—$DispEn$ focuses on the class sequence that maps the elements of time series into positive integers. According to the mapping rule of dispersion entropy, the same dispersion pattern results from multiple forms of sample vectors. The time series $\{x_i\}_{i=1}^N$ is reduced with the standard distribution function to normalized series $y_j^m = [y_j, y_{j+\tau}, \ldots, y_{j+(m-1)\tau}]$:

$$y_i = \frac{1}{\sigma\sqrt{2\pi}} \int_{-\inf}^{x_i} \exp \frac{-(s-\mu)^2}{2\sigma^2} ds \tag{10}$$

where $y_i \in (0,1)$. The phase space is restructured in c class number as $z_i^c = round(c \cdot y_i + 0.5)$ and $z_j^{m,c} = [z_j^c, z_{j+\tau}^c, \ldots, z_{j+(m-1)\tau}^c]$. Each z_i^c corresponds to the dispersion pattern v. The frequency of v can be deduced as:

$$p = \frac{Number\{j|j \le n - (m-1)\tau, v\}}{n - (m-1)\tau} \tag{11}$$

where $Number\{j|j \le n - (m-1)\tau, v\}$ is the number of dispersion patterns v corresponding to $z_j^{m,c}$. Dispersion entropy can be defined according to information entropy theory:

$$DispEn(m, c, \tau) = - \sum_{v=1}^{c^m} p \cdot log(p). \tag{12}$$

- Cosine similarity entropy [37]—$CoSiEn$ evaluates the angle between two embedding vectors instead of the Chebyshev distance. The global probability of occurrence of similar patterns using the local probability of occurrence of similar patterns is used to estimate entropy. The angular distance for all pairwise embedding vectors is:

$$AngDist_{i,j}^m = \frac{1}{\pi} cos^{-1} \left(\frac{x_i^m \cdot x_j^m}{|x_i^m| \cdot |x_j^m|} \right), i \ne j, \tag{13}$$

where $x_i^m = [x_i, x_{i+\tau}, \ldots, x_{i+(m-1)\tau}]$ is the embedding vector of $\{x_i\}_{i=1}^N$. When $AngDist_{i,j}^m \le r$, the number of similar patterns $P_i^m(r)$ is obtained. The local and global probabilities of occurrence are:

$$B_i^m(r) = \frac{1}{N-m-1} P_i^m(r) \quad and \quad B^m(r) = \frac{1}{N-m} \sum_{i=1}^{N-m} B_i^m(r). \tag{14}$$

Finally, cosine similarity entropy is defined by:

$$CoSiEn(m, \tau, r, N) = -B^m(r) \cdot log_2 B^m(r) - (1 - B^m(r)) \cdot log_2(1 - B^m(r)). \quad (15)$$

- Bubble entropy [38,39]—*BubbEn* reduces the significance of the parameters employed to obtain an estimated entropy. Based on permutation entropy, the *BubbEn* vectors are ranked in the embedding space. The bubble sort algorithm is used for the ordering procedure and counts the number of swaps performed for each vector. More coarse-grained distributions are created and then compute the entropy of this distribution. *BubbEn* reduces the dependence on input parameters (such as N and m) by counting the number of sample swaps necessary to achieve the ordered subsequences instead of counting order patterns. *BubbEn* embeds a given time series $\{x_i\}_{i=1}^N$ into an m dimensional space, producing a series of vectors of size $N - m + 1$: $X_1, X_2, \ldots,$ X_N, where $X_i = (x_i, x_{i+1}, \ldots, x_{i+m-1})$. The number of swaps required for sorting is counted for each vector X_i. The probability p_i of having i swaps is used to evaluate Renyi entropy:

$$B_2^m(x) = -\log \sum_{i=0}^{\frac{m(m-1)}{2}} p_i^2. \quad (16)$$

Increasing by one the embedding dimension m, the procedure is repeated to obtain a new entropy value B_2^{m+1}. Finally, *BubbEntropy* is obtained as for *ApEntropy*:

$$BubbEn(x, m, N) = \frac{B_2^{m+1} - B_2^m}{log \frac{m+1}{m-1}}. \quad (17)$$

- Fuzzy entropy [40,41]—*FuzzEn* employs the fuzzy membership functions as triangular, trapezoidal, bell-shaped, Z-shaped, Gaussian, constant-Gaussian and exponential functions. *FuzzEn* has less dependence on N and uses the same step as in the *SampEn* approach. Firstly, the zero-mean embedding vectors (centered using their own means) are constructed $q_i^m = x_i^m - \mu_i^m$, where:

$$x_i^m = [x_i, x_{i+\tau}, \ldots, x_{i+(m-1)\tau}] \quad \text{and} \quad \mu_i^m = \frac{1}{m} \sum_{k=1}^m x_i^m[k]. \quad (18)$$

FuzzEn calculates the $S_i^m(r, \eta)$ fuzzy similarity:

$$S_i^m(r, \eta) = e^{\left(ChebDist_{i,j}^m\right)^\eta / r} \quad (19)$$

obtained from a fuzzy membership function, where η is the order of the Gaussian function. The Chebyshev distance is:

$$ChebDist_{i,j}^m = max_{k=\overline{1,m}}\{q_i^m[k] - q_j^m[k]\}, i \neq j. \quad (20)$$

As in the *SampEn* approach, the local and global probabilities of occurrence are computed, obtaining a subsequent fuzzy entropy:

$$FuzzEn(m, \tau, r, N) = ln\frac{B^m(r)}{B^{m+1}(r)}. \quad (21)$$

- Increment entropy [42]: the *IncrEn* approach (similar to the permutation entropy) encodes the time series in the form of symbol sequences. For a time series $\{x_i\}_{i=1}^N$, an increment series $v(i) = x(i+1) - x(i), (1 \leq i \leq N)$ is constructed and then divided into vectors of m length $V(l) = [v(l), \ldots, v(l+m-1)], 1 \leq l \leq N - m$. Each element in each vector is mapped to a word consisting of the sign $s_k = sgn(v(j))$ and the size q_k, which is:

$$q_k = min\left(q, \left| \frac{v(i) \cdot q}{std(v(i))} \right| \right). \tag{22}$$

However, the sign indicates the direction of the volatility between the corresponding neighboring elements in the original time series. The pattern vector w is a combination of all corresponding s_k and q_k pairs. The relative frequency of each word w_n is defined as $P(w_n) = Q(w_n)/(N - m)$, where $Q(w_n)$ is the total number of instances of the nth word. Finally, *IncrEn* is defined as:

$$IncrEn = -\frac{1}{m-1} \sum_{n=1}^{(2q+1)^m} P(w_n) log P(w_n). \tag{23}$$

- *PhasEntropy* [43] quantifies the distribution of the time series $\{x_i\}$ in a two-dimensional phase space. First, the time-delayed time series $Y[n]$ and $X[n]$ are calculated as follows:

$$Y[n] = x[n+2] - x[n+1] \tag{24}$$

$$X[n] = x[n+1] - x[n] \tag{25}$$

The second-order difference plot of x is constructed as a scatter plot of $Y[n]$ against $X[n]$. The slope angle of $\theta[n]$ of each point $(X[n], Y[n])$ is measured from the origin $(0, 0)$. The plot is split into k sectors serving as a coarse-graining parameter. For each k, the sector slope angle $S_\theta[i]$ is the addition of the slope angle of points as follows:

$$S_\theta[i] = \sum_{j=1}^{N_i} \theta[n] \tag{26}$$

where $i = 1, 2, \dots , k$ and N_i is the points number of the ith sector. The probability distribution $p(i)$ of the sector is:

$$p(i) = \frac{S_\theta}{\sum_{j=1}^{i} S_\theta} \tag{27}$$

The estimation of the Shannon entropy of the probability distribution $p(i)$ leads to *PhasEn*, computed as:

$$PhasEn = -\frac{1}{log(k)} \sum_{i=1}^{k} p(i) \cdot log p(i). \tag{28}$$

- Slope entropy [44]—*SlopEn* includes amplitude information in a symbolic representation of the input time series $\{x_i\}_{i=1}^{N}$. Thus, each subsequence of length m drawn from $\{x_i\}_{i=1}^{N}$, can be transformed into another subsequence of length $m - 1$ with the differences of $x_i - x_{i-1}$. In order to find the corresponding symbols, a threshold is added to these differences. Then, *SlopEn* uses 0, 1 and 2 symbols with positive and negative versions of the last two. Each symbol covers a range of slopes for the segment joining two consecutive samples of the input data. The frequency of each pattern found is mapped into a value using a Shannon entropy approach: it is applied with the factor corresponding to the number of slope patterns found.

- Entropy of entropy [45]—*EnofEn*: the time series $\{x_i\}_{i=1}^N$ is divided into consecutive non-overlapping windows w_j^τ of length τ: $w_j^\tau = \left\{x_{(j-1)\tau+1}, \ldots, x_{(j-1)\tau+\tau}\right\}$. The probability p_{jk} for the interval x_i over w_j^τ to occur in state k is:

$$p_{jk} = \frac{\text{total number of } x_i \text{ over } w_j^\tau \text{ in state } k}{\tau}. \tag{29}$$

Shannon entropy is used now to characterize the system state inside each window. Consequently:

$$y_j^\tau = \sum_{k=1} p_{jk} \cdot \log p_{jk}. \tag{30}$$

In the second step, the probability p_l for the interval y_j to occur in state l is:

$$p_l = \frac{\text{total number of } y_j^\tau \text{ in level } l}{N/\tau}. \tag{31}$$

Shannon entropy is used for the second time instead of the Sample entropy, to characterize the degree of the state change.

$$EnofEn(\tau) = -\sum_{l=1} p_l \cdot \log p_l. \tag{32}$$

- Attention entropy [46]—*AttEn*; traditional entropy methods focus on the frequency distribution of all the observations in a time-series, while attention entropy only uses the key patterns. Instead of counting the frequency of all observations, it analyzes the frequency distribution of the intervals between the key patterns in a time-series. The last calculus is the Shannon entropy of intervals. The advantages of attention entropy are that it does not need any parameter to tune, is robust to the time-series length and requires only a linear time to compute.
- Multiscale entropy [5,47,48]—*MSEn* extends entropy to multiple time scales by calculating the entropy values for each coarse-grained time series. The multiple time scales are constructed from the original time series $\{x_1, x_2, \ldots, x_N\}$ of length N by averaging the data points within non-overlapping windows of increasing length. The coarse-grained time series $\{y^{(s)}\}$ is:

$$y_j^{(s)} = \frac{1}{\tau} \sum_{i=(j-1)s+1}^{js} x_i, \quad 1 \leq j \leq [N/s]. \tag{33}$$

MSEntropy is:

$$MSEn(m, r, s) = -\ln \frac{A_s^m(r)}{B_s^m(r)} \tag{34}$$

where $A_s^m(r)$ and $B_s^m(r)$ represent the probability that two sequences match for $m + 1$ points and m points, respectively, calculated from the coarse-grained time series at the scale factor s. Multiscale entropy reduces the accuracy of entropy estimation and is often undefined as the data length becomes shorter with an increase in scale s. This is true in the case of *SampEn*, which is sensitive to parameters (data length N, embedding dimension m, time lag τ, tolerance r) of short signals. To avoid this, many variants of the traditional multiscale entropy method, such as composite multiscale entropy [49,50] and refined multiscale entropy [51,52], are proposed. In the classical multiscale entropy method, there is only one coarse-grained time series derived from a non-overlapping coarse-grained procedure at scale s. However, s is the number of coarse-grained time series in the composite multiscale entropy method. The sliding windows of all coarse-grained procedures overlap. The mean of entropy values for all coarse-grained time series is defined as the composite multiscale entropy value at the scale s to improve the multiscale entropy accuracy. At a scale factor s, the $cMSEntropy$ is defined as:

$$cMSEn(m, r, s) = \frac{1}{s} \sum_{k=1}^{s} \left(-ln \frac{n_{k,s}^{m+1}}{n_{k,s}^{m}} \right),$$

(35)

where $n_{k,s}^{m}$ is the total number of m-dimensional matched vector pairs and is calculated from the kth coarse-grained time series at a scale factor s.

The refined multiscale entropy [51,52], based on the multiscale entropy approach, applies different entropies as a function of time scale in order to perform a multiscale irregularity assessment. $rMSEn$ prevents the influence of the reduced variance on the complexity evaluation and removes the fast temporal scales. Thus, an $rMSEn$ method improves the coarse-grained process.

3. System Description

Many industrial applications require precise regulation of the speed of the drive motors. A brushless motor operates under various speed and load conditions and the knowledge of some physical parameters (speed, torque, current) for proper speed regulation is essential. Figure 1 shows a system implementation for brushless motor control as a Permanent Magnet Synchronous Machine in Matlab/Simulink. A three-phase inverter is used to feed the motor phases, thereby injecting currents in the coils to create the necessary magnetic fields for three phases. The three-phase inverter is modeled as an universal bridge in Matlab, with three arms and MOSFET/ Diode as power electronic devices (T_i and B_i, i = a, b, c), controlled by pulse width modulation.

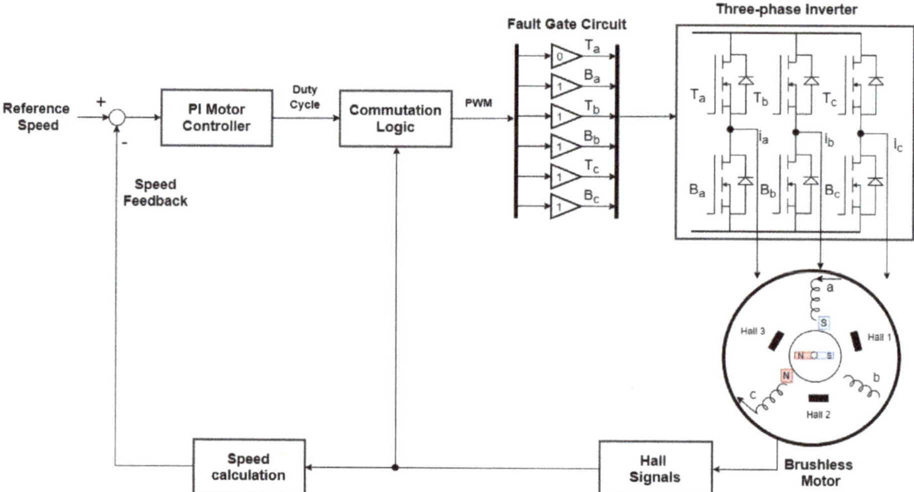

Figure 1. Power circuit structure of a brushless motor.

A simplified model of stator consists of three coils arranged in a, b and c directions. To ensure the brushless motor movement, the a, b and c stator windings are powered according to the rotor's position. The rotor magnetic field position is detected by three Hall sensors (placed every 120°) and provides the corresponding winding excitation through the commutation logic circuit. Table 1 summarizes the main specifications of this brushless machine.

In permanent magnet synchronous motors, a physical phenomenon can appear: the electromagnetic torque oscillations. These oscillations are named the Cogging effect and are taken in consideration by [53,54]. The Cogging phenomenon is the interaction of the magnetic field produced by the permanent magnet rotor with the stator teeth. This interaction can be reduced by the physical modification of the rotor and stator internal structure. Another modality to reduce the phenomenon is the control technique introducing this knowledge directly in the controller design, as in [53,54]. Cogging torque is a significant problem for high-precision applications where position control is required. In order to simplify the analysis, the previously mentioned Cogging phenomenon can be neglected.

The brushless motor design and the analysis of various control techniques are discussed in [55], where double closed loops (speed and current) are considered. The current loop is used to improve the dynamic performance of the controlled system. Wu [56] used only one speed control loop as [57]. Nga [58] assumed that the current control loop is ideal, meaning that the transfer function of the closed current control loop is equal to 1. In the cascaded control structure, the inner loops are designed to achieve fast response and outer loop is designed to achieve optimum regulation and stability.

The inner loop also keeps the torque output below a safe limit. Moreover, the controller should be developed in such a manner that it produces less torque ripple. Torque ripple is developed from motor control through inefficient commutation strategies and internal gate control schemes. Ideally, the torque ripple is constant due to the in-phase back electromotive force and quasi-square wave stator current. In this paper, we consider the dynamics of the current control loop much faster than that of the speed control loop in order to decouple both dynamics. Satisfying this condition, the reference value of the inner loop, which is the output of the outer controller can be considered nearly constant (a simple current limit closed control loop). To achieve the regulation objective, we are interested by the steady-state phase currents and not by their dynamic performances. This paper proposes a simple control structure using only the speed control loop.

The outer loop helps to control the speed of the motor. The speed feedback comes from the Hall sensor positions. The three-phase control technique for brushless motor uses a proportional integral (PI) controller. The controller receives the error signal and generates the control signal to regulate the speed response (referred to the target speed). The PI controls the duty cycle of the PWM pulses to maintain the desired speed. The proportional and integral gains of the controller are described in Table 1. The inner loop synchronizes the inverter gate states of the brushless motor and stator winding excitation as in Table 2. With this table, the commutation logic can easily control the commutation. The fault gate circuit is implemented by gains destined to each MOSFET. The gain is zero for an open-circuit fault, and one if not faulty.

Both single-switch and multi-switch open-circuit faults are classified and studied.

- One open-circuit fault may occur in the switch: T_a or B_a of the first phase a, T_b or B_b of the second phase b, T_c or B_c of the phase c.
- Open-circuit phase fault can be detected in T_a and B_a, T_b and B_b or T_c and B_c.
- If two upper *MOSs* faults are detected, the two open-circuit faults can be T_a and T_b, T_a and T_c or T_b and T_c. The two open-circuit faults, B_a and B_b, B_a and B_c or B_b and B_c, are the symmetrical faults of the lower arms.
- The cases of two open-circuit faults on the upper and lower arms are T_a and B_b, T_a and B_c, T_b and B_a, T_b and B_c, T_c and B_a and T_c and B_b.
- The brushless motor is still running even in three fault cases: T_a, B_a, T_b; T_a, B_a, B_b; T_a, B_a, T_c; T_a, B_a, B_c; T_b, B_b, T_a; T_b, B_b, B_a; T_b, B_b, T_c; T_b, B_b, B_c; T_c, B_c, T_a; T_c, B_c, B_a; T_c, B_c, T_b; T_c, B_c, B_b; T_a, B_b, T_c ; T_a, B_b, B_c; T_a, T_b, B_c; B_a, B_b, T_c; B_a, T_b, B_c; B_a, T_b, T_c.
- If the upper and lower arms are affected by multiple open-circuits, the open-circuit faults can be: T_a, B_a, T_b, B_c; T_a, B_a, B_b, T_c; T_a, T_b, B_b, B_c; B_a, T_b, B_b, T_c; T_a, B_b, T_c, B_c; B_a, T_b, T_c, B_c.

With no loss of generality, this work focuses on the open-circuit fault on the first switch T_a of the first phase a. Two open switch faults are also considered: on the second switch B_a of the first phase and on the first switch T_b of the second phase; then, two open-circuit faults on the first phases T_a, B_a, followed by the case T_b, T_c. The cases of multiple open-circuits are: B_a, B_b, T_c, then B_a, T_b, B_b and finally B_a, T_b, B_b, T_c.

Table 1. Mean specifications of the brushless machine.

Specification	Parameter	Value
Brushless motor	Stator phase resistance	2.8 Ω
	Stator phase inductance	$8.5 \cdot 10^{-3}$ H
	Flux linkage	0.175
	Inertia	0.8^{-3} kg/m^2
	Viscous damping	0.001 Nms
	Pole pairs	4
	Rotor flux position	90°
Speed controller	Proportional	0.015
	Integral	16
	Min output	−500
	Max output	500

Table 2. Truth table of Hall effect sensors and gate state of the brushless motor.

Hall 1	Hall 2	Hall 3	T_a	B_a	T_b	B_b	T_c	B_c
1	0	1	1	0	0	0	0	1
0	0	1	0	1	0	0	1	0
0	1	1	0	0	0	1	1	0
0	1	0	0	0	1	0	0	1
1	1	0	0	1	1	0	0	0
1	0	0	1	0	0	1	0	0

4. Datasets

Each inverter phase has two arms, i.e., the upper arm and the lower arm, whose currents are denoted as i_a, i_b and i_c. The three-phase currents of the brushless motor are recorded as a one-dimensional time series.

Let us observe the current of phase a under normal operating conditions (i.e., without any fault on the switches T_a, B_a, T_b, B_b, T_c or B_c of the inverter). Figure 2a shows this time-series data, sampled with sampling time $T = 5$ μs and composed of $N = 6000$ samples. The currents of phase b and c are similar to phase a's current. Under healthy conditions, Table 3 (first line) shows the zero average of the phase currents.

Then, an open-circuit fault occurs in phase a on the T_a switch. The output currents of phases a, b and c corresponding to an open-circuit fault are shown in Figures 2b and 3a,b. When an open-circuit fault occurs in phase a, the positive phase current gets distorted and that phase average current becomes negative; it is positive for the two others. The DC side of phase a output current can be observed in Table 3 (line 2). The current amplitude of phases b and c change; their means change too. Similarly, when an open-circuit fault occurs in phase a on switch B_a, the negative phase current gets distorted and the average current of that phase becomes positive when it is negative for the two others.

Considering a phase fault in switches T_a and B_a, the phase current waveforms are illustrated in Figures 4a,b and 5a. Consequently to these faults, the mean of the phase current i_a has a very low amplitude. The currents of the other phases recover the alternating waveforms. Line 8 of Table 3 presents the current means.

For instance, when two upper open-circuit faults simultaneously occur in T_a and T_b, the currents in the upper half-bridges are only able to flow in T_c. Figures 5b and 6a,b show the abnormal distortions of the currents of phases a, b and c, which differ from normal operating conditions. During this process, the open-circuit faults degrade the system's performances, but do not cause a shutdown.

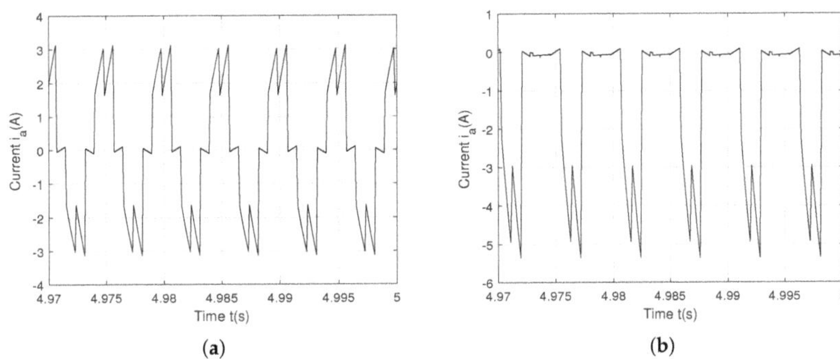

Figure 2. (**a**) Current i_a during a no$-$fault case; (**b**) Current i_a during open$-$circuit faults on T_a.

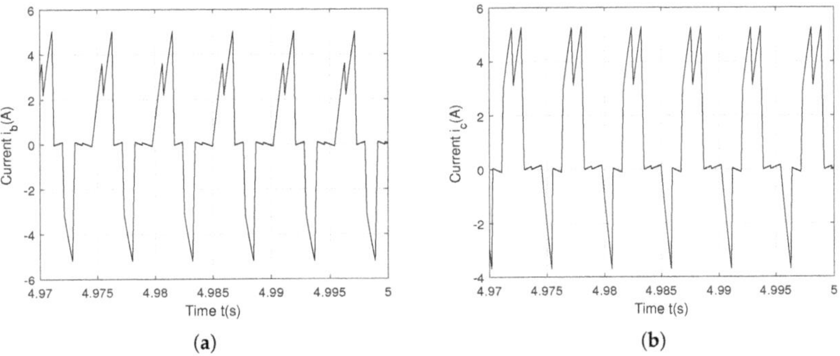

Figure 3. (**a**) Current i_b during open$-$circuit faults on T_a; (**b**) Current i_c during open$-$circuit faults on T_a.

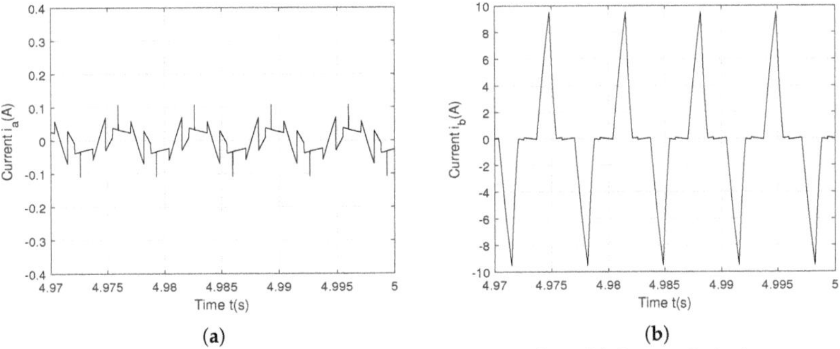

Figure 4. (**a**) Current i_a during open$-$circuit faults on T_a and B_a; (**b**) Current i_b during open$-$circuit faults on T_a and B_a.

Table 3. Current means of phases *a*, *b* and *c* with no−fault, one open−circuit fault, two open−circuit faults, three open−circuit faults and four open−circuit faults on the switches for a couple = 3 Nm and a reference speed = 3000 tr/min.

No.	Open-Circuit Fault	i_a Current Mean	i_b Current Mean	i_c Current Mean	$xyzw$
1.	No fault	0.00033	−0.00001	−0.00034	
2.	T_a	−1.4164	0.2167	1.1994	*abac*
3.	T_b	1.2017	−1.4161	0.2144	*abbc*
4.	T_c	0.2885	1.1260	−1.4145	*acbc*
5.	B_a	1.4158	−0.2652	−1.1506	*abac*
6.	B_b	−1.1234	1.2713	−0.1479	*abbc*
7.	B_c	−0.2702	−1.1422	1.4125	*acbc*
8.	T_a, B_a	0.0006	0.2708	−0.2714	*abac*
9.	T_b, B_b	-0.2761	0.0007	0.2754	*abbc*
10.	T_c, B_c	−0.173	0.1757	−0.0027	*acbc*
11.	T_a, T_b	-0.8987	-1.6516	2.5503	*acbc*
12.	T_b, T_c	2.465	−0.8106	−1.6545	*abac*
13.	T_a, T_c	−1.6474	2.5753	−0.9278	*abbc*
14.	B_a, B_b	0.9219	1.6456	−2.5675	*acbc*
15.	B_b, B_c	−2.4387	0.9258	1.5129	*abac*
16.	B_a, B_c	1.6466	−2.5087	0.8621	*abbc*
17.	T_a, B_b	−1.7775	1.3916	0.3858	*acbc*
18.	T_a, B_c	−1.3516	−0.4263	1.7779	*abbc*
19.	T_b, B_a	1.9577	−1.5853	−0.3724	*acbc*
20.	T_b, B_c	0.5589	−1.8893	1.3304	*abac*
21.	T_c, B_a	1.5627	0.5331	−2.0958	*abbc*
22.	T_c, B_b	−0.6330	2.0175	−1.3846	*abac*
23.	T_a, B_a, T_b	0.0004	−2.6371	2.6367	*acbc*
24.	T_a, B_a, B_b	0.0044	2.9920	−2.9964	*acbc*
25.	T_a, B_a, T_c	0.0025	2.4690	−2.4715	*abbc*
26.	T_a, B_a, B_c	−0.0032	−2.8272	2.8304	*abbc*
27.	T_b, B_b, T_a	−2.9702	0.0025	2.9678	*acbc*
28.	T_b, B_b, B_a	2.3538	0.0007	−2.3546	*acbc*
29.	T_b, B_b, T_c	2.9561	−0.0022	−2.9538	*abac*
30.	T_b, B_b, B_c	−2.3683	−0.0019	2.3702	*abac*
31.	T_c, B_c, T_a	−2.5577	2.5564	0.0014	*abbc*
32.	T_c, B_c, B_a	2.2754	−2.2745	−0.0009	*abbc*
33.	T_c, B_c, T_b	2.6707	−2.6692	−0.0015	*abac*
34.	T_c, B_c, B_b	−2.9748	2.9760	−0.0012	*abac*
35.	T_a, B_b, T_c	−1.6465	2.4144	−0.7679	*abac*
36.	T_a, B_b, B_c	−2.5268	0.9357	1.5911	*abac*
37.	T_a, T_b, B_c	−0.7666	−1.6462	2.4128	*abac*
38.	B_a, B_b, T_c	0.9229	1.6451	−2.5680	*abac*
39.	B_a, T_b, B_c	1.6455	−2.5318	0.8864	*abac*
40.	B_a, T_b, T_c	2.3986	−0.7597	−1.6389	*abac*
41.	T_a, B_a, T_b, B_c	0.0016	−2.6055	2.6038	*abac*
42.	T_a, B_a, B_b, T_c	0.0041	3.0170	−3.0211	*abac*
43.	T_a, T_b, B_b, B_c	−2.8549	0.0023	2.8526	*abac*
44.	B_a, T_b, B_b, T_c	2.6972	−0.0011	−2.6961	*abac*
45.	T_a, B_b, T_c, B_c	−2.4336	2.4316	0.002	*abac*
46.	B_a, T_b, T_c, B_c	2.3253	−2.3255	0.0001	*abac*

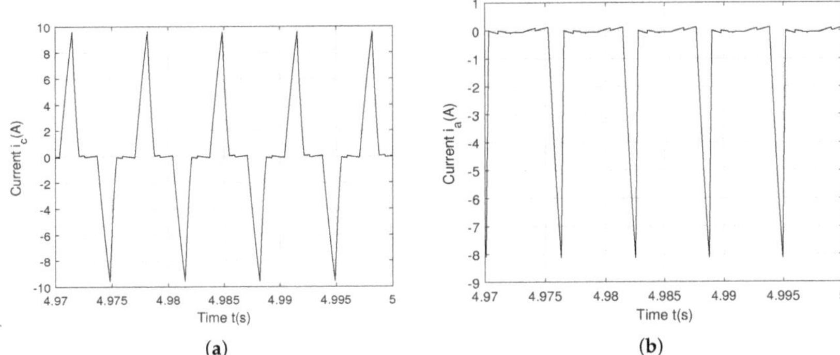

Figure 5. (**a**) Current i_c during open−circuit faults on T_a and B_a; (**b**) Current i_a during open−circuit faults on T_a and T_b.

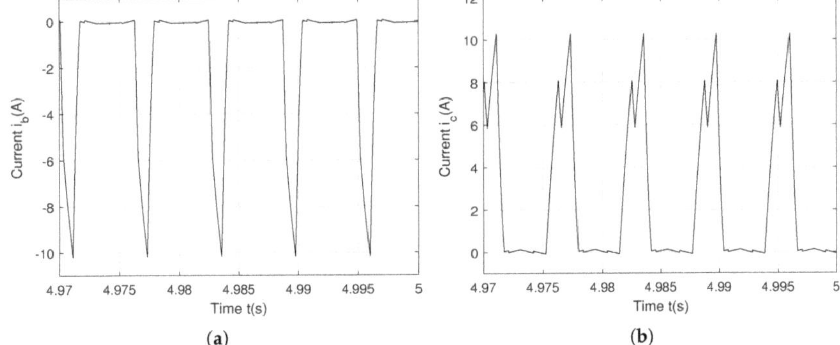

Figure 6. (**a**) Current i_b during open−circuit faults on T_a and T_b; (**b**) Current i_c during open−circuit faults on T_a and T_b.

If two open-circuit faults occur in T_a and B_b, the phase a current remains positive and the phase b current remains negative, as shown in Figure 7a,b . The other phase current (i_c) is also affected with unbalance during these fault conditions, as shown in Figure 8a.

Considering three faults in T_a, B_a, T_b, the phase a current is near zero and the phase b current during the positive cycle is eliminated as shown in Figures 8b and 9a. Consequently, the phase c current only remains during the positive cycles, as shown in Figure 9b.

Similarly, the effects of B_a, T_b and B_c faults on the phase currents are easy to find out. When the lower switch B_a is faulty, the current i_a flows only T_a, having a positive mean (Figure 10a). With T_b fault, the positive cycle of phase current i_b vanishes, as shown in Figure 10b. Figure 11a shows i_c: when the open-circuit fault of B_c occurs, current i_c has a positive waveform.

In the case of multiple open-circuit faults in several switches (B_a, T_b, B_b, T_c), phase current waveforms are seriously affected as shown in Figures 11b and 12a,b: phase b current is near zero according to phase fault, while phase a current and phase c current have a positive mean and a negative mean, respectively.

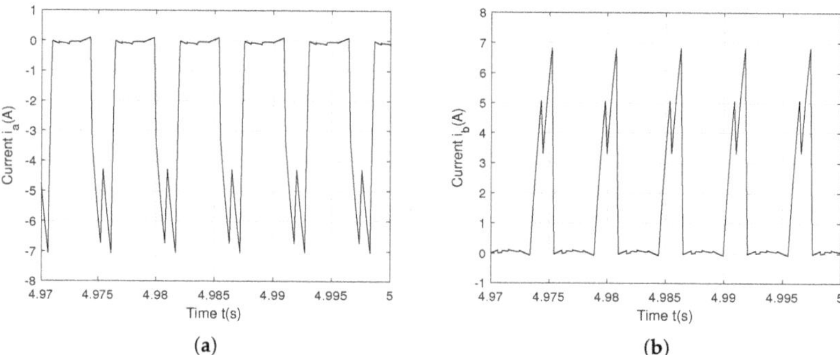

Figure 7. (**a**) Current i_a during open−circuit faults on T_a and B_b; (**b**) Current i_b during open−circuit faults on T_a and B_b.

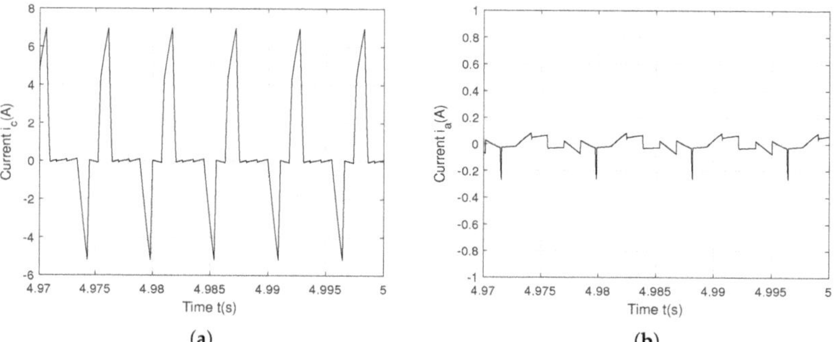

Figure 8. (**a**) Current i_c during open−circuit faults on T_a and B_b; (**b**) Current i_a during open−circuit faults on T_a, B_a and T_b.

The faults are divided into eight categories: no fault, 6 single-switch faults, 3 double-switch faults in the same bridge arm, 3 two upper-switch faults and 3 two lower-switch faults, 6 double faults from crossed half-bridges, 12 triple-switch faults with phase failure, 6 triple-switch faults in different bridge arms and, finally, 6 multiple faults with phase failure. For a typical three-phase inverter, there are 45 possible open-circuit faults, as shown in Table 3. For these cases, the mean of the phase currents are calculated and shown in Table 3.

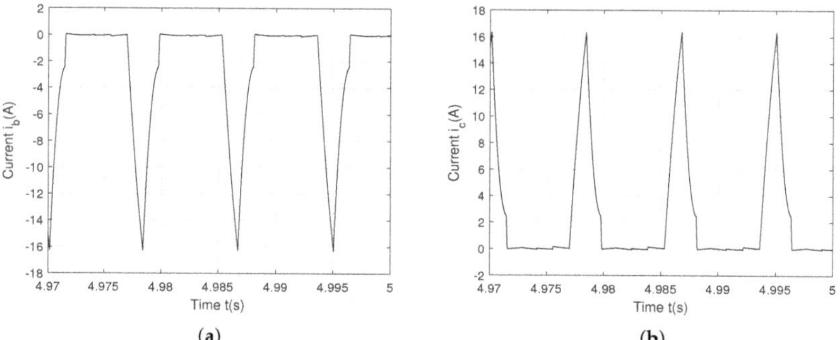

Figure 9. (**a**) Current i_b during open−circuit faults on T_a, B_a and T_b; (**b**) Current i_c during open−circuit faults on T_a, B_a and T_b.

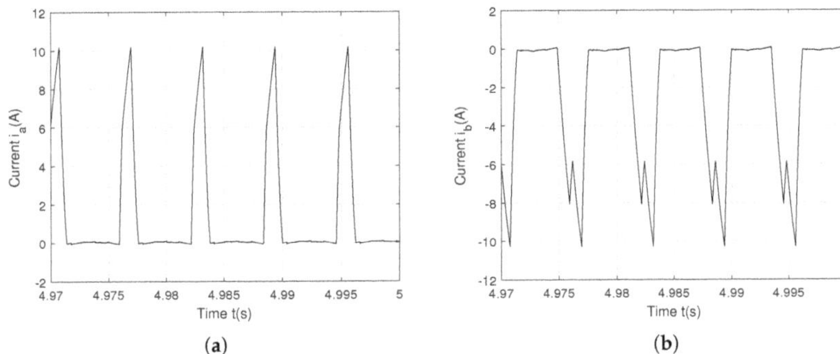

Figure 10. (**a**) Current i_a during open−circuit faults on B_a, T_b and B_c; (**b**) Current i_b during open−circuit faults on B_a, T_b and B_c.

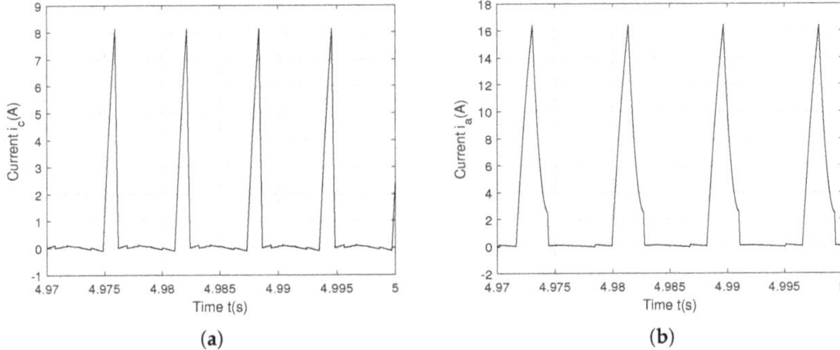

Figure 11. (**a**) Current i_c during open−circuit faults on B_a, T_b and B_c; (**b**) Current i_a during open−circuit faults on B_a, T_b, B_b and T_c.

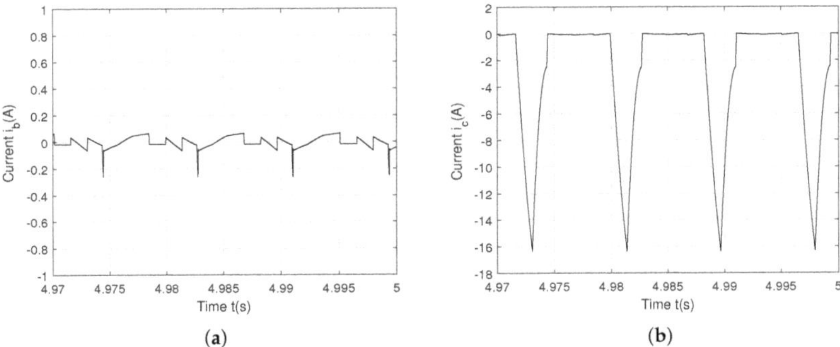

Figure 12. (**a**) Current i_b during open−circuit faults on B_a, T_b, B_b and T_c; (**b**) Current i_c during open−circuit faults on B_a, T_b, B_b and T_c.

For the first fault case (T_a), the signs of the phase currents (i_a, i_b and i_c) are negative, positive and positive. Positive, negative and negative values are found, respectively, on lines 15 (B_b and B_c faults), 17 (T_a and B_c faults) and 36 (T_a, B_b and B_c faults). If a phase is faulty, for example phase b, the line 9 of Table 3 presents a negative, zero and positive mean values. However, negative, zero and positive mean values can also be on the other fault cases: line 27 (T_b, B_b and T_a faults), line 30 (T_b, B_b and B_c faults) and line 43 (T_a, T_b, B_b and

B_c faults). For the last example with negative, negative and positive mean current values, there are four fault cases: line 7 (B_c fault), line 11 (T_a and T_b faults), line 18 (T_a and B_c faults) and line 37 (T_a, T_b and B_c faults).

5. Selection of Entropy Functions

In this part, entropy is employed to characterize the complexity of signals in the open-circuit case, such as the healthy and faulty waveforms, as in Figures 2a–12b. Phase currents are directly used as information. A fault is detected, based on the average current. When the fault occurs in the inverter, the current waveforms vary. *MSEn*, *cMSEn* and *rMSEn* algorithms can use *SampEn*, *K2En*, *CondEn*, *DispEn*, *CoSiEn*, *BubbEn*, *ApEn*, *FuzzEn*, *IncrEn*, *PhasEn*, *SlopEn* and *EnofEn* approaches or *AttEn*, giving 52 entropy functions to evaluate the signals complexity. For the ease of comparison, the entropy of phases a, b and c when one open-circuit fault occurs on T_a, is divided by the entropy of phase a under healthy conditions. Similarly, the entropy of phases a, b and c when multiple open-circuit faults occur on T_i and B_i $i = a, b, c$, is divided by the entropy of phase a under healthy conditions.

5.1. One Open-Circuit Fault on T_a on the Phase a

This study investigates the efficiency of different entropies with several parameters, such as data length N = 6000 samples, embedding dimension m = 2, time delay τ = 1, tolerance r = 0.2 and scale s = 3. The entropies of the 6000 samples are shown in Figure 13. The *BubbEn* entropy of phase a samples (represented in red), where the open-circuit fault occurs, has larger value than the entropy of phases b and c (represented in black). Incontestably, they are clearly separated. Even the entropy of phase a is lower than the entropy of phases b and c for *SampEn*, *K2En*, *DispEn*, *ApEn*, *SlopEn* and *AttEn*. The separation of the three phases a, b and c is shown in Figure 13: phases b and c have an entropy very close to each other, and different from that of phase a. Each of these entropies is able to detect the faulty phase. Figure 13 represents the larger difference between the entropy of phase a; the entropy of phases b and c is given by *BubbEn*.

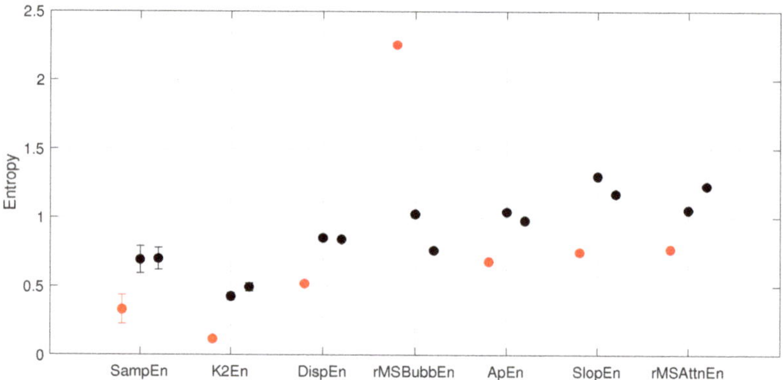

Figure 13. Entropy evaluation using *SampEn* (mean of four sample entropies), *K2En* (mean of four Kolmokov entropies), *DispEn* (mean of four dispersion entropies), *rMSBubbEn*, *ApEn*, *SlopEn* (mean of four slope entropies) and *rMSAttEn* for one open−circuit fault on T_a: phase a entropy with an open−circuit fault in red and phases b and c entropies in black.

Many values represented in Figure 13 are an average of two, three or four entropies. Relevant values of *SampEn*, *MSSampEn*, *cMSSampEn* and *rMSSampEn* are averaged to give a mean entropy of phases a, b and c. In the same way, for slope entropy, the same entropy value is obtained with *SlopEn*, *MSSlopEn*, *cMSSlopEn* and *rMSSlopEn* functions. With dispersion entropy also, for *DispEn*, *MSDispEn*, *cMSDispEn* and *rMSDispEn*, the same value is obtained. *K2En*, *MSK2En*, *cMSK2En* and *rMSK2En* give similar entropy values. For bubble entropy, a pertinent value is obtained only with $rMSBubbEn$: unfortunately, $BubbEn$, $MSBubbEn$ and $cMSBubbEn$ do not distinguish the open-circuit fault on phase a from phases b and c. Figure 13 presents the approximation entropy using only $ApEn$. The other values of $MSApEn$, $cMSApEn$ and $rMSApEn$ do not distinguish the open-circuit fault on phase a. A relevant value of attention entropy is obtained with $rMSAttEn$. For the other entropies, such as $CondEn$, $CoSiEn$, $FuzzEn$, $IncrEn$, $PhasEn$ and $EnofEn$, the distance between faulty phase a and phases b or c is nearly zero: the open-circuit fault is not detected.

The optimal entropy should be searched from all possible combinations, according to the following rules:

$$max\left(\underset{j=1}{\overset{52}{distance}} \left| entropy_{phase-a} - entropy_{phase-b} \right| \right) \tag{36}$$

and

$$max\left(\underset{j=1}{\overset{52}{distance}} \left| entropy_{phase-a} - entropy_{phase-c} \right| \right). \tag{37}$$

The objective is to maximise the distance between the phase a entropy, where the open-circuit fault occurs and the entropy of phases b and c. For a typical brushless motor, 52 possible open-circuit faults can be diagnosed using Equations (36) and (37) according to the principle shown in Table 4. Normally, the entropy is able to detect the faulty phase: it is denoted with '✓'. Otherwise, if the open-circuit fault is not detected (the distance between faulty phase a and phases b or c is nearly zero), it is pointed out by '✗'. The neutral mark '-' is employed if the distance between phase a and phases b or c is not zero but not enough to detect the open-circuit fault. However, the distance is only an approximate measure on the characteristic plot.

Table 4. Several entropies' fault-detection capability with one open-circuit fault, two open-circuit faults, three open-circuit faults and four open-circuit switches faults for a couple = 3 Nm and a reference speed = 3000 tr/min.

Entropies	T_a	B_aT_b	T_aB_a	T_bT_c	$B_aB_bT_c$	$B_aT_bB_b$	$B_aT_bB_bT_c$
$SampEn$	✓	✗	✓	✓	✗	✗	✗
$MSSampEn$	✓	✗	✓	✗	✗	✗	✗
$cMSSampEn$	✓	✗	✓	✗	✗	✗	✗
$rMSSampEn$	✓	✓	✓	✗	✗	✗	✗
$K2En$	✓	✗	✓	✓	✗	✗	✗
$MSK2En$	✓	✗	✓	✓	✗	✗	✗
$cMSK2En$	✓	✗	✓	✓	✗	✗	✗
$rMSK2En$	✓	✗	✓	✓	✗	✗	✗
$CondEn$	-	✗	-	✓	-	✗	✓
$MSCondEn$	✗	✗	-	✗	-	✗	✓
$cMSCondEn$	✗	✗	-	✗	-	✗	✓
$rMSCondEn$	✗	✗	-	✗	-	✗	-
$DispEn$	✓	✗	-	✓	✗	✗	✗
$MSDispEn$	✓	✗	-	✓	✗	✗	✗
$cMSDispEn$	✓	✗	-	✓	✗	✗	✗
$rMSDispEn$	✓	✗	-	✓	✗	✗	✗
$CoSiEn$	-	✓	✗	✗	✗	✗	✗
$MSCoSiEn$	-	-	✗	✗	✗	✗	✗
$cMSCoSiEn$	-	✓	✗	✗	✗	✗	✗
$rMSCoSiEn$	-	✓	✗	✗	✗	✗	✗
$BubbEn$	✗	✗	-	✗	✓	✗	✗
$MSBubbEn$	✗	✗	-	✗	✓	✗	✗
$cMSBubbEn$	✗	✗	-	✗	✓	✗	✗
$rMSBubbEn$	✓	-	✓	✓	-	✗	✗

Table 4. *Cont.*

Entropies	T_a	B_aT_b	T_aB_a	T_bT_c	$B_aB_bT_c$	$B_aT_bB_b$	$B_aT_bB_bT_c$
ApEn	✓	✗	-	-	✗	✗	✗
MSApEn	✗	✗	✓	✗	✗	✗	✗
cMSApEn	✗	✗	✓	✗	✗	✗	✗
rMSApEn	✗	✗	✓	✗	✗	✗	✗
FuzzEn	-	✓	✓	✗	✗	✗	✗
MSFuzzEn	✗	✓	✓	✗	✗	✗	✗
cMSFuzzEn	✗	✓	✓	✗	✗	✗	✗
rMSFuzzEn	✗	✓	✓	✗	✗	✗	✗
IncrEn	✗	✗	✗	✗	✓	✓	✓
MSIncrEn	✗	✗	✗	✗	✓	✓	✓
cMSIncrEn	✗	✗	✗	✗	✓	✓	✓
rMSIncrEn	-	✗	✗	✗	-	✗	-
PhasEn	✗	-	✗	-	✗	✓	✗
MSPhasEn	✗	-	✗	-	✗	✓	✗
cMSPhasEn	✗	-	✗	-	✗	✓	✗
rMSPhasEn	✗	-	✗	-	✗	✗	✗
SlopEn	✓	✓	✓	✗	✗	✗	✗
MSSlopEn	✓	✓	✓	✗	✗	✗	✗
cMSSlopEn	✓	✓	✓	✗	✗	✗	✗
rMSSlopEn	✓	✓	✓	✗	✗	✗	✗
EnofEn	-	✓	✗	✗	✗	✗	✗
MSEnEn	-	✓	✗	✗	✗	✗	✗
cMSEnEn	-	✓	✗	✗	✗	✗	✗
rMSEn	✗	✓	✗	✗	✗	✗	✗
AttEn	✗	✗	✗	✗	✗	✗	✗
MSAttEn	✗	✗	✗	✗	✗	✗	✗
cMSAttEn	✗	✗	✗	✗	✗	✗	✗
rMSAttEn	✓	-	✗	-	✗	✗	✗

5.2. Two Open-Circuit Faults on B_a—Phase a and on T_b—Phase b

The embedding dimension m, data length N, time delay τ and the choice of tolerance r remain unchanged. Figure 14 shows the performance of several entropies with two open-circuit faults: on B_a—phase a and on T_b—phase b. The entropies of faulty phases a and b are in red, the entropy of phase c is in black.

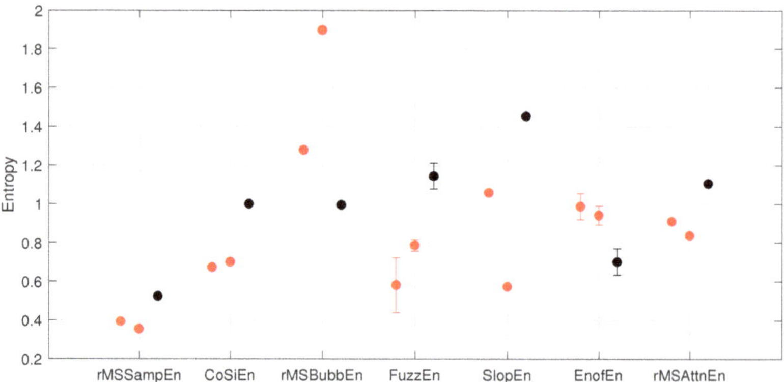

Figure 14. *rMSSampEn, CosiEn, FuzzEn, EnofEn, ApEn, rMSSlopEn* and *rMSAttEn* for one open−circuit fault on B_a and T_b: entropy of faulty phases a and b in red and phase c in black.

The optimal entropy should be searched from all possible combinations, according to the following rules:

$$max\left(\underset{j=1}{\overset{52}{distance}}\left|entropy_{phase-a} - entropy_{phase-c}\right|\right) \tag{38}$$

and

$$max\left(\underset{j=1}{\overset{52}{distance}}\left|entropy_{phase-b} - entropy_{phase-c}\right|\right). \tag{39}$$

The objective is to maximise the distance between the entropy of phase a and this of phase c, as distance between the entropy of phase b and phase c. Relevant results are obtained with $rMSSampEn$, $CoSiEn$, $FuzzEn$, $EnofEn$ and $rMSAttEn$. For following explanations, $SlopEn$ and $rMSBubbEn$ are also represented even if they do not distinguish as well the open-circuits on phases a and b.

5.3. Two Open-Circuit Faults on T_a and on B_a—Phase a

The entropy parameters are unchanged. We investigate a phase fault: on T_a and on B_a—phase a. Figure 15 shows the investigation of different entropies: the phase a entropy with open-circuit is in red, the entropies of phases b and c are in black. The largest distance between phase a entropy and those of phases b and c are obtained with $SampEn$, $ApEn$ and $rMSBubbEn$. The entropies are selected using Equations (36) and (37). The values of $SampEn$, $ApEn$ are for the particular form of the phase current i_a. As shown in Figure 4a, this current has a regular shape with a very small amplitude.

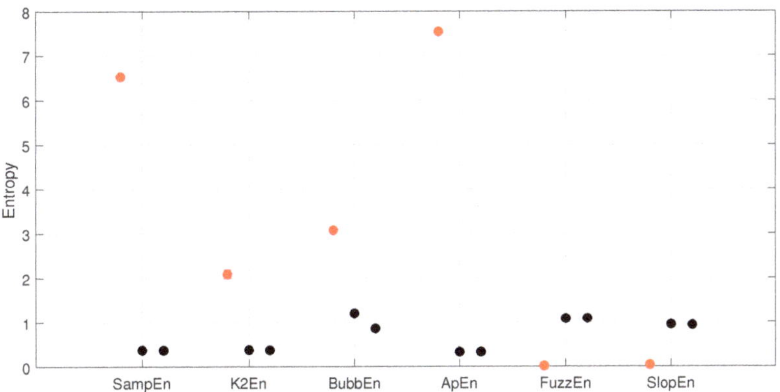

Figure 15. $SampEn$, $K2En$, $ApEn$, $FuzzEn$ and $SlopEn$ for two open−circuit faults on T_a and B_a: phase a entropy with open−circuit fault in red and phases b and c entropies in black.

5.4. Two Open-Circuit Faults on T_b—Phase b and on T_c—Phase c

The two open-circuit faults considered in this subsection are on T_b—phase b and on T_c—phase c. Figure 16 shows the different entropies: the entropy of phase a is in black, phases b and c entropies are in red. The biggest distance between the phase a entropy and those of phases b and c is obtained with $rMSBubbEn$. The entropies are selected using Equations (36) and (37). For the following explanations, $SlopEn$ is also represented even if it does not distinguish as well the open-circuits on phases b and c compared with phase a.

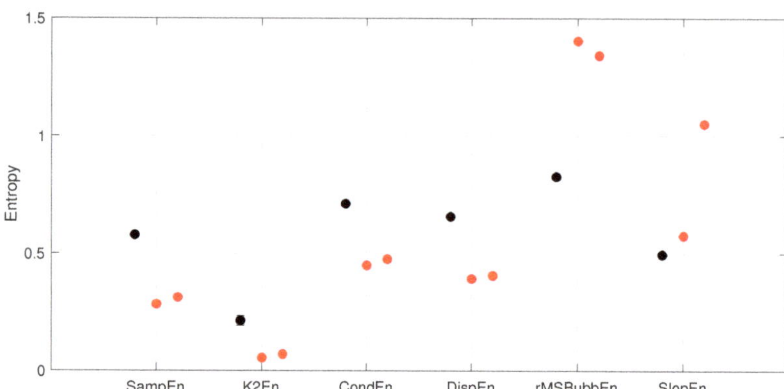

Figure 16. *SampEn*, *K2En*, *CondEn*, *DispEn*, *rMSBubbEn* and *SlopEn* for two open−circuit faults on T_b and T_c: phase *a* entropy in black and phases *b* and *c* entropies with open−circuits in red.

5.5. Three Open-Circuit Faults on B_a—Phase a, T_b—Phase b and on T_c—Phase c

Three open-circuit faults occur on B_a—phase *a*, on T_b—phase *b* and T_c—phase *c*. Figure 17 shows the entropies of phases *a*, *b* and *c* with open-circuit faults, in red. The optimal entropy should be searched from all possible combinations, according to the following rules:

$$min\left(\underset{j=1}{\overset{52}{distance}} \left| entropy_{phase-a} - entropy_{phase-c} \right| \right) \tag{40}$$

and

$$min\left(\underset{j=1}{\overset{52}{distance}} \left| entropy_{phase-b} - entropy_{phase-c} \right| \right). \tag{41}$$

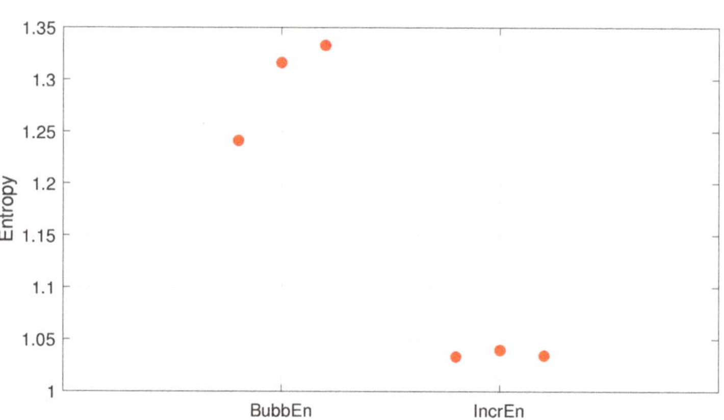

Figure 17. *BubbEn* and *IncrEn* with three open−circuit faults on B_a, T_b and T_c: phase *a*, *b* and *c* entropies in red.

According to Equations (40) and (41), Figure 17 presents *BubbEn* and *IncrEn*. The entropies of phases *a*, *b* and *c* are very closed.

5.6. Three Open-Circuit Faults on B_a—Phase a, T_b and B_b—Phase b

Three open-circuit faults occur on B_a—phase a, on T_b and B_b—phase b. As we can see in Figure 18, the entropies of phases a and b with open-circuit faults is represented in red, and the entropy of phase c is in black. The optimal entropy should be searched from all possible combinations, according to Equations (38) and (39). This time, only *PhasEn* is able to detect the phases where the open-circuit faults occur. For example, this is not the case with *SlopEn*. Phase a entropy is too close to phase c entropy.

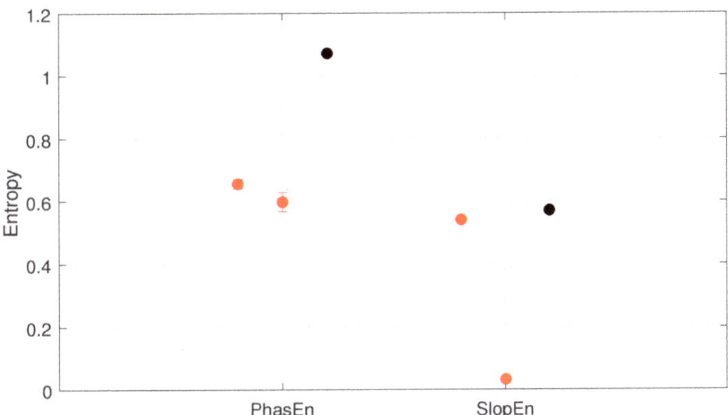

Figure 18. *PhasEn* and *SlopEn* for three open−circuit faults on B_a, T_b and B_b: phases a and b entropies with open−circuit faults in red and entropy of phase c in black.

5.7. Four Open-Circuit Faults on B_a—Phase a, T_b and B_b—Phase b and T_c—Phase c

Four open-circuit faults occur on B_a—phase a, on T_b and B_b—phase b and T_c—phase c. Figure 19 shows the entropies of phases a, b and c with open-circuit faults, in red according to Equations (40) and (41). Once again, *IncrEn* presents very good results.

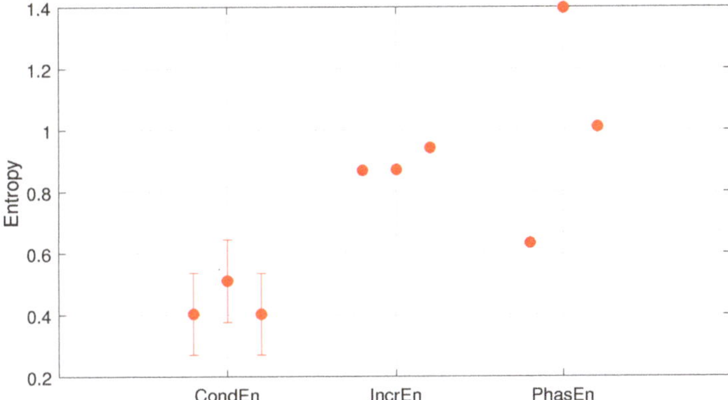

Figure 19. *CondEn*, *IncrEn* and *PhaseEn* for four open−circuit faults on B_a, T_b, B_b and T_c: entropies of phase a, b and c with open−circuit faults in red.

6. Optimization of Parameters L, m, r, τ and s

The parameters, data length N, embedding dimension m, time lag τ and tolerance r, are discussed in the next subsections. The calculated values of entropy depend on the parameters as embedded dimension m and tolerance r. The scale s may also affect the performance of our fault detection method. Finding an optimum set is a major challenge. The parameter optimization is carried out by the maximization of the distance:

$$max\left(\underset{j=1}{\overset{52}{distance}}\left|entropy(L,m,r,\tau,s)_{phase-x} - entropy(L,m,r,\tau,s)_{phase-y}\right|\right) \qquad (42)$$

and

$$max\left(\underset{j=1}{\overset{52}{distance}}\left|entropy(L,m,r,\tau,s)_{phase-z} - entropy(L,m,r,\tau,s)_{phase-w}\right|\right). \qquad (43)$$

in the cases of single or two open-circuit faults and three open-circuit faults (two faults on the same phase). For three and four open-circuit faults on the three phases, the parameter optimization is carried out by the minimization of the distance:

$$min\left(\underset{j=1}{\overset{52}{distance}}\left|entropy(L,m,r,\tau,s)_{phase-x} - entropy(L,m,r,\tau,s)_{phase-y}\right|\right) \qquad (44)$$

and

$$min\left(\underset{j=1}{\overset{52}{distance}}\left|entropy(L,m,r,\tau,s)_{phase-z} - entropy(L,m,r,\tau,s)_{phase-w}\right|\right). \qquad (45)$$

All x, y, z and w cases are presented in Table 3. There are five main parameters for the entropy methods, including length L, embedding dimension m, threshold r, time delay τ and scale s. The optimal combination of L, m, r, τ and s should be searched. In order to check the incidence of the parameters variation on the entropy, we present $DispEn$ $rMSBubbEn$ and $rMSAttnEn$ with one open-circuit on T_a (Figure 13). The values of $CoSiEn$ and $EnofEn$ are evaluated if two open-circuits occur on B_a and T_b (Figure 14). $rMSSampEn$ $K2En$, $MSApEn$, $FuzzEn$ and $SlopEn$ are studied considering the phase fault as in Figure 15. Then, we examine $CondEn$ with two other open-circuit faults: T_b and T_c, as in Figure 14. If there are three open-circuit faults on B_a, T_b and T_c, $IncrEn$ is evaluated (Figure 17). The last entropy, $PhasEn$, is considered with three open-circuit fauts B_a, T_b and B_b, as in Figure 18.

6.1. Varied Data Length (L)

Figure 20 show the analysis for the data length L. The parameter we used are: $m = 2$, $\tau = 1$, $r = 0.2$ and $s = 2$.

When changing from 1000 samples to 6000 samples, the data lengths are: $L_1 = 1000$ points; then, the length increases up to $L_2 = 2000$ samples, approximately two periods of the signal, followed by $L_3 = 3000$ points; $L_4 = 4000$ represents four signal periods; then, 5000 points are saved of L_5 data; and, finally, $L_6 = 6000$ samples, i.e., six signal periods, as in Figures 2b–12b.

$rMSSampEn$, $K2En$, $CondEn$, $DispEn$, $CoSiEn$, $rMSBubbEn$, $MSApEn$, $FuzzEn$, $PhasEn$, $SlopEn$, $EnofEn$ and $rMSAttEn$ are performed with a specifier open-circuit fault. Then, the distance between the healthy phase (represented by a red curve) and the open-circuit phases (represented by a black curve) is maximal, except for one case: the distance between the three red curves is minimal for $IncrEn$, performed with three open-circuit faults B_a, T_b and T_c.

Figure 20a shows $rMSSampEn$ as function of the data length. In Figure 20a, $rMSSampEn$, increases for L_1–L_2 in the sample range (1000, 2000), decreases for $[L_2$-$L_3]$ in the sample range (2000, 3000) and is followed by an increase in the range L_3–L_4 in the sample range (3000, 4000). Then, it slowly decreases to a constant value for L_6. $K2En$ $rMSBubbEn$ are unchanged as L increases, keeping a constant entropy value as in Figure 20b,f. In Figure 20c, $CondEn$ increases, decreases and increases slowly, keeping a constant distance between the entropies curves. $DispEn$ of the healthy phase gradually decreases when the data length increases, as shown in Figure 20d. For $DispEn$, it is appropriate to choose L_1 because the entropy values are length independent. To ensure a large difference between phases *a* and *b* entropies and phase *c* entropy (Figure 20e), it is appropriate to choose L_6 for *CoSiEn*.

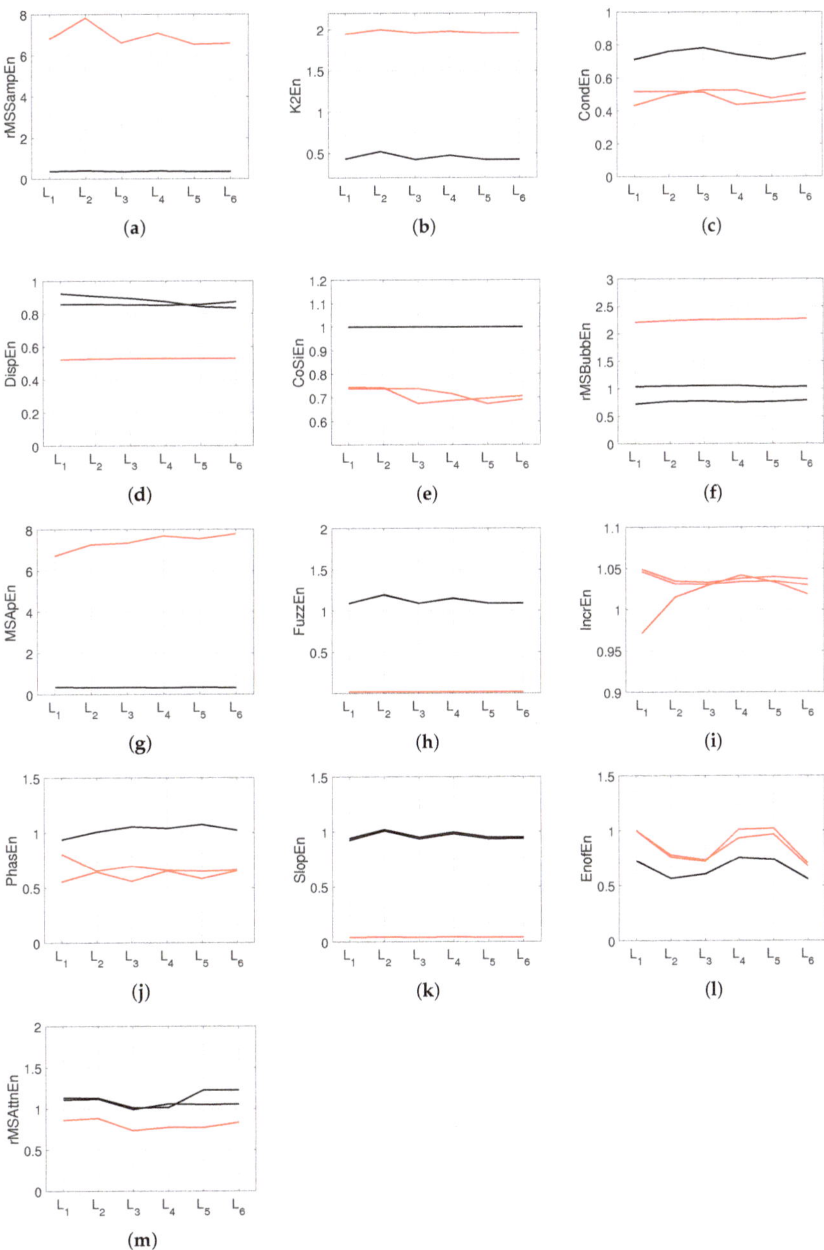

Figure 20. Entropies computed with (**a**) *rMSSampEn*, (**b**) *K2En*, (**c**) *CondEn*, (**d**) *DispEn*, (**e**) *CoSiEn*, (**f**) *rMSBubbEn*, (**g**) *MSApEn*, (**h**) *FuzzEn*, (**i**) *IncrEn*, (**j**) *PhasEn*, (**k**) *SlopEn*, (**l**) *EnofEn* and (**m**) *rMSAttEn* for the data length *L*: healthy phase represented by a black curve and the open−circuit phase represented by a red curve.

In Figure 20g, $MSApEn$ increases slowly for $[L_1-L_6]$ in the sample range (1000, 6000). With regards to the entropy shape, a large length of data ensures a maximum distance between the healthy and faulty phases. Figure 20h,k show $FuzzEn$ and $SlopEn$. After an insignificant variation, the entropies are nearly constant when L is in the range $[L5, L6]$. A suitable value of data length is L_3 for $IncrEn$: the distance between the three entropies is minimal. A maximal distance between the healthy phase (red curve) and the open-circuit phases (black curve) of $PhasEn$ is for L_6, as in Figure 20j. The results of $EnofEn$ and $rMSAttEn$ are shown in Figure 20l,m. Even if these entropies vary (increase or decrease), the distance between the healthy and faulty phases is constant. It seems better to choose L_5 as the data length.

6.2. Varied Embedding Dimension (m)

Let us change m from 2 to 8 to study the effect of m on these approaches ($rMSSampEn$, $K2En$, $CondEn$, $DispEn$, $CoSiEn$, $rMSBubbEn$, $MSApEn$, $FuzzEn$, $IncrEn$ and $SlopEn$) as in Figures 21a–j. The functions $PhasEn$, $EnofEn$ and $AttEn$ do not have an embedding dimension m.

Results of $SampEn$, $rMSBubbEn$ and $ApEn$ are shown in Figure 21a,f,g. These entropies decrease faster, ensuring a large difference between phase a entropy and phases b and c entropies for $m = 2$. $K2En$, $CondEn$, $FuzzEn$ and $CoSiEn$ are constant when m is in the range $[2, 8]$, as in Figure 21b,c,e,h. In Figure 21d, $DispEn$ gradually decreases when the embedding dimension m increases, keeping a constant difference between the entropy of phase a and the entropies of phases b and c, as in Figure 21d.

Figure 21i shows the entropies $IncrEn$ of phases a, b and c (in red), which are very close to each other. $m = 2$ is chosen in order to minimise the distance between these entropies. For the last entropy, $SlopEn$ (Figure 21j): the entropies of the healthy phases b and c decrease when m increases; the entropy of the open-circuit phase a increases when m increases. It is appropriate to choose $m = 2$ for $SlopEn$.

6.3. Varied Time Lag (τ)

Here, we deepen the influence of another indicator, such as time lag τ on some entropies. Time lag τ varies from 1 to 7. We already illustrated the influence of data length L and embedding dimension m on the entropy; let us now examine the performance of $rMSSampEn$, $K2En$, $CondEn$, $DispEn$, $CoSiEn$, $rMSBubbEn$, $MSApEn$, $FuzzEn$, $IncrEn$, $PhasEn$, $SlopEn$ and $EnofEn$ with the variation of time lag τ. The function $AttnEn$ does not require a time lag τ. The data length L, embedding dimension m, scale s and tolerance r were fixed at $N = 6000$, $m = 2$, $s = 2$ and $r = 0.2$ in the following analysis.

Figure 22a,f,g, show the impact of different values of τ on $rMSSampEn$, $rMSBubbEn$ and $rMSApEn$: a steep decrease of these entropies of phase a and a nearly constant value for phases b and c can be observed when τ increases. The major difference between the curves is for a smaller $\tau = 1$. We find that the difference between the $CondEn$, $DispEn$, $CoSiEn$, $PhasEn$ and $EnofEn$ of healthy phase and open-circuit phase is nearly constant, suggesting correlation, as plotted in Figure 22c–e,j,l.

In Figure 22b the shape of $rMSK2En$ in function of τ, decreases at the beginning of the interval $\tau = [1, 2]$, followed by a slow increase for $\tau = [2, 4]$, ending with an abrupt increase of open-circuit phase entropy. Figure suggests that $\tau = 7$ suits well for the calculation of $rMSK2En$. $FuzzEn$ entropy of phases a and b is shown in Figure 22h: only larger time-lag entropies have a relevant significance. For τ equals to 1, $FuzzEn$ is 1.1, and exceeds 1.7 for $\tau = 7$.

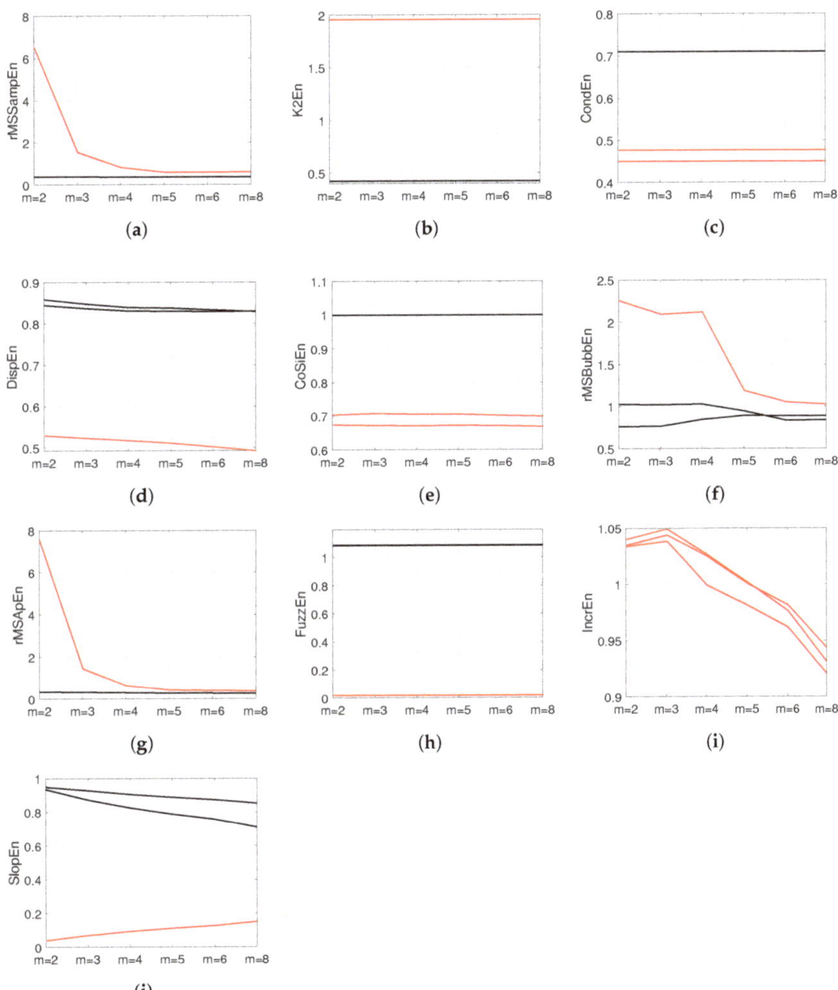

Figure 21. Entropies computed with (**a**) *rMSSampEn*, (**b**) *K2En*, (**c**) *CondEn*, (**d**) *DispEn*, (**e**) *CoSiEn*, (**f**) *rMSBubbEn*, (**g**) *rMSApEn*, (**h**) *FuzzEn*, (**i**) *IncrEn* and (**j**) *SlopEn* for the embedding dimension *m*: healthy phase represented by a black curve and the open−circuit phase represented by a red curve.

Figure 22i shows the entropies *IncrEn* of phases *a*, *b* and *c*, which are very close to each other. $\tau = 1$ is chosen in order to minimize the distance between these entropies. *SlopEn* presents a peak for $\tau = 3$: the difference between *FuzzEn* of phase *a* and of phase *b* is then maximal. However, as the time-lag increases, the difference between black and red curves become smaller. Only a lower time-lag ($\tau = 3$) entropy has a relevant significance, as in Figure 22k.

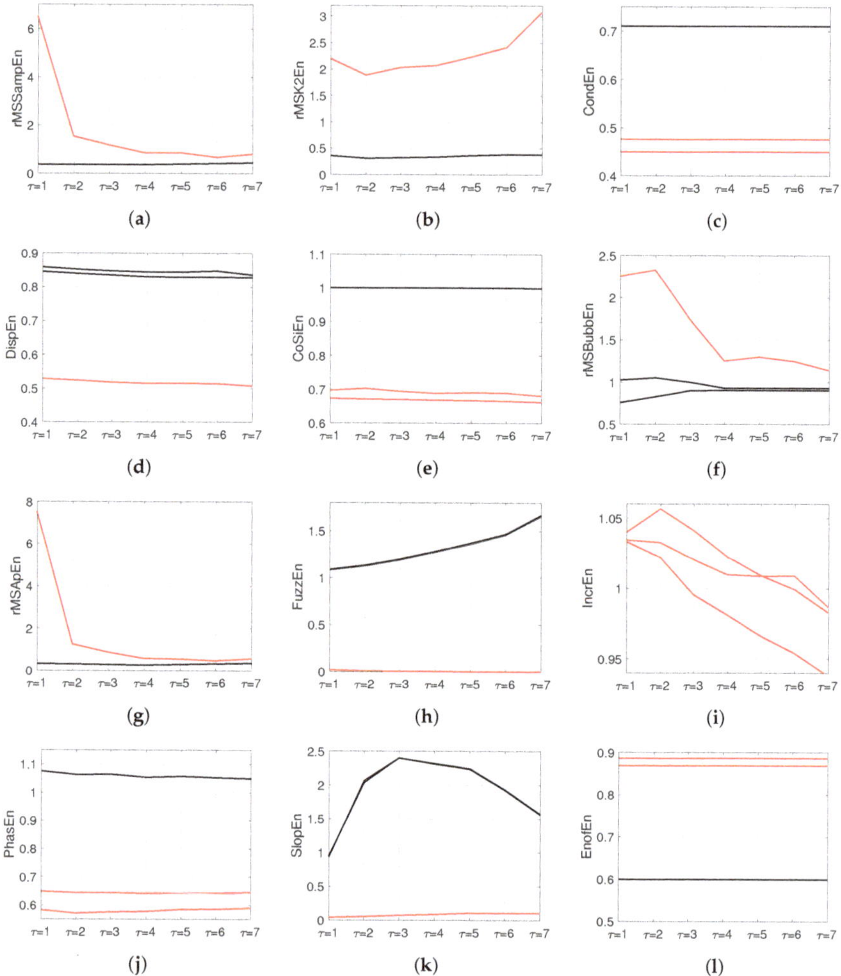

Figure 22. Entropies computed with (**a**) *rMSSampEn*, (**b**) *rMSK2En*, (**c**) *CondEn*, (**d**) *DispEn*, (**e**) *CoSiEn*, (**f**) *rMSBubbEn*, (**g**) *rMSApEn*, (**h**) *FuzzEn*, (**i**) *IncrEn*, (**j**) *PhasEn*, (**k**) *SlopEn* and (**l**) *EnofEn* for the time lag τ: healthy phase represented by a black curve and the open–circuit phase represented by a red curve.

6.4. Varied Tolerance (r)

The analysis of the tolerance r, changing from 0.2 to 0.7, was only performed on *rMSSampEn*, *rMSK2En*, *CoSiEn*, *rMSApEn* and *FuzzEn*. The data length, time lag and embedding dimension are $N = 6000$, $\tau = 2$, $s = 2$ and $m = 2$.

Figure 23a,c present the impact of several r values on, respectively, *rMSSampEn* and *rMSApEn*. The increase of r results in a monotone increase of *rMSSampEn* and *rMSApEn* of faulty phase a (Figure 23a) except for the constant values for $r = [0.4, 0.5]$. *rMSSampEn* and *rMSApEn* both of phase b, with no-fault, are nearly constant. The largest difference between the two curves is for a large r. The figures suggest that $r = 0.7$ is suitable for the calculation of *rMSSampEn* and *rMSApEn* values. Figure 23b shows *rMSK2En*: the entropy of the healthy phases b and c increases when the embedding dimension m increases; the entropy of the open-circuit phase a decreases when the embedding dimension m increases.

It is appropriate to choose $r = 0.2$ for $rMSK2En$. For the last entropy, the difference between phases a and $bFuzzEn$ is nearly constant, as plotted in Figure 23d. The entropy $FuzzEn$ is valid for any value of r.

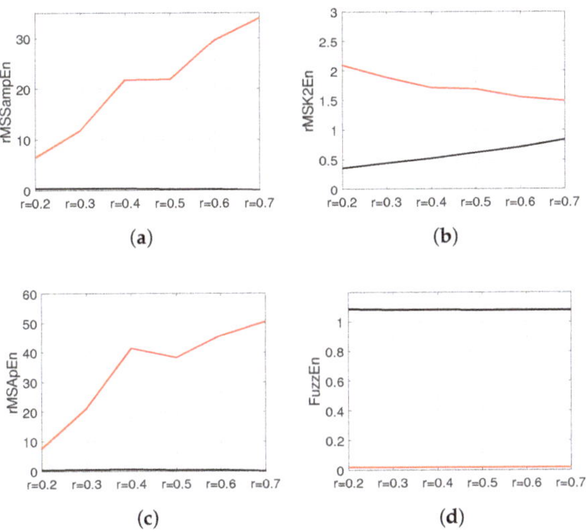

Figure 23. Entropies computed with (**a**) $rMSSampEn$, (**b**) $rMSK2En$, (**c**) $rMSApEn$ and (**d**) $FuzzEn$ for the tolerance r: healthy phase represented by a black curve and the open−circuit phase represented by a red curve.

6.5. Varied Scale (s)

Figure 24a–l illustrate the performance of all entropies: $rMSSampEn$, $K2En$, $CondEn$, $DispEn$, $CoSiEn$, $rMSBubbEn$, $rMSApEn$, $FuzzEn$, $IncrEn$, $PhasEn$, $SlopEn$, $EnofEn$ and $rMSAttEn$. To investigate the effects of scale s on these entropies, we used: $L = 6000$ points, $m = 2$, $\tau = 1$ and $r = 0.2$.

Figure 24a shows $rMSSampEn$ as a functions of scale. The entropy of the open-circuit phase a decreases for $s = [1, 2]$, is nearly constant in the range $s = [3, 7]$, followed by an increase in the range of $s = [7, 9]$, ending with a decrease for $s = [9, 10]$. In the meantime, the entropy of the healthy phase is nearly constant for all ranges of s. The scale $s = 2$ is appropriate. As for $rMSSampEn$, $rMSK2En$ is nearly constant for a healthy case. After a very slow variation, $rMSK2En$ of the open-circuit phase a increases in the range $s = [5, 9]$, decreasing at the end for the last scale. To ensure a large difference between phase a entropy and phases b and c entropies, it is appropriate to choose $s = 9$ for $rMSK2En$, as in Figure 24b.

In Figure 24c–e,h,j, $CondEn$, $DispEn$, $CoSiEn$, $FuzzEn$ and $PhasEn$ are represented. The differences between the healty phase entropy and the open-circuit phase entropyare nearly constant over the range $s = [2, 10]$.

Results of $rMSBubbEn$ are shown in Figure 24f. The entropy of phase a decreases gradually with an increase of s. Meanwhile, entropy of phase b undergoes slight variations. The first scale $s = 2$ gives the largest distance between the entropies of phases a and b. The same result is obtained for $rMSApEn$, as in Figure 24g. At the end of the s interval, the two curves merge and the open-circuit fault on phase a cannot be detected any more. Only a lower scale ($s = 2$) entropy has relevant significance, as in Figure 24g. Scale $s = 4$ or 5 gives the smallest distance between the faulty phases a, b and c for $IncrEn$, as shown in Figure 24i.

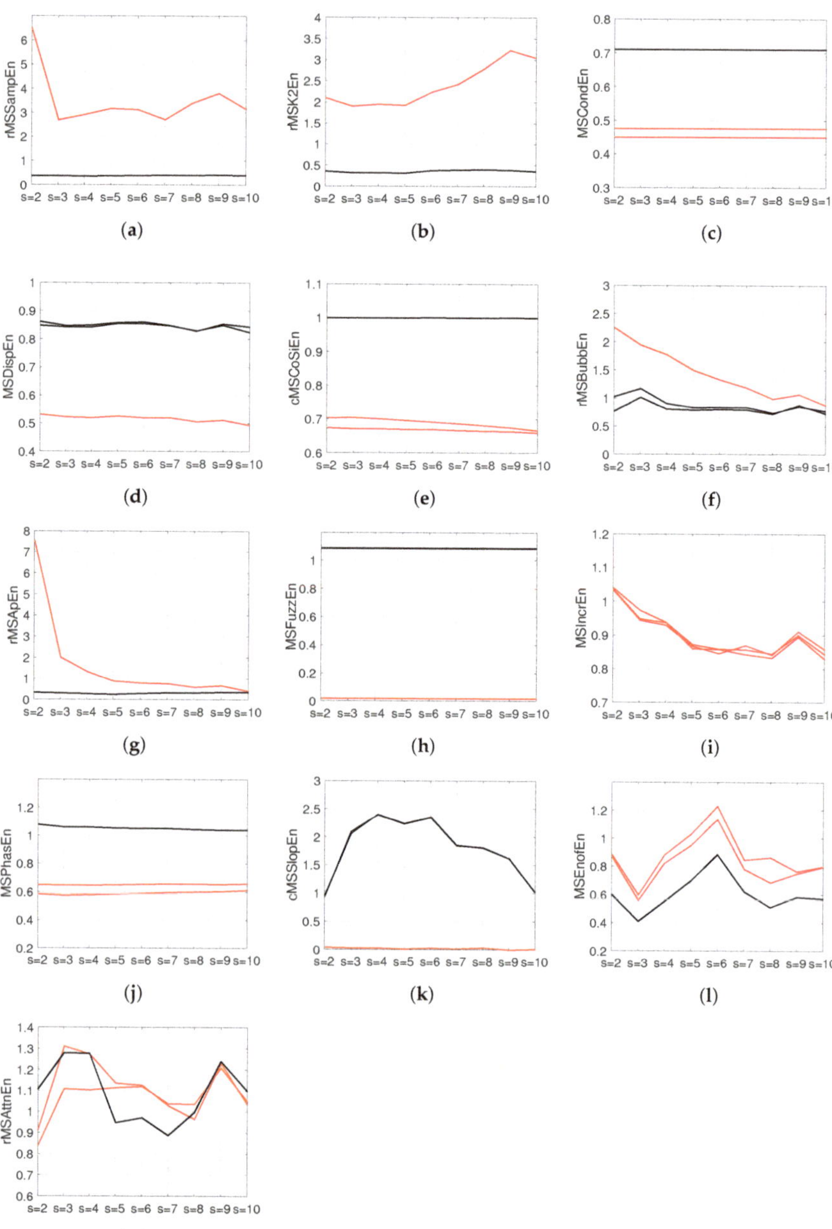

Figure 24. Entropies computed with (**a**) *rMSSampEn*, (**b**) *rMSK2En*, (**c**) *MSCondEn*, (**d**) *MSDispEn*, (**e**) *cMSCoSiEn*, (**f**) *rMSBubbEn*, (**g**) *rMSApEn*, (**h**) *MSFuzzEn*, (**i**) *MSIncrEn*, (**j**) *MSPhasEn*, (**k**) *cMSSlopEn*, (**l**) *MSEnofEn* and (**m**) *rMSAttEn* for the scale *s*: healthy phase represented by a black curve and the open−circuit phase represented by a red curve.

cMSSlopEn entropy of phases *a* and *b* is shown in Figure 24k: only lower scale entropies show a relevant significance. For *s* = 2, *cMSSlopEn* is 1, exceeding 2.4 for *s* = 4. Furthermore, the scale analysis reveals additional entropy information not previously observed at scale $s = 2$. $cMSSlopEn$ clearly presents two peaks for $s = 4$ and 6: the difference between phase a and phase b $cMSSlopEn$ is maximal for $s = 4$. However, the difference between the black and red curves becomes smaller as the scale increases.

The results of $MSEnofEn$ are shown in Figure 24l. Even if these entropies vary (increase or decrease), the distance between the healty and faulty phases is constant. It seems better to choose the scale $s = 2$.

Figure 24m shows that $rMSAttEn$ with repeated up and down, where phase a entropy and phases b and c entropies are interlaced. As long as the two curves (phase a entropy and other phases entropy) merge, the phase a open-circuit fault cannot be detected. Only lower scale ($s = 2$) and middle scale ($s = 5$, 6 or 7) entropies have relevant significance.

6.6. New Setting of Parameters

Parameters are now set to: $N = 2000$, $m = 2$, $\tau = 1$, $r = 0.7$, $s = 2$ for $rMSSampEn$; $N = 2000$, $m = 2$, $\tau = 7$, $r = 0.2$, $s = 9$ for $K2En$; $N = 2000$, $m = 2$, $\tau = 1$, $s = 2$ for $CondEn$; $N = 5000$, $m = 8$, $\tau = 7$, $s = 9$ for $DispEn$; $N = 5000$, $m = 2$, $\tau = 1$, $r = 0.2$, $s = 10$ for $CoSiEn$; $N = 2000$, $m = 2$, $\tau = 1$, $s = 2$ for $rMSBubbEn$; $N = 6000$, $m = 2$, $\tau = 1$, $r = 0.7$, $s = 2$ for $rMSApEn$; $N = 2000$, $m = 2$, $\tau = 7$, $r = 0.2$, $s = 2$ for $FuzzEn$; $N = 3000$, $m = 2$, $\tau = 1$, $s = 2$ for $IncrEn$; $N = 2000$, $\tau = 1$ and $s = 2$ for $PhasEn$; $N = 2000$, $m = 2$, $\tau = 3$, $s = 4$ for $SlopEn$; $N = 5000$, $\tau = 1$ and $s = 4$ for $EnofEn$; $N = 3000$ and $s = 6$ for $rMSAttnEn$. These new settings of parameters are able to increase the distance between faulty and non-faulty phases or to decrease this distance in the $IncrEn$ case. Some entropy functions will be applied to distinguish the healthy state and an open-circuit faulty state and to fault classification, considering the new setting parameters.

7. Conclusions

In this paper, we provide a systematic overview of many known entropy measures, highlighting their applicability to inverter fault detection. Several usual entropies (sample entropy, Kolmogorov entropy, dispersion entropy, cosine entropy, bubble entropy, approximation entropy, fuzzy entropy, incremental entropy, phase entropy, slope entropy, entropy of entropy, attention entropy) and multiscale entropies (and also refined multiscale entropy, composite multiscale entropy) are proposed to quantify the complexity of the brushless motor currents. Their roles in fault detection are summarized into the entropy distance between a healthy phase and an open-circuit faulty phase. Moreover, this paper reveals the great ability of some entropies to distinguish between a healthy and an open-circuit faulty phase. Finally, the simulation results show that these entropies are able to detect and locate the arms of the bridge with one, two, three or even four open-circuit faults.

Author Contributions: Formulation: C.M.; Problem solving: C.M.; Contribution to the numerical computation and results: C.M., S.R., B.L.G. and J.P.; Writing of the manuscript: C.M.; Discussion: S.C.; Revision: C.M. All authors have read and agreed to the published version of the manuscript.

Funding: This research received no external funding.

Data Availability Statement: Data sharing is not applicable to this article.

Conflicts of Interest: The authors declare no conflict of interest.

References

1. Wu, S.D.; Wu, P.H.; Wu, C.W.; Ding, J.J.; Wang, C.C. Bearing fault diagnosis based on multiscale permutation entropy and support vector machine. *Entropy* **2012**, *14*, 1343–1356. [CrossRef]
2. Shang, Y.; Lu, G.; Kang, Y.; Zhou, Z.; Duan, B.; Zhang, C. A multi-fault diagnosis method based on modified Sample Entropy for lithium-ion battery strings. *J. Power Sources* **2020**, *446*, 227275. [CrossRef]
3. Guzman-Vargas, L.; Ramirez-Rojas, A.; Angulo-Brown, F. Multiscale entropy analysis of electroseismic time series. *Nat. Hazards Earth Syst. Sci.* **2008**, *8*, 855–860. [CrossRef]
4. Niu, H.; Wang, J. Quantifying complexity of financial short-term time series by composite multiscale entropy measure. *Commun. Nonlinear Sci. Numer. Simul.* **2015**, *22*, 375–382. [CrossRef]

5. Morel, C.; Humeau-Heurtier, A. Multiscale permutation entropy for two-dimensional patterns. *Pattern Recognit. Lett.* **2021**, *150*, 139–146. [CrossRef]
6. Richman, J.; Randall Moorman, J. Physiological time-series analysis using approximate entropy and sample entropy. *Am. J. Physiol. Heart Circ. Physiol.* **2000**, *278*, 2039–2049. [CrossRef]
7. Jana, S.; Dutta, G. Wavelet entropy and neural network based fault detection on a non radial power system network. *J. Electr. Electron. Eng.* **2012**, *2*, 26–31. [CrossRef]
8. Dasgupta, A.; Nath, S.; Das, A. Transmission line fault classification and location using wavelet entropy and neural network. *Electr. Power Compon. Syst.* **2012**, *40*, 1676–1689. [CrossRef]
9. Yang, Q.; Wang, J. Multi-Level Wavelet Shannon Entropy-Based Method for Single-Sensor Fault Location. *Entropy* **2015**, *17*, 7101–7117. [CrossRef]
10. Guan, Z.Y.; Liao, Z.Q.; Li, K.; Chen, P. A precise diagnosis method of structural faults of rotating machinery based on combination of empirical mode decomposition, sample entropy, and deep belief network. *Sensors* **2019**, *19*, 591. [CrossRef]
11. Huo, Z.; Miguel Martínez-García, M.; Zhang, Y.; Yan, R.; Shu, L. Entropy Measures in Machine Fault Diagnosis: Insights and Applications. *IEEE Trans. Instrum. Meas.* **2020**, *69*, 2607–2620. [CrossRef]
12. Li, B.; Jia, S. Research on diagnosis method of series arc fault of three-phase load based on SSA-ELM. *Sci. Rep.* **2022**, *12*, 592. [CrossRef]
13. Wei, H.; Zhang, Y.; Yu, L.; Zhang, M.; Teffah, K. A new diagnostic algorithm for multiple IGBTs open-circuit faults by the phase currents for power inverter in electric vehicles. *Energies* **2018**, *11*, 1508. [CrossRef]
14. Lu, S.; Ye, W.; Xue, Y.; Tang, Y.; Guo, M. Dynamic feature information extraction using the special empirical mode decomposition entropy value and index energy. *Energy* **2020**, *193*, 116610. [CrossRef]
15. Sarita, K.; Kumar, S.; Saket, R.K. OC fault diagnosis of multilevel inverter using SVM technique and detection algorithm. *Comput. Electr. Eng.* **2021**, *96*, 107481. [CrossRef]
16. Patel, B.; Bera, P.; Saha, B. Wavelet packet entropy and rbfnn based fault detection, classification and localization on HVAC transmission line. *Electr. Power Compon. Syst.* **2018**, *46*, 15–26. [CrossRef]
17. Ahmadi, S.; Poure, P.; Saadate, S.; Khaburi, D.A. A Real-Time Fault Diagnosis for Neutral-Point-Clamped Inverters Based on Failure-Mode Algorithm. *IEEE Trans. Ind. Inform.* **2021**, *17*, 1100–1110. [CrossRef]
18. Nsaif, Y.; Lipu, M.S.H.; Hussain, A.; Ayob, A.; Yusof, Y.; Zainuri, M.A. A New Voltage Based Fault Detection Technique for Distribution Network Connected to Photovoltaic Sources Using Variational Mode Decomposition Integrated Ensemble Bagged Trees Approach. *Energies* **2022**, *15*, 7762. [CrossRef]
19. Estima, J.O.; Cordoso, A.J.M. A new approach for real-time multiple open-circuit fault diagnosis in voltage-source inverters. *IEEE Trans. Ind. Appl.* **2011**, *47*, 2487–2494. [CrossRef]
20. Lee, J.S.; Lee, K.B.; Blaabjerg, F. Open-switch fault detection method of a back-to-back converter using NPC topology for wind turbine systems. *IEEE Trans. Ind. Appl.* **2014**, *51*, 325–335. [CrossRef]
21. Yu, L.; Zhang, Y.; Huang, W.; Teffah, K. A fast-acting diagnostic algorithm of insulated gate bipolar transistor open-circuit faults for power inverters in electric vehicles. *Energies* **2017**, *10*, 552. [CrossRef]
22. Park, G.B.; Lee, K.J.; Kim, R.Y.; Kim, T.S.; Ryu, J.S.; Hyun, D.S. Simple fault diagnosis based on operating characteristic of brushless direct-current motor drives. *IEEE Trans. Ind. Electron.* **2011**, *58*, 1586–1593. [CrossRef]
23. Wu, Y.; Zhang, Z.; Li, Y.; Sun, Q. Open-Circuit Fault Diagnosis of Six-Phase Permanent Magnet Synchronous Motor Drive System Based on Empirical Mode Decomposition Energy Entropy. *IEEE Access* **2021**, *9*, 91137–91147. [CrossRef]
24. Estima, J.O.; Freire, N.M.A.; Cardoso, A.J.M. Recent advances in fault diagnosis by Park's vector approach. *IEEE Workshop Electr. Mach. Des. Control Diagn.* **2013**, *2*, 279–288.
25. Yan, H.; Xu, Y.; Cai, F.; Zhang, H.; Zhao, W.; Gerada, C. PWM-VSI fault diagnosis for PMSM drive based on fuzzy logic approach. *IEEE Trans. Power Electron.* **2018**, *34*, 759–768. [CrossRef]
26. Faraz, G.; Majid, A.; Khan, B.; Saleem, J.; Rehman, N. An Integral Sliding Mode Observer Based Fault Diagnosis Approach for Modular Multilevel Converter. In Proceedings of the 2019 International Conference on Electrical, Communication, and Computer Engineering (ICECCE), Swat, Pakistan, 24–25 July 2019; pp. 1–6.
27. Wang, Q.; Yu, Y.; Hoa, A. Fault detection and classification in MMC-HVDC systems using learning methods. *Sensors* **2020**, *20*, 4438. [CrossRef]
28. Xing, W. An open-circuit fault detection and location strategy for MMC with feature extraction and random forest. In Proceedings of the 2021 IEEE Applied Power Electronics Conference and Exposition—APEC, Virtual, 14–17 June 2021; Volume 52007166, pp. 1111–1116.
29. Ke, L.; Liu, Z.; Zhang, Y. Fault Diagnosis of Modular Multilevel Converter Based on Optimized Support Vector Machine. In Proceedings of the 2020 39th Chinese Control Conference, Shenyang, China, 27–29 July 2020; pp. 4204–4209.
30. Wang, C.; Lizana, F.; Li, Z. Submodule short-circuit fault diagnosis based on wavelet transform and support vector machines for a modular multi-level converter with series and parallel connectivity. In Proceedings of the IECON 2017—43rd Annual Conference of the IEEE Industrial Electronics Society, Beijing, China, 29 October–1 November 2017; pp. 3239–3244.
31. Morel, C.; Akrad, A.; Sehab, R.; Azib, T.; Larouci, C. IGBT Open-Circuit Fault-Tolerant Strategy for Interleaved Boost Converters via Filippov Method. *Energies* **2022**, *15*, 352. [CrossRef]

32. Morel, C.; Akrad, A. Open-Circuit Fault Detection and Location in AC-DC-AC Converters Based on Entropy Analysis. *Energies* **2023**, *16*, 1959. [CrossRef]
33. Unakafova, V.; Unakafov, A.; Keller, K. An approach to comparing Kolmogorov-Sinai and permutation entropy. *Eur. Phys. J. ST* **2013**, *222*, 353–361. [CrossRef]
34. Porta, A.; Baselli, G.; Liberati, D.; Montano, N.; Cogliati, C.; Gnecchi-Ruscone, T.; Malliani, A.; Cerutti, S. Measuring regularity by means of a corrected conditional entropy in sympathetic outflow. *Biol. Cybern.* **1998**, *78*, 71–78. [CrossRef]
35. Fu, W.; Tan, J.; Xu, Y.; Wang, K.; Chen, T. Fault Diagnosis for Rolling Bearings Based on Fine-Sorted Dispersion Entropy and SVM Optimized with Mutation SCA-PSO. *Entropy* **2019**, *21*, 404. [CrossRef]
36. Rostaghi, M.; Azami, H. Dispersion entropy: A measure for time-series analysis. *IEEE Signal Process. Lett.* **2016**, *23*, 610–614. [CrossRef]
37. Theerasak, C.; Mandic, D. Cosine Similarity Entropy: Self-Correlation-Based Complexity Analysis of Dynamical Systems. *Entropy* **2017**, *19*, 652.
38. Cuesta-Frau, D.; Vargas, B. Permutation Entropy and Bubble Entropy: Possible interactions and synergies between order and sorting relations. *Math. Biosci. Eng.* **2019**, *17*, 1637–1658. [CrossRef]
39. Manis, G.; Bodini, M.; Rivolta, M.W.; Sassi, R. A Two-Steps-Ahead Estimator for Bubble Entropy. *Entropy* **2021**, *23*, 761. [CrossRef]
40. Chen, W.; Wang, Z.; Xie, H.; Yu, W. Characterization of surface EMG signal based on fuzzy entropy, Neural Systems and Rehabilitation Engineering. *IEEE Trans. Neural Syst. Rehabil. Eng.* **2007**, *15*, 266–272. [CrossRef]
41. Azami, H.; Li, P.; Arnold, S.; Escuder, J.; Humeau-Heurtier, A. Fuzzy Entropy Metrics for the Analysis of Biomedical Signals: Assessment and Comparison. *IEEE Access* **2019**, *7*, 104833–104847. [CrossRef]
42. Liu, X.; Wang, X.; Zhou, X.; Jiang, A. Appropriate use of the increment entropy for electrophysiological time series. *Comput. Biol. Med.* **2018**, *95*, 13–23. [CrossRef]
43. Reyes-Lagos, J.; Pliego-Carrillo, A.C.; Ledesma-Ramírez, C.I.; Peña-Castillo, M.A.; García-González, M.T.; Pacheco-López, G.; Echeverría, J.C. Phase Entropy Analysis of Electrohysterographic Data at the Third Trimester of Human Pregnancy and Active Parturition. *Entropy* **2020**, *22*, 798. [CrossRef]
44. Cuesta-Frau, D. Slope Entropy: A New Time Series Complexity Estimator Based on Both Symbolic Patterns and Amplitude Information. *Entropy* **2019**, *21*, 1167. [CrossRef]
45. Chang, F.; Wei, S.-Y.; Huang, H.P.; Hsu, L.; Chi, S.; Peng, C.K. Entropy of Entropy: Measurement of Dynamical Complexity for Biological Systems. *Entropy* **2017**, *19*, 550.
46. Yang, J.; Choudhary, G.; Rahardja, S. Classification of Interbeat Interval Time-series Using Attention Entropy. *IEEE Trans. Affect. Comput.* **2020**, *14*, 321–330. [CrossRef]
47. Liu, T.; Cui, L.; Zhang, J.; Zhang, C. Research on fault diagnosis of planetary gearbox based on Variable Multi-Scale Morphological Filtering and improved Symbol Dynamic Entropy. *Int. J. Adv. Manuf. Technol.* **2022**, *124*, 3947–3961.. [CrossRef]
48. Silva, L.; Duque, J.; Felipe, J.; Murta, L.; Humeau-Heurtier, A. Twodimensional multiscale entropy analysis: Applications to image texture evaluation. *Signal Process.* **2018**, *147*, 224–232. [CrossRef]
49. Wu, S.; Wu, C.; Lin, S.; Wang, C.; Lee, K. Time series analysis using composite multiscale entropy. *Entropy* **2013**, *15*, 1069–1084. [CrossRef]
50. Humeau-Heurtier, A. The Multiscale Entropy Algorithm and Its Variants: A Review. *Entropy* **2015**, *17*, 3110–3123. [CrossRef]
51. Valencia, J.; Porta, A.; Vallverdu, M.; Claria, F.; Baranovski, R.; Orlowska-Baranovska, E.; Caminal, P. Refined multiscale entropy: Application to 24-h Holter records of heart period variability in hearthy and aortic stenosis subjects. *IEEE Trans. Biomed. Eng.* **2009**, *56*, 2202–2213. [CrossRef]
52. Wu, S.; Wu, C.; Lin, S.; Lee, K.; Peng, C. Analysis of complex time series using refined composite multiscale entropy. *Phys. Lett. A* **2014**, *378*, 1369–1374. [CrossRef]
53. Dini, P.; Saponara, S. Model-Based Design of an Improved Electric Drive Controller for High-Precision Applications Based on Feedback Linearization Technique. *Electronics* **2021**, *10*, 2954. [CrossRef]
54. Dini, P.; Saponara, S. Design of an observer-based architecture and non-linear control algorithm for cogging torque reduction in synchronous motors. *Energies* **2020**, *13*, 2077. [CrossRef]
55. Mohanraj, D.; Aruldavid, R.; Verma, R.; Sathyasekar, K.; Barnawi, A.B.; Chokkalingam, B.; Mihet-Popa, L. A Review of BLDC Motor: State of Art, Advanced Control Techniques, and Applications. *IEEE Access* **2017**, *2*, 40. [CrossRef]
56. Wu, H.; Wen, M.-Y.; Wong, C.-C. Speed control of BLDC motors using hall effect sensors based on DSP. In Proceedings of the IEEE International Conference on System Science and Engineering, Puli, Taiwan, 7–9 July 2016.
57. Akin, B.; Bhardwaj, M.; Warriner, J. Trapezoidal Control of BLDC Motors Using Hall Effect Sensors. *Tex. Instrum.—SPRAB4*, **2011**, *2954*, 34.
58. Nga, N.T.T.N.; Chi, N.T.P.; Quang, N.H. Study on Controlling Brushless DC Motor in Current Control Loop Using DC-Link Current. *Am. J. Eng. Res.* **2016**, *7*, 522–528.

Article

Currents Analysis of a Brushless Motor with Inverter Faults—Part II: Diagnostic Method for Open-Circuit Fault Isolation

Cristina Morel [1,*], Baptiste Le Gueux [2], Sébastien Rivero [2] and Saad Chahba [1]

[1] Ecole Supérieure des Techniques Aéronautiques et de Construction Automobile, ESTACA'Lab Paris-Saclay, 12 Avenue Paul Delouvrier—RD10, 78180 Montigny-le-Bretonneux, France; saad.chahba@estaca.fr
[2] ESTACA Campus Ouest, Rue Georges Charpak—BP 76121, 53009 Laval, France; baptiste.legueux@estaca.eu (B.L.G.); sebastien.rivero@estaca.eu (S.R.)
* Correspondence: cristina.morel@estaca.fr

Abstract: In this paper, a brushless motor with a three-phase inverter is investigated under healthy and multiple open-circuit faults. The occurrence of faults in an inverter will lead to atypical characteristics in the current measurements. This is why many usual entropies and multiscale entropies have been proposed to evaluate the complexity of the output currents by quantifying such dynamic changes. Among this multitude of entropies, only some are able to differentiate between healthy and faulty open-circuit conditions. In addition, another selection is made between these entropies in order to improve diagnostic speed. After the fault detection based on the mean values, the open-circuit faults are localized based on the fault diagnostic method. The simulation results ensure the ability of these entropies to detect and locate open-circuit faults. Moreover, they are able to achieve fault diagnostics for a single switch, double switches, three switches, and even four switches. The diagnostic time to detect and to isolate faults is between 10.85 ms and 13.67 ms. Then, in order to prove the ability of the fault diagnostic method, a load variation is performed under the rated speed conditions of the brushless motor. The validity of the method is analyzed under different speed values for a constant torque. Finally, the fault diagnostic method is independent from power levels.

Keywords: open-circuit switch fault; fault detection; fault isolation; fault diagnostic method

Citation: Morel, C.; Gueux, B.L.; Rivero, S.; Chahba, S. Currents Analysis of a Brushless Motor with Inverter Faults—Part II: Diagnostic Method for Open-Circuit Fault Isolation. *Actuators* **2023**, *12*, 230. https://doi.org/10.3390/act12060230

Academic Editors: Qinfen Lu, Xinyu Fan, Cao Tan and Jiayu Lu

Received: 4 May 2023
Revised: 26 May 2023
Accepted: 31 May 2023
Published: 2 June 2023

1. Introduction

In recent years, numerous open-switch fault diagnostic methods applied in multilevel inverters have been presented [1–3]. Fault detection and isolation methods for Mosfet or IGBT open-circuit faults have been studied in previous works, particularly focusing on arm voltages or current measurements. If open-circuit faults are not detected as quickly as possible, they may cause secondary damage. Thus, it is very important to identify and detect faulty switches as soon as possible.

Recently, many current-type methods based on the direct analysis of current waveforms have gained attention. These methods have the ability to detect and locate one or multiple open-circuit faults. Twenty-one methods for open-circuit faults were evaluated and summarized in [4], based on their performance and implementation efforts. In [5,6], a current-based method was used for fault diagnostics. The average current Park's vector strategy [5,7] was applied to diagnose the faulty upper or lower half-leg of a three-level inverter. However, the proposed strategy was not able to identify the faulty switch in the defected half-leg. In [8], a diagnostic method based on empirical mode decomposition energy entropy and normalized average current was proposed to identify one or two faults: an open-circuit fault on the upper or lower bridge arm; two open-circuit faults in the cross side bridge arm of the double phase, in the same side bridge arm of the double phase or in the double-bridge arm of the single phase. In paper [9], one open-transistor fault was detected analyzing the normalized load current. Nevertheless, this method has a higher complexity and larger detection times. The authors of [10] proposed a fast fault detection and isolation approach to identify a single open-circuit fault in power electronics sub-modules for modular multilevel converters with model-predictive control. In [11], a comparison method of fault diagnostic variables with threshold values was presented for a three-level three-phase NPC inverter. Once a faulty leg is detected, the localization of the faulty switch is based on the value of the average current. This method can detect only single-switch faults in one time period. An implementation of open-transistor fault detection and diagnostic method based on the current trajectory of phase was presented in [12]. This method is load-independent, but detects only one open circuit. A sliding mode observer was proposed in [13–15]. The current form factors of the estimated current and measured current are needed for faulty phase detection. However, the complexity of the method is medium, and it can only detect a single switch fault. Several faults were detected in [16] using the ratio of the theoretical and the practical voltage values on the capacitor of each inverter's sub-module. The authors of [17] proposed a novel diagnostic algorithm for single and multiple IGBT open-circuit faults.

When algorithms are presented, an analysis of complexity [18] is very important. To better understand the capabilities of the algorithm, the results (under several speed values and load variations in the brushless motor) are presented in terms of computational time. The authors of [19] used a residual observer-based fault detection algorithm, detecting open-circuit faults in 15 ms to 20 ms. The authors of [20,21] proposed a Kalman-filter-based approach: the comparison of measured and estimated voltages and currents obtained by their filter enables the detection of the open-circuit faults of a modular multilevel converter sub-module. However, the diagnostic time was longer than 100 ms. In [22–24], a sliding-mode observer was proposed as a fault diagnostic method. Detecting and locating faults require at least 50 ms. In [25,26], only a single fault could be diagnosed by the state observer in more than 30 ms. A clustering algorithm and calculated capacitance methods are proposed in [27]. The algorithm is complicated and requires a large amount of calculation: both methods need at least 13 ms to detect and locate faults. An adaptive linear neuron-recursive least squares algorithm to estimate capacitor voltages is proposed to detect and locate different types of sub-module faults in [28], in more than 30 ms. The authors used simple sub-modules in [29] and integral sub-modules in [30] to detect and isolate the multiple open-switch faults in modular multi-level converters. Nevertheless, this method diagnoses and isolates multiple open-switch faults within 20 ms. The algorithm developed by [31] is based on the instant voltage error in the converter and requires only signals already available to the control system, avoiding the use of additional hardware. The algorithm is independent from the load and from the used control strategy and provides very fast detection and identification of the fault, with diagnostic times as low as two sample periods. Kiranyaz [32] used 1D CNN to detect and locate one switch open-circuit fault using circulating current, load current signals, and four-cell capacitor voltage. This method achieved a detection probability of 0.989 and an average identification probability of 0.997 in less than 100 ms. Two deep-learning methods and a stand-alone SoftMax classifier [33] were used with raw data collected by current sensors to improve classification accuracy and reduce computation time. In the method proposed by [34], the three-phase currents were used to calculate the fault diagnostic variables with the average current Park's Vector method for the identification of only one open-circuit fault. Then, these variables were processed with a fuzzy logic method [35], and the faulty information of the PWM-VSI could be obtained. Faulty power switches can be identified in less than 90 ms after fault occurrence (approximately two motor phase current fundamental periods).

Many usual entropies and multiscale entropies were proposed to evaluate the complexity of phase currents in the previous article "Current Analysis of a Brushless Motor with Inverter Faults-Part I: Parameters of Entropy Functions and Open-Circuit Faults Detection"— [36]. We are now able to select the appropriate entropy functions with an appropriate combination of parameters (as the data length, N; the embedding dimension,

m; the time lag, τ; the tolerance, r; and the scale, s), which differentiate between healthy and faulty open-circuit conditions. The main goal of this paper is to present a fault diagnostic method to detect and locate the open-circuit switch faults of a brushless motor with a three-phase inverter. Among the multitude of entropies, only some are able to improve diagnostic speeds. Starting from the brushless inverter under healthy and faulty conditions, the various possible switching fault states are discussed. After the fault detection based on the mean values, the open-circuit faults are localized, based on the fault diagnostic method. The simulation results ensure the ability of these entropies to detect and locate open-circuit faults. Moreover, they are able to achieve fault diagnostics for a single switch, double switches, three switches and even four switches. Generally, the fault detection technique with more than two parameters leads to a complex detection system. That is not the case here. Fault localization is performed using a combination of entropy functions incorporating a threshold variable. Synthesis of the total computation time to detect faults in single, double, three and even four switches is realized. Then, in order to prove the ability of the fault diagnostic method, load variation (from minimum to maximum) is performed under the rated speed condition of the brushless motor. Finally, the validity of the fault diagnostic method is analyzed under different speed conditions for a constant torque.

The paper is organized as follows: Section 2 illustrates the fault location with the fault diagnostic method, together with the total computation time, to detect and locate single or multiple open-circuit switch faults. Then, the influence of torque and speed variations on the detection and location of faults is introduced in Section 3. Section 4 ends with the Conclusion.

2. Fault Location with the Fault Diagnostic Method

In the first part [36], many entropies were used to identify one or multiple open-circuit faults: *SampEn, K2En, CondEn, DispEn, CoSiEn, BubbEn, ApEn, FuzzEn, IncrEn, PhasEn, SlopEn, EnofEn* and *AttEn* functions, each one with multiscale, composite multiscale and refined multiscale approaches, providing 52 entropy functions to evaluate the complexity of the phase currents. However, among these entropies, some are more sensitive to fault conditions. Such sensitivity is able to increase the distance between the faulty and the non-faulty phases or decrease this distance in the *IncrEn* case. Some entropies will be applied to distinguish between healthy conditions and faulty open-circuit conditions, considering the following parameter settings: $N = 2000$, $m = 2$, $\tau = 1$ and $s = 2$ for $rMSBubbEn$, *PhasEn* and *CondEn*; $N = 2000$, $m = 2$, $\tau = 3$ and $s = 4$ for *SlopEn*.

Fault detection is essential in a fault diagnostic approach. Phase currents are used in a fault diagnostic procedure to detect and isolate open-circuit switches. A block diagram is presented in Figure 1. Three sensors are added to the circuit to measure the currents of the inverter (i_a, i_b and i_c of phases a, b and c), which can be used to identify faults in switches. In terms of algorithm complexity, the dSpace platform is suggested to implement the entropy functions due to the memory space. Furthermore, with a MATLAB Function block, we can write a MATLAB function (a functionality programmed in the M language) into a Simulink model and execute it for simulation. We specify the inputs to the MATLAB Function block in the function header as the arguments (the currents i_a, i_b and i_c) and return the output data (the entropy values). In addition, after the detection of open-switch fault occurrence (Figure 2), the fault localization strategy is initiated, based on the entropy analysis of Figures 2 and 3. Moreover, the localization of the faulty switches is necessary to isolate the faults.

Fault localization is performed using one or a combination of entropy functions by incorporating threshold variables, $\epsilon_1 = 0.01$, $\epsilon_2 = 1.5$ and $\epsilon_3 = 0.2$. The selected values of ϵ_1, ϵ_2 and ϵ_3 depend on the simulation results of the first part [36]. ϵ_3 is 20 times bigger than ϵ_1 and 7.5 times smaller than ϵ_2.

The means of the three currents are calculated as in Figure 2. If the means of i_a and i_b are nearly zero, then the mean of i_c is also zero. This is the normal operating condition (without faults). The phase currents of the brushless motor are alternative waves, with zero mean. Under healthy conditions, the average value of the positive half cycle of the phase currents i_a, i_b and i_c is equal to the negative half cycle of the phase currents i_a, i_b and i_c.

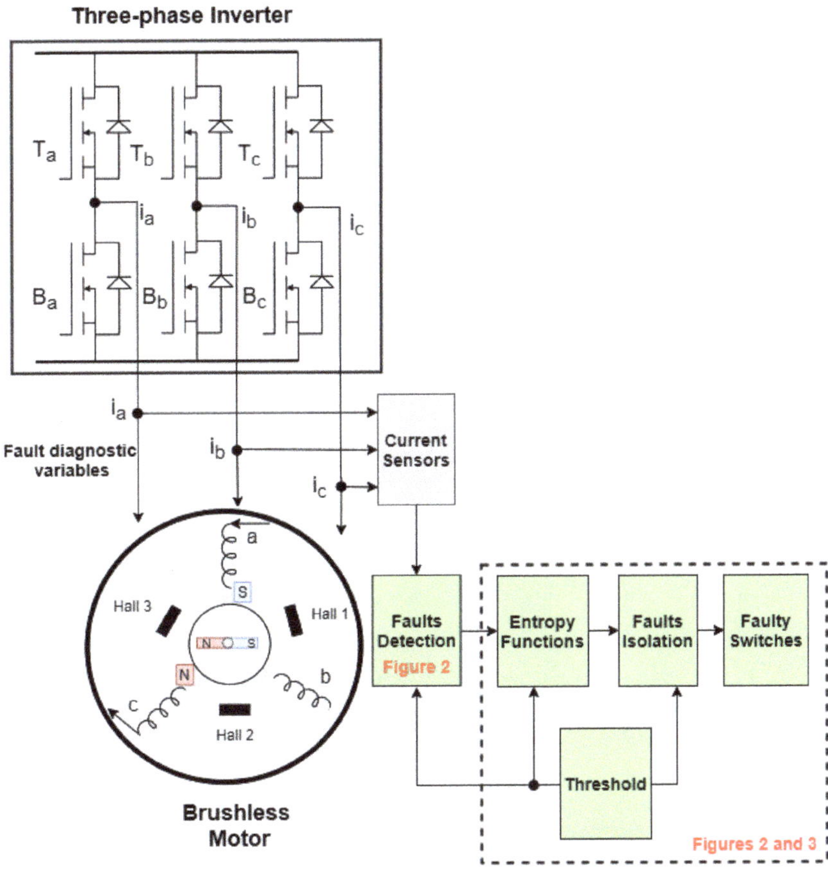

Figure 1. Schematic representation of fault diagnostic method.

The average of the phase currents i_a, i_b and i_c is nearly zero (when no fault occurs in any switch of the inverter), as in the equations

$$mean(i_a) \approx 0, \tag{1}$$

$$mean(i_b) \approx 0, \tag{2}$$

$$mean(i_c) \approx 0. \tag{3}$$

When a fault occurs in the inverter, the current waveforms are distorted. At least two of the equations are not respected: an open-switch fault in upper pair switches determines a negative current wave, with a negative mean; an open-switch fault in lower pair switches determine a positive current wave, with a positive mean. This change in the waveform carries the fault information that can be extracted using different entropy functions.

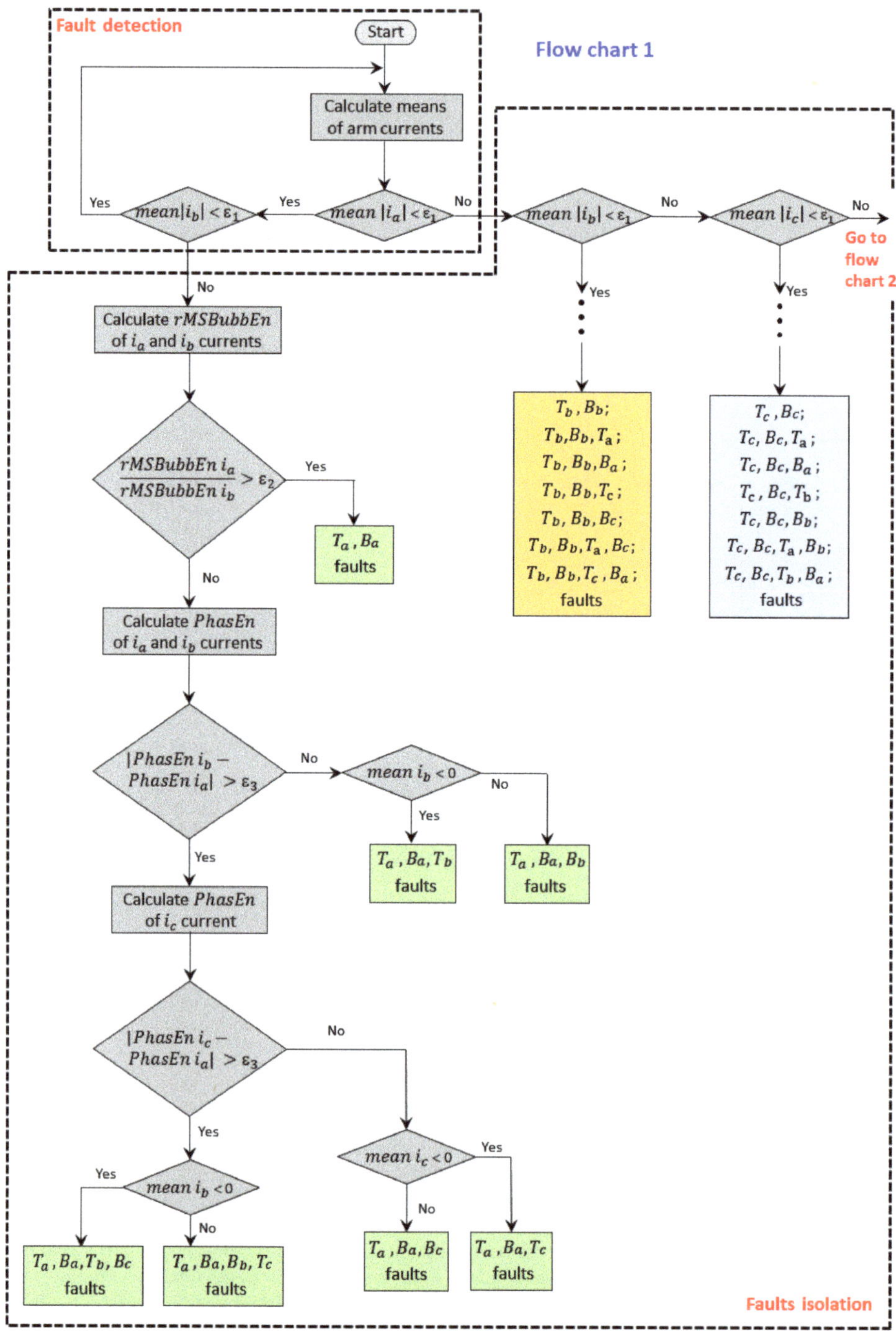

Figure 2. First part of the flow chart of the fault diagnostic approach.

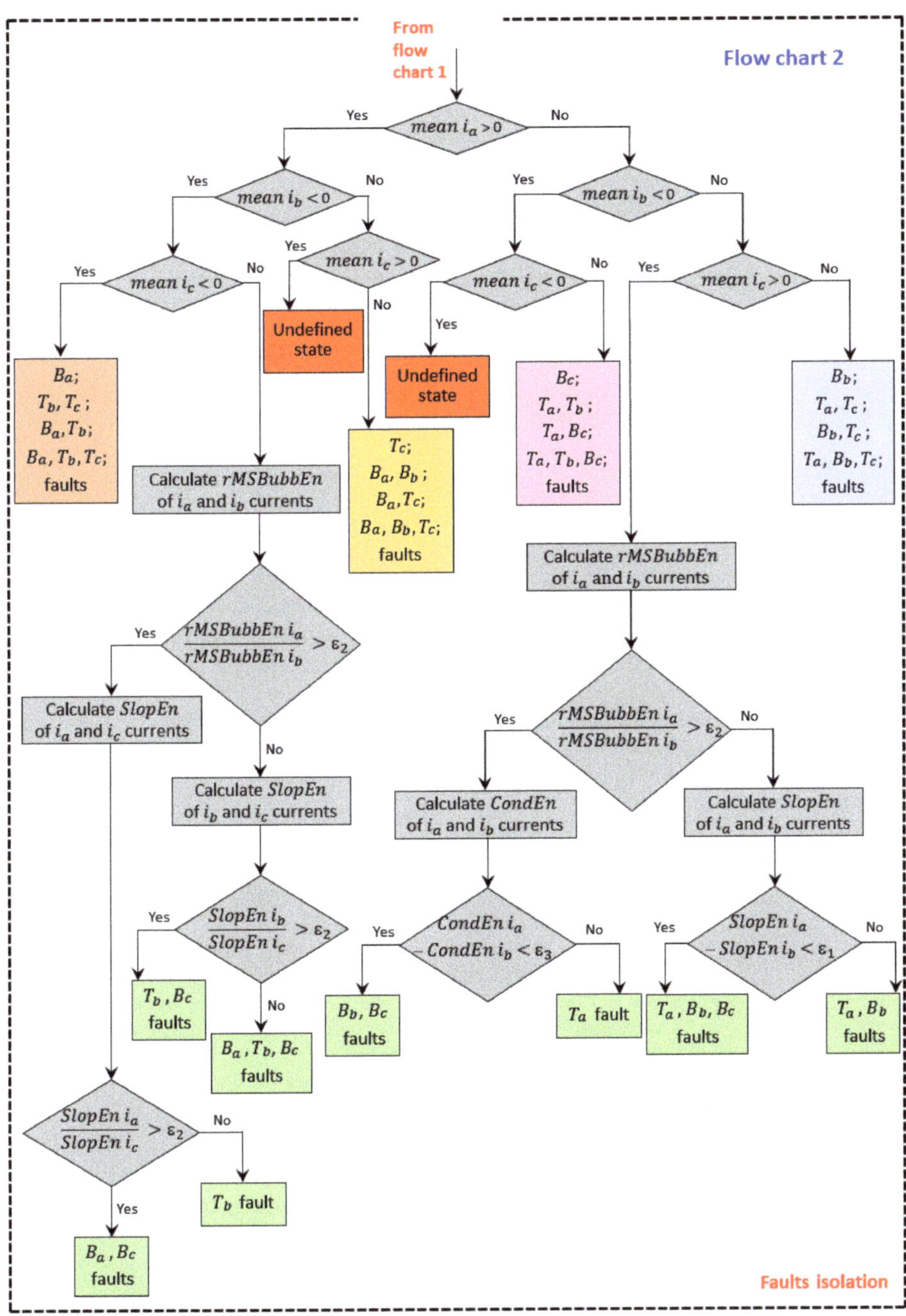

Figure 3. Second part of the flow chart of the proposed fault diagnostic approach.

Let us suppose that the average of the phase current i_a is zero, but not for i_b or i_c. This means that, for the moment, there are two open-circuit faults on the upper and lower switches of phase a. All the possibilities of open-circuit faults together with the phase current mean values are given in [36], as:

- T_a, B_a: mean of i_a = 0.0006, mean of i_b = 0.2708, mean of i_c = −0.2714;

- T_a, B_a, T_b: mean of i_a = 0.0004, mean of i_b = −2.6371, mean of i_c = 2.6367;

- T_a, B_a, B_b: mean of i_a = 0.0044, mean of i_b = 2.9920, mean of i_c = −2.9964;

- T_a, B_a, T_c: mean of i_a = 0.0025, mean of i_b = 2.4690, mean of i_c = −2.4715;

- T_a, B_a, B_c: mean of i_a = 0.0032, mean of i_b = −2.8272, mean of i_c = 2.8304;

- T_a, B_a, T_b, B_c: mean of i_a = 0.0016, mean of i_b = −2.6055, mean of i_c = 2.6038;

- T_a, B_a, B_b, T_c: mean of i_a = 0.0041, mean of i_b = 3.0170, mean of i_c = −3.0211.

Once a fault is detected by the average method, the faulty switches are localized with the diagnostic method we propose, which combines the multiple entropy functions of the phase currents. The growing interest in entropy approaches relies on their ability to analyze and to provide information related to signal complexity. The entropy of the phase currents is directly used as fault information. A larger difference between the 52 entropies of phase a and of phases b and c is given by $SampEn$, $K2En$, $MSApEn$, $rMSBubbEn$, $FuzzEn$ and $SlopEn$, as can be seen in [36]. The largest distance is obtained with $ApEn$, followed by $SampEn$. The smallest distance can be obtained with $FuzzEn$ and $SlopEn$.

Another important issue for the fault detection and isolation is the assessment of computation time. Table 1 shows the computation time of all entropies. This time depends on the length of the phase current, but is not proportional to it. For a length of 6000 samples, the $SampEn$ computation time is 1.44 s and 0.1862 s for a 2000-sample length. In order to reduce the fault diagnostic cost and to improve its speed, $rMSBubbEn$ is chosen. This entropy is insensitive to the length of the phase current.

The $rMSBubbEn$ of current i_a is compared with the $rMSBubbEn$ of current i_b. If

$$\frac{rMSBubbEn\ i_a}{rMSBubbEn\ i_b} > \epsilon_2 \tag{4}$$

is greater than the threshold ϵ_2, then T_a and B_a switches' faults are isolated.

The computation time to calculate a mean of a 2000-sample wave is 0.2537 ms and 0.0964 ms for an if condition. The total computing time to detect and isolate T_a and B_a is 10.85 ms, for which it is necessary to calculate three means of arm currents, to apply three if conditions and to calculate two $rMSBubbEn$ of i_a and i_b, as in Table 2.

If Condition (4) is not respected, there is another fault on phase b or c or two faults on phases b and c. Ref. [36] shows that $PhasEn$ is the right entropy to isolate the faults thereafter. $PhasEn$ of the phase currents i_a and i_b are compared. If

$$|PhasEn\ i_b - PhasEn\ i_a| < \epsilon_3, \tag{5}$$

and the mean of current i_b is negative, then there is also an open-circuit fault on T_b. In all, T_a, B_a, T_b. If the mean of the current i_b is positive, then three open-circuit faults are on T_a, B_a, B_b. The total computing time to detect and isolate T_a, B_a, T_b or T_a, B_a, B_b is 12.67 ms, for which it is necessary to calculate three means of arm currents, to apply five if conditions and to calculate two $rMSBubbEn$ and two $PhasEn$ of i_a and i_b.

Table 1. Entropy computation time for two different lengths of phase current.

Entropy Type	Entropies	Computation Time for Length L_6	Computation Time for Length L_2
Sample Entropy	*SampEn*	1.4441	0.1862
	MSSampEn	2.3840	0.2977
	cMSSampEn	4.1252	0.5084
	rMSSampEn	2.1296	0.3278
Kolmokov Entropy	*K2En*	1.5930	0.2009
	MSK2En	2.9755	0.3438
	cMSK2En	3.9457	0.4508
	rMSK2En	2.2020	0.2643
Conditional Entropy	*CondEn*	1.241×10^{-3}	7.548×10^{-4}
	MSCondEn	4.153×10^{-3}	2.916×10^{-3}
	cMSCondEn	18.61×10^{-3}	18.11×10^{-3}
	rMSCondEn	6.309×10^{-3}	4.643×10^{-3}
Dispersion Entropy	*DispEn*	2.204×10^{-3}	1.501×10^{-3}
	MSDispEn	7.120×10^{-3}	4.324×10^{-3}
	cMSDispEn	21.66×10^{-3}	19.28×10^{-3}
	rMSDispEn	9.220×10^{-3}	7.827×10^{-3}
Cosine Similarity Entropy	*CoSiEn*	1.801	0.215
	MSCoSiEn	2.952	0.327
	cMSCoSiEn	4.107	0.489
	rMSCoSiEn	2.467	0.282
Bubble Entropy	*BubbEn*	1.546×10^{-3}	9.429×10^{-4}
	MSBubbEn	10.91×10^{-3}	5.180×10^{-3}
	cMSBubbEn	18.81×10^{-3}	16.80×10^{-3}
	rMSBubbEn	7.978×10^{-3}	4.905×10^{-3}
Approximation Entropy	*ApEn*	2.096	0.260
	MSApEn	4.381	0.512
	cMSApEn	5.977	0.723
	rMSApEn	3.081	0.373
Fuzzy Entropy	*FuzzEn*	1.213	0.156
	MSFuzzEn	2.278	0.281
	cMSFuzzEn	3.196	0.393
	rMSFuzzEn	1.993	0.238
Increment Entropy	*IncrEn*	1.924×10^{-3}	1.058×10^{-3}
	MSIncrEn	8.639×10^{-3}	5.763×10^{-3}
	cMSIncrEn	21.42×10^{-3}	16.69×10^{-3}
	rMSIncrEn	8.931×10^{-3}	7.047×10^{-3}
Phase Entropy	*PhasEn*	8.168×10^{-4}	8.143×10^{-4}
	MSPhasEn	3.928×10^{-3}	3.159×10^{-3}
	cMSPhasEn	15.64×10^{-3}	16.62×10^{-3}
	rMSPhasEn	6.090×10^{-3}	5.229×10^{-3}

Table 1. *Cont.*

Entropy Type	Entropies	Computation Time for Length L_6	Computation Time for Length L_2
Slope Entropy	*SlopEn*	1.084×10^{-3}	8.216×10^{-4}
	MSSlopEn	5.801×10^{-3}	4.128×10^{-3}
	cMSSlopEn	18.56×10^{-3}	18.40×10^{-3}
	rMSSlopEn	7.102×10^{-3}	5.716×10^{-3}
Entropy of Entropy	*EnofEn*	55.20×10^{-3}	20.72×10^{-3}
	MSEnofEn	139.1×10^{-3}	49.65×10^{-3}
	cMSEnofEn	247.6×10^{-3}	71.19×10^{-3}
	rMSEnofEn	118.3×10^{-3}	41.33×10^{-3}
Attention Entropy	*AttEn*	7.306×10^{-4}	7.317×10^{-4}
	MSAttEn	3.879×10^{-3}	2.768×10^{-3}
	cMSAttEn	14.25×10^{-3}	14.04×10^{-3}
	rMSAttEn	6.108×10^{-3}	5.036×10^{-3}

If Condition (5) is not met, and if the phase entropy of i_a and the phase entropy of i_c are nearby

$$|PhasEn\ i_c - PhasEn\ i_a| < \epsilon_3, \tag{6}$$

then T_a, B_a, B_c faults can be isolated if the mean of current i_c is positive. For negative values of i_c, the fault diagnostic method identifies T_a, B_a, B_c as faulty. If Conditions (5) and (6) are not valid, the distances between the three *PhasEn* of i_a, i_b and i_c are greater than ϵ_3, as in [36]. A positive average of i_b allows the detection and isolation of T_a, B_a, B_b, T_c faults. For a negative average of i_b, T_a, B_a, T_b, B_c faults can be detected and isolated. After the calculation of three means of arm currents, applying six *if* conditions, calculation two *rMSBubbEn* and three *PhasEn*, a total computing time of 13.58 ms is obtained to detect and isolate the following cases: T_a, B_a, T_c or T_a, B_a, B_c or T_a, B_a, T_b, B_c or T_a, B_a, B_b, T_c.

According to the previous analysis, when the average of i_b is equal to zero and the means of i_b and i_c are close to zero represents a case similar to the previous one. We can isolate the following faults: T_b, B_b or T_b, B_b, T_a or T_b, B_b, B_a or T_b, B_b, T_c or T_b, B_b, B_c or T_b, B_b, T_a, B_c or T_b, B_b, B_a, T_c, with a total computing time as in the previous case (respecting Condition (1)).

If the mean of i_c is zero and Conditions (1) and (2) are not respected, the faults T_c, B_c or T_c, B_c, T_a or T_c, B_c, B_a or T_c, B_c, T_b or T_c, B_c, B_b or T_c, B_c, T_a, B_b or T_c, B_c, B_a, T_b are detected and isolated. In this case, an additional *if* condition is presented in the first flow chart of Figure 2, increasing the total computing time by 0.0964 ms.

To clearly understand the fault diagnostic method, when the averages of the currents i_a, i_b and i_c are not zero, the first flow chart of Figure 2 is extended by the second flow chart of Figure 3. Eight cases stand out in the function of i_a, i_b, i_c phase current signs. If the means of these three currents are positive, negative and negative, the four cases of faults are detected and isolated as B_a or T_b, T_c or B_a, T_b or B_a, T_b, T_c.

The second flow chart details the case when the averages are positive, negative and positive, highlighting T_b or T_b, B_c or B_a, B_c or B_a, T_b, B_c faults. It is impossible to have the three mean values positive, positive and positive or negative, negative and negative. If the means of these three currents are positive, positive and negative, the four faults are T_c or B_a, B_b or B_a, T_c or B_a, B_b, T_c. Another possibility is negative, negative and positive, leading to the isolation of B_c or T_a, T_b or T_a, B_c or T_a, T_b, B_c faults. Details of the proposed diagnostic approach are presented in the second flow chart isolating the following faults: T_a or T_a, B_b or B_a, B_c or T_a, B_b, B_c. The last case is the isolation of B_b or T_a, T_c or B_b, T_c or T_a, B_b, T_c.

Let us detail the case when the averages are positive, negative and positive. The *rMSBubbEn* of current i_a and i_b are calculated and compared. If Condition (4) is respected, the $SlopEn$ of i_a and i_c is studied. In order to detect T_b or T_b, B_c or B_a, B_c or B_a, T_b, B_c faults, the following condition is applied:

$$\frac{SlopEn\ i_a}{SlopEn\ i_c} > \epsilon_2. \tag{7}$$

If this condition is true, T_b is detected. Otherwise, B_a and B_c are isolated. Furthermore, after the calculation of three means of arm currents, applying eight *if* conditions and calculating two $rMSBubbEn$ and two $SlopEn$, the total computing time is 12.84 ms for the detection and isolation of T_b or B_a, B_c. If the $rMSBubbEn$ of i_a and i_b do not fulfill Condition (4), the $SlopEn$ of i_b and i_c are determined and compared as

$$\frac{SlopEn\ i_b}{SlopEn\ i_c} > \epsilon_2 \tag{8}$$

Providing T_b, B_c or B_a, T_b, B_c faults' isolation. The total computing time is 12.84 ms, as in the previous case.

Table 2. Computation time to locate faults: current phase of L_2 length.

No.	Open-Circuit Fault	Number of Operation Type	Total Computation Time
1.	No fault	3 mean, 2 if	8.25×10^{-4}
2.	T_a	3 mean, 8 if, 2 $CondEn$, 2 $rMSBubbEn$	12.84×10^{-3}
3.	T_b	3 mean, 8 if, 2 $SlopEn$, 2 $rMSBubbEn$	12.97×10^{-3}
4.	T_c	3 mean, 8 if, 2 $SlopEn$, 2 $rMSBubbEn$	12.97×10^{-3}
5.	B_a	3 mean, 8 if, 2 $CondEn$, 2 $rMSBubbEn$	12.84×10^{-3}
6.	B_b	3 mean, 8 if, 2 $SlopEn$, 2 $rMSBubbEn$	12.97×10^{-3}
7.	B_c	3 mean, 8 if, 2 $CondEn$, 2 $rMSBubbEn$	12.84×10^{-3}
8.	T_a, B_a	3 mean, 3 if, 2 $rMSBubbEn$	10.85×10^{-3}
9.	T_b, B_b	3 mean, 3 if, 2 $rMSBubbEn$	10.85×10^{-3}
10.	T_c, B_c	3 mean, 4 if, 2 $rMSBubbEn$	10.94×10^{-3}
11.	T_a, T_b	3 mean, 8 if, 2 $CondEn$, 2 $rMSBubbEn$	12.84×10^{-3}
12.	T_b, T_c	3 mean, 8 if, 2 $CondEn$, 2 $rMSBubbEn$	12.84×10^{-3}
13.	T_a, T_c	3 mean, 8 if, 2 $SlopEn$, 2 $rMSBubbEn$	12.97×10^{-3}
14.	B_a, B_b	3 mean, 8 if, 2 $SlopEn$, 2 $rMSBubbEn$	12.97×10^{-3}
15.	B_b, B_c	3 mean, 8 if, 2 $CondEn$, 2 $rMSBubbEn$	12.84×10^{-3}
16.	B_a, B_c	3 mean, 8 if, 2 $SlopEn$, 2 $rMSBubbEn$	12.97×10^{-3}
17.	T_a, B_b	3 mean, 8 if, 2 $SlopEn$, 2 $rMSBubbEn$	12.97×10^{-3}
18.	T_a, B_c	3 mean, 8 if, 2 $SlopEn$, 2 $rMSBubbEn$	12.97×10^{-3}
19.	T_b, B_a	3 mean, 8 if, 2 $SlopEn$, 2 $rMSBubbEn$	12.97×10^{-3}
20.	T_b, B_c	3 mean, 8 if, 2 $SlopEn$, 2 $rMSBubbEn$	12.97×10^{-3}
21.	T_c, B_a	3 mean, 8 if, 2 $SlopEn$, 2 $rMSBubbEn$	12.97×10^{-3}
22.	T_c, B_b	3 mean, 8 if, 2 $SlopEn$, 2 $rMSBubbEn$	12.97×10^{-3}
23.	T_a, B_a, T_b	3 mean, 5 if, 2 $rMSBubbEn$, 2 $PhasEn$	12.67×10^{-3}
24.	T_a, B_a, B_b	3 mean, 5 if, 2 $rMSBubbEn$, 2 $PhasEn$	12.67×10^{-3}
25.	T_a, B_a, T_c	3 mean, 6 if, 2 $rMSBubbEn$, 3 $PhasEn$	13.58×10^{-3}
26.	T_a, B_a, B_c	3 mean, 6 if, 2 $rMSBubbEn$, 3 $PhasEn$	13.58×10^{-3}
27.	T_b, B_b, T_a	3 mean, 5 if, 2 $rMSBubbEn$, 2 $PhasEn$	12.67×10^{-3}
28.	T_b, B_b, B_a	3 mean, 5 if, 2 $rMSBubbEn$, 2 $PhasEn$	12.67×10^{-3}
29.	T_b, B_b, T_c	3 mean, 6 if, 2 $rMSBubbEn$, 3 $PhasEn$	13.58×10^{-3}
30.	T_b, B_b, B_c	3 mean, 6 if, 2 $rMSBubbEn$, 3 $PhasEn$	13.58×10^{-3}

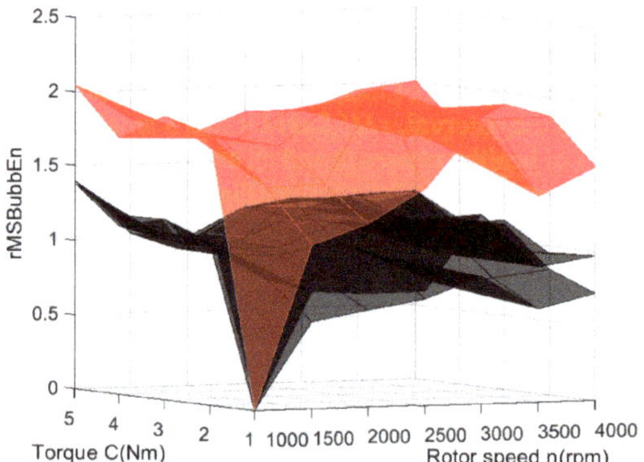

Figure 5. *rMSBubbEn* with one open-circuit fault on T_a with different combinations of torque and speed: healthy phases are represented by two black surfaces and the open-circuit phase is represented by a red surface.

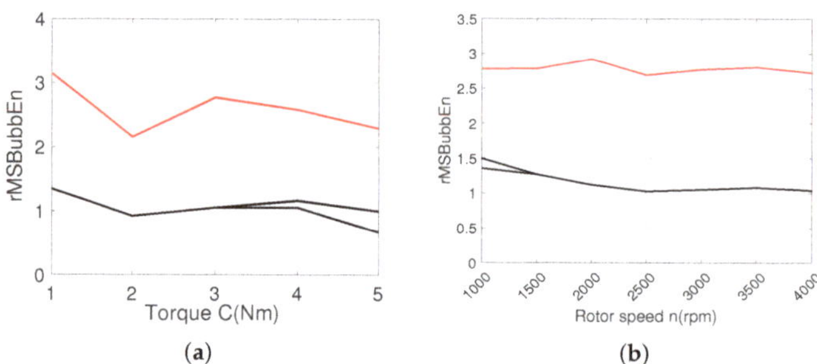

(a) **(b)**

Figure 6. *rMSBubbEn* for (**a**) torque and (**b**) speed variations with two open-circuit faults on T_a and B_a: healthy phases are represented by two black curves and the open-circuit phase is represented by a red curve.

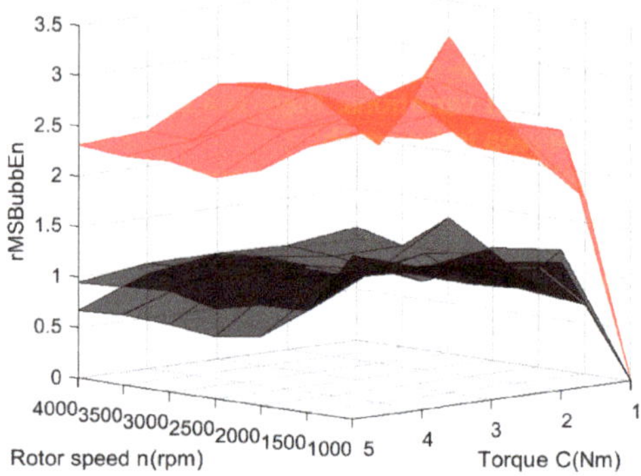

Figure 7. *rMSBubbEn* for torque and speed variations with two open-circuit faults on T_a and B_a: healthy phases are represented by two black surfaces and the open-circuit phase is represented by a red surface.

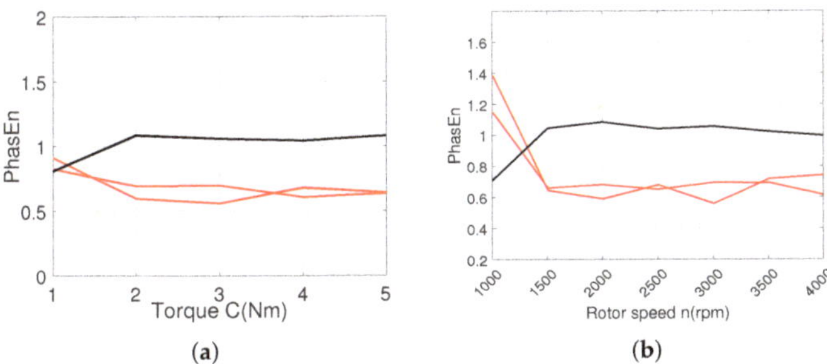

(a)   (b)

Figure 8. *PhasEn* for (**a**) torque and (**b**) speed variations with three open-circuit faults on B_a, T_b and T_b: healthy phase is represented by a black curve and the open-circuit phases are represented by two red curves.

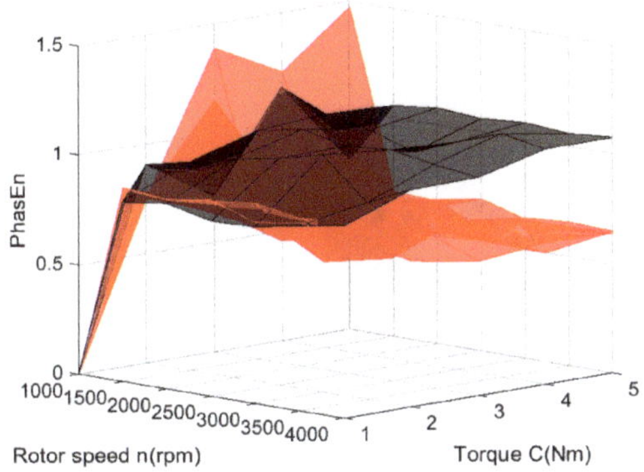

Figure 9. *PhasEn* for torque and speed variations with three open-circuit faults on B_a, T_b and T_b: healthy phase is represented by a black surface and the open-circuit phases are represented by two red surfaces.

In order to check the incidence of the speed and load variations on *SlopEn*, Figure 10 illustrates the case of one open-circuit fault on T_a.

The distance between *SlopEn* of healthy phases *b* and *c* and *PhasEn* of open-circuit phase *a* is rather constant, as in Figure 10a. This distance presents a little narrowing for the load in the range of (4 Nm to 5 Nm). The simulation results underline the effectiveness of the proposed algorithm in the case of load variation in a large domain. According to Figure 10b, at the beginning of the interval (for small values of speed *n*), the three *SlopEn* of phases *a*, *b* and *c* tend to interweave.

Figure 10b shows a whole picture of *PhasEn* with various *n* and *C*. The fault diagnostic method is efficient for the range of (2000 rpm to 4000 rpm) for speed variations. As in the case above for *PhasEn* with three open-circuit faults on B_a, T_b and T_b, *SlopEn* fails to isolate a T_a open-circuit fault under lower speed conditions and lower torques. This is a limitation of this approach. The open-circuit faults are detected based on the mean values of phase currents, but they cannot be located only for lower values of speed and torque. On the other hand, the simulation results in Figure 11 demonstrate the effectiveness of the proposed method for high speeds (2000 rpm to 4000 rpm and high torques (2 Nm to 5 Nm).

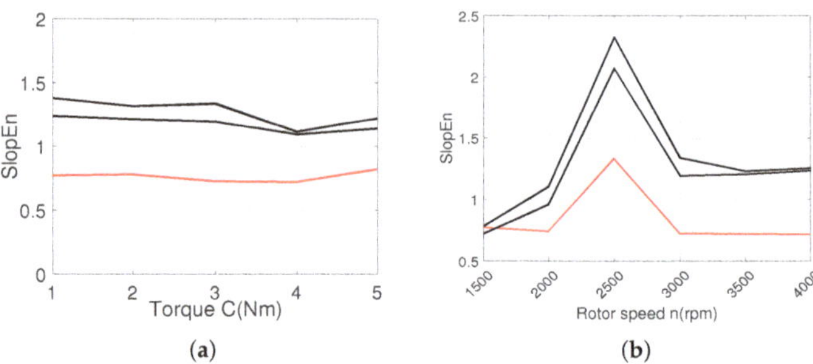

Figure 10. *SlopEn* for (**a**) torque and (**b**) speed variations with one open-circuit fault on T_a: healthy phases are represented by two black curves and the open-circuit phase represented by a red curve.

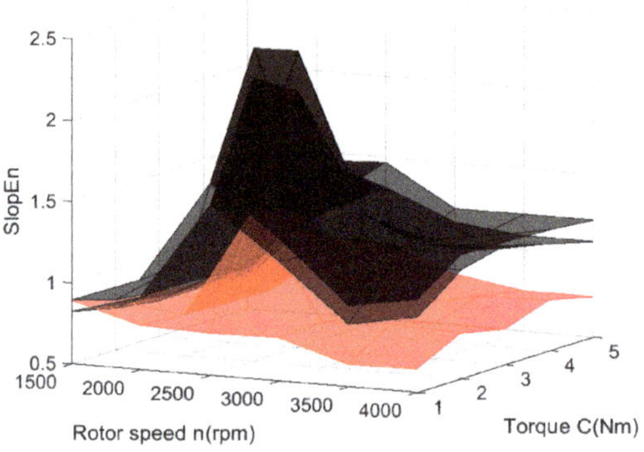

Figure 11. *SlopEn* for torque and speed variations with one open-circuit fault on T_a: healthy phases are represented by two black surfaces and the open-circuit phase is represented by a red surface.

4. Conclusions

Fault detection and identification are becoming increasingly important for industrial applications. This paper propose a diagnostic method for open-circuit faults of an inverter connected to a brushless motor. This algorithm requires only phase inverter currents and computing operations to generate different entropies. Some entropies are able to differentiate between healthy and unhealthy open-circuit conditions. Among these entropies, another selection is made in order to speed up the diagnostic. The simulation results ensure that these entropies are able to detect and locate open-circuit faults and, moreover, are able to achieve fault diagnostics for a single switch, double switches, three switches and even four switches.

The fault detection method is based on the average phase currents and the detection time is rather short. Then, the work in this paper deals with the localization of multiple open-circuit faults by a rapid and robust fault diagnostic method using three threshold values. The simulation results also confirm that the proposed fault diagnostic method can detect and locate multiple faults within (10.85 ms to 13.8 ms). This is much faster than many other diagnostic methods that usually require several fundamental periods. Then, in order to prove the robustness and ability of fault detection, a load variation is performed under the rated speed conditions of the brushless motor. The validity of the method is analyzed under several speed values for a constant torque.

Nevertheless, as mentioned above, there is a limit to this approach: open-circuit faults are detected, but their isolation fails under lower speed conditions and lower torques.

Consequently, the simulation demonstrates the feasibility and effectiveness of the proposed approach for large speeds of (2000 rpm to 4000 rpm) and large torques of (2 Nm to 5 Nm).

In the future, another interesting extension to our work may be to increase the brushless motor number of phases up to five (this requires an inverter with ten switches) and to compare the proposed solution with the current solution. The simulation shows that the proposed method can effectively diagnose and locate faults. Future research will focus on the proposed diagnosis method under slow variations in speed and torque (increasing and decreasing profiles), not only constant values. It will be interesting to add a fault-tolerant control strategy to our future work to ensure that the inverter and the motor work normally.

Author Contributions: Formulation was conducted by C.M.; problems were solved by C.M.; B.L.G. and S.R. contributed to the numerical computation and results; manuscript writing was conducted by C.M.; discussion: S.C.; revision: C.M. All authors have read and agreed to the published version of the manuscript.

Funding: This research received no external funding.

Institutional Review Board Statement: Not applicable.

Informed Consent Statement: Not applicable.

Data Availability Statement: Not applicable.

Conflicts of Interest: The authors declare no conflict of interest.

References

1. Puthiyapurayil, M.R.M.K.; Nasirudeen, M.N.; Saywan, Y.A.; Ahmad, W.; Malik, H. A Review of Open-Circuit Switch Fault Diagnostic Methods for Neutral Point Clamped Inverter. *Electronics* **2022**, *11*, 3169. [CrossRef]
2. Choi, U.M.; Blaabjerg, F.; Lee, K.B. Study and handling methods of power IGBT module failures in power electronic converter systems. *IEEE Trans. Power Electron.* **2015**, *30*, 2517–2533. [CrossRef]
3. An, Q.; Sun, L.K.; Zhao, K.; Sun, L. Switching function model-based fast-diagnostic method of open-switch faults in inverters without sensors. *IEEE Trans. Power Electron.* **2011**, *26*, 119–126. [CrossRef]
4. Lu, B.; Sharma, S. A literature review of IGBT fault diagnostic and protection methods for power inverters. *IEEE Trans. Ind. Appl.* **2009**, *45*, 1770–1777.
5. Mendes, A.M.S.; Abadi, M.B.; Cruz, S.M.A. Fault diagnostic algorithm for three-level neutral point clamped AC motor drives based on average current Park's vector. *IET Power Electron.* **2014**, *7*, 1127–1137. [CrossRef]
6. Jung, F.; Zhao, J. A real-time multiple open-circuit fault diagnostic method in voltage-source-inverter fed vector controlled drives. *IEEE Trans. Power Electron.* **2016**, *31*, 1425–1437.
7. Estima, J.O.; Freire, N.M.A.; Cardoso, A.J.M. Recent advances in fault diagnostic by Park's vector approach. *IEEE Workshop Electr. Mach. Des. Control Diagn.* **2013**, *2*, 279–288.
8. Wu, Y.; Zhang, Z.; Li, Y.; Sun, Q. Open-Circuit Fault diagnostic of Six-Phase Permanent Magnet Synchronous Motor Drive System Based on Empirical Mode Decomposition Energy Entropy. *IEEE Access* **2021**, *9*, 91137–91147. [CrossRef]
9. Sleszynski, W.; Nieznanski, J.; Cichowski, A. Open-transistor fault diagnostics in voltage-source inverters by analyzing the load currents. *IEEE Trans. Ind. Electron.* **2009**, *56*, 4681–4688. [CrossRef]
10. Zhou, D.; Yang, S.; Tang, Y. A Voltage-Based Open-Circuit Fault Detection and Isolation Approach for Modular Multilevel Converters With Model-Predictive Control. *IEEE Trans. Power Electron.* **2018**, *33*, 9866–9874. [CrossRef]
11. Patel, K.; Borole, S.; Manikandan, R.; Singh, R.R. Short Circuit and open-circuit Fault Identification Strategy for 3-Level Neutral Point Clamped Inverter. In Proceedings of the 2021 Innovations in Power and Advanced Computing Technologies (i-PACT), Kuala Lumpur, Malaysia, 27–29 November 2021; pp. 1–6.
12. Raj, N.; Mathewb, J.; Ga, J.; George, S. Open-Transistor Fault Detection and diagnostic Based on Current Trajectory in a Two-level Voltage Source Inverter. *Procedia Technol.* **2016**, *25*, 669–675. [CrossRef]
13. Xu, S.; Huang, W.; Wang, H.; Zheng, W.; Wang, J.; Chai, Y.; Ma, M. A Simultaneous diagnostic Method for Power Switch and Current Sensor Faults in Grid-Connected Three-Level NPC Inverters. *IEEE Trans. Power Electron.* **2022**, *26*, 1–15.
14. Oubabas, H.; Djennoune, S.; Bettayeb, M. Interval sliding mode observer design for linear and nonlinear systems. *J. Process. Control* **2018**, *26*, 959–1524. [CrossRef]
15. Song, B.; Qi, G.; Xu, L. A new approach to open-circuit fault diagnostic of MMC sub-module. *Syst. Sci. Control Eng.* **2020**, *8*, 119–127. [CrossRef]
16. Li, T.; Zhao, C.; Li, L.; Zhang, F.; Zhai, X. Sub-module fault diagnostic and local protection scheme for MMC-HVDC system. *Proceeding CSEE* **2014**, *34*, 1641–1649.
17. Wei, H.; Zhang, Y.; Yu, L.; Zhang, M.; Teffah, K. A new diagnostic algorithm for multiple IGBTs open-circuit faults by the phase currents for power inverter in electric vehicles. *Energies* **2018**, *11*, 1508. [CrossRef]

18. Cosimi, F. Analysis and design of a non-linear MPC algorithm for vehicle trajectory tracking and obstacle avoidance. In *Applications in Electronics Pervading Industry, Environment and Society: APPLEPIES 2020*; Springer International Publishing: Berlin, Germany, 2021; Volume 8.

19. Campos-Delgado, D.U.; Pecina-Sánchez, J.A.; Espinoza-Trejo, D.R.; Arce-Santana, E.R. Diagnostic of open-switch faults in variable speed drives by stator current analysis and pattern recognition. *IET Electr. Power Appl.* **2013**, *7*, 509–522. [CrossRef]

20. Deng, F.; Chen, Z.; Khan, M.; Zhu, R. Fault Detection and Localization Method for Modular Multilevel Converters. *IEEE Trans. Power Electron.* **2015**, *30*, 2721–2732. [CrossRef]

21. Zhang, Y.; Hu, H.; Liu, Z. Concurrent fault diagnostic of a modular multi-level converter with kalman filter and optimized support vector machine. *Syst. Sci. Control Eng.* **2019**, *7*, 43–53. [CrossRef]

22. Faraz, G.; Majid, A.; Khan, B.; Saleem, J.; Rehman, N. An Integral Sliding Mode Observer Based Fault diagnostic Approach for Modular Multilevel Converter. In Proceedings of the 2019 International Conference on Electrical, Communication, and Computer Engineering (ICECCE), Swat, Pakistan, 24–25 July 2019; pp. 1–6.

23. Shao, S.; Wheeler, P.W.; Clare, J.C.; Watson, A.J. Fault detection for modular multilevel converters based on sliding mode observer. *IEEE Trans. Power Electron.* **2013**, *28*, 4867–4872. [CrossRef]

24. Shao, S.; Watson, A.J.; Clare, J.C.; Wheeler, P.W. Robustness analysis and experimental validation of a fault detection and isolation method for the modular multilevel converter. *IEEE Trans. Power Electron.* **2016**, *31*, 3794–3805. [CrossRef]

25. Li, B.; Shi, S.; Wang, B.; Wang, G.; Wang, W.; Xu, D. Fault diagnostic and tolerant control of single IGBT open-circuit failure in modular multilevel converters. *IEEE Trans. Power Electron.* **2016**, *31*, 3165–3176. [CrossRef]

26. Li, B.; Zhou, S.; Xu, D.; Yang, R.; Xu, D.; Buccella, C.; Cecati, C. An improved circulating current injection method for modular. multilevel converters in variable-speed drives. *IEEE Trans. Ind. Electron.* **2016**, *63*, 7215–7225. [CrossRef]

27. Yang, Q.; Qin, J.; Saeedifard, M. Analysis, detection, and location of open-switch submodule failures in a modular multilevel converter. *IEEE Trans. Power Del.* **2016**, *31*, 155–164. [CrossRef]

28. Abdelsalam, M.; Marei, M.I.; Tennakoon, S. An integrated control strategy with fault detection and tolerant control capability based on capacitor voltage estimation for modular multilevel converters. *IEEE Trans. Ind. Appl.* **2017**, *53*, 2840–2851. [CrossRef]

29. Li, W.; Li, G.; Zeng, R. The fault detection, localization, and tolerant operation of modular multi-level converters with an insulated gate bipolar transistor (IGBT) open-circuit fault. *Energies* **2018**, *11*, 837. [CrossRef]

30. Chakraborty, R.; Samantaray, J.; Chakrabarty, S.; Variable gain observer-based estimation of capacitor voltages in modular multi-level converters. In Proceedings of the IEEE International Conference on Power Electronics, Drives & Energy Systems: PEDES, Jaipur, India, 16–19 December 2020, pp. 7–12.

31. Caseiro, L.M.A.; Mendes, A.M.S. Real-time IGBT open-circuit fault diagnostic in three-level neutral-point-clamped voltage-source rectifiers based on instant voltage error. *IEEE Trans. Ind. Electron.* **2015**, *62*, 1669–1678. [CrossRef]

32. Kiranyaz, S.; Gastli, A.; Ben-Brahim, L.; Al-Emadi, N.; Gabbouj, M. Real-time fault detection and identification. for MMC Using 1-D convolutional neural networks. *IEEE Trans. Ind. Electron.* **2019**, *66*, 8760–8771. [CrossRef]

33. Wang, Q.; Yu, Y.; Hoa, A. Fault detection and classification in MMC-HVDC systems using learning methods. *Sensors* **2020**, *20*, 4438. [CrossRef]

34. Yan, H.; Xu, Y.; Cai, F.; Zhang, H.; Zhao, W.; Gerada, C. PWM-VSI fault diagnostic for PMSM drive based on fuzzy logic approach. *IEEE Trans. Power Electron.* **2018**, *34*, 759–768. [CrossRef]

35. Bissey, S.; Jacques, S.; Le Bunetel, J.-C. The Fuzzy Logic Method to Efficiently Optimize Electricity Consumption in Individual Housing. *Energies* **2017**, *10*, 1701. [CrossRef]

36. Morel, C.; Rivero, S.; Le Gueux, B.; Portal, J; Chahba, S. Currents Analysis of a Brushless Motor with Inverter Faults—Part I: Parameters of Entropy Functions and Open-Circuit Faults Detection. *Actuators* **2023**, *12*, 228. [CrossRef]

 *actuators*

MDPI

Article

Evaluation and Simulation Analysis of Mixing Performance for Gas Fuel Direct Injection Engine under Multiple Working Conditions

Hongchen Wang, Tianbo Wang *, Jing Chen, Lanchun Zhang, Yan Zheng, Li Li and Yanyun Sun

School of Automotive and Traffic Engineering, Jiangsu University of Technology, Changzhou 213001, China; 2021655039@smail.jsut.edu.cn (H.W.); 2021655263@smail.jsut.edu.cn (J.C.); zlc@jsut.edu.cn (L.Z.); zhengyan@jsut.edu.cn (Y.Z.); liliorigin@jsut.edu.cn (L.L.); sunyanyun@jsut.edu.cn (Y.S.)
* Correspondence: wangtianbo@jsut.edu.cn

Abstract: Gas fuel direct injection (DI) technology can improve the control precision of the in-cylinder mixing and combustion process and effectively avoid volumetric efficiency reduction in a compressed natural gas (CNG) engine, which has been a tendency. However, compared with the port fuel injection (PFI) method, the former's mixing path and duration are shortened greatly, which often leads to poor mixing uniformity. What is worse, the in-cylinder mixing performance would be seriously affected by engine working conditions, such as engine speed and load. Based on this situation, the fluid mechanics software FLUENT is used in this article, and the computational fluid dynamics (CFD) model of the injection and mixing process in a gas-fueled direct injection engine is established. A quantitative evaluation mechanism of the in-cylinder mixing performance of the CNG engine is proposed to explore the influencing rule of different engine speeds and loads on the mixing process and performance. The results indicate that phase space analysis can accurately reflect the characteristics of the mixture mixing process. The gas fuel mixture rapidly occupies the cylinder volume in the injection stage. During the transition stage, the gas fuel mixture is in a highly transient state. The diffusion stage is characterized by the continuous homogenization of the mixture. The in-cylinder mixing performance is linearly dependent on the engine's working condition in the phase space.

Keywords: gas fuel; gas fuel direct injection; mixing performance; multiple working conditions; evaluation mechanism

Citation: Wang, H.; Wang, T.; Chen, J.; Zhang, L.; Zheng, Y.; Li, L.; Sun, Y. Evaluation and Simulation Analysis of Mixing Performance for Gas Fuel Direct Injection Engine under Multiple Working Conditions. *Actuators* **2023**, *12*, 239. https://doi.org/10.3390/act12060239

Academic Editor: Luigi de Luca

Received: 23 May 2023
Revised: 6 June 2023
Accepted: 7 June 2023
Published: 8 June 2023

1. Introduction

The energy crisis and environmental pollution are the two major challenges facing current social development. The transportation industry is one of the main contributors to the energy consumption and emission growth, and its share continues to grow [1]. Vigorously promoting orthodox car energy conservation and emission reductions, as well as the industrialization of new energy vehicles, will become major challenges which urgently need to be addressed in the current transportation industry and pressing tasks for promoting the sustainable development of the transportation industry [2]. Compared to traditional diesel fuel, the use of natural gas (NG) engines can significantly reduce pollution from sulfur dioxide, nitrogen oxide and CO_2 in emissions. CNG engines have been widely used in key areas of the national economy, including automobiles (including sedans, buses, trucks, etc.), ships and distributed power stations [3,4].

Currently, the fuel supply system of CNG engines can be classified as in-cylinder direct injection (DI) and port fuel injection (PFI) [5]. The PFI mode reduces the CNG engine's volumetric efficiency, with engines having lower power output than gasoline and diesel engines with the same displacement [6]. Moreover, it is difficult to achieve accurate control of the mixing and combustion process for PFI, which limits the development of

CNG engines. With the maturity of DI technology, the combination of gas fuel and DI technology will become a development trend in future [7–9]. The gas fuel DI technology can eliminate the volumetric efficiency loss due to the PFI mode, avoid fuel loss during the scavenging process and facilitate the suppression of knocking. The DI technology enables the enhanced control of the mixing and combustion process by optimizing the matching of injection timing, air–fuel ratio and ignition advance angle [7].

The mixing performance for CNG engines can be divided into homogeneous mixtures and stratified mixtures [8,9]. At high loads and steady speeds, the gas fuel needs to be sufficiently mixed with air to increase the engine power and meet the vehicle's driving performance. The homogeneous mixing method can ensure that the fuel is fully mixed, which makes the fuel easier to burn, thus improving thermal efficiency and dynamic performance [10]. However, under a partial load, incomplete combustion and the wastage of energy may occur with the homogeneous mixing method, where some exhaust is not fully burned, which can lead to environmental pollution. In this case, a stratified mixing method is necessary [11]. A stratified mixing method can make the engine run smoothly by reducing the noise and vibration [12].

The energy power and economy performance of engines are affected by the in-cylinder mixing performance of the gas fuel DI technology. Compared with the PFI mode, the mixing path and duration are shortened greatly for the DI mode [13]. In contrast to liquid fuels, gas fuel injection and mixing processes do not require complex physical changes such as phase transitions. However, due to the lower gas density, it is more difficult for the gas fuel to fully mix with the air. What is worse, the mixing performance is greatly affected by variable factors, such as engine speed and load [14]. The formation of the mixture of in-cylinder DI engines was studied by Karri Keskinen et al. through computational fluid dynamics (CFD) [15]. They designed three mixing performance metrics based on injection pressure and nozzle type to describe the mixing performance of the gas fuel throughout the compression stroke. An objective classification method of mixture distribution in the combustion chamber of a DI engine was proposed by S.K et al. [16]. On this basis, four types of mixture distribution, random, linear, Gaussian and parabolic, were simulated in the combustion chamber of a DI engine by using the CFD method. The effects of different engines' intake pressure, intake temperature and engine speed on mixture uniformity were studied by Mohammad et al. [17], and it was found that intake pressure and speed had greater effects on the mixture gas fuel. The numerical simulation of the formation mechanism of the mixture in-cylinder was carried out by Baratta et al. [18]. The research results showed that the mixing rate of gas fuel and air was dominated by the fuel jet and the tumble and vortex flow, and the mixing duration had an influence on the mixing performance of gas fuel.

In view of this, a CFD model of the DI mixing process for CNG engines is established. A quantitative evaluation mechanism for the mixing performance of CNG engines is proposed. The influence of different engine speeds and loads on the mixing process and mixing performance is investigated in this paper. This study has significant theoretical implications for improving the power and emission performance of CNG DI engines and increasing engine efficiency.

2. Model and Research Method

2.1. Model

This study is based on a certain type of CNG DI engine. The main engine parameters are shown in Table 1. Based on the power requirements of the prototype engine, the authors have previously designed a high-pressure CNG injection device based on a moving-coil electromagnetic linear actuator and mushroom-type poppet valve [19], with the main parameters shown in Table 2. Building upon the high-pressure CNG injection device studied in the author's previous research, this article further investigates the in-cylinder mixing performance of gas-fueled engines using this type of injection device. To facilitate the calibration of gas fuel duration of injection (DOI), the inlet pressure of the injection device was set to 1.0 MPa, which was based on the one-dimensional isentropic flow calculation formula and the in-cylinder pressure without gas injection (sourced from CFD model calculation results) [19]. Under this injection pressure, the choking phenomenon of the nozzle throat near the field could be ensured. A lower injection pressure can reduce production costs and improve the gas fuel utilization efficiency of the upstream gas storage tanks in the gas supply system.

Table 1. Specifications of CNG engine.

Parameter	Value
Bore (mm) × Stroke (mm)	131 × 155
Displacement volume(L)	12.53
Compression ratio	11.5
Rated power (kW)/speed (rpm)	255/2000
IVO/IVC(CA)	30° BTDC/46° ATDC
EVO/EVC(CA)	78° BTDC/30° ATDC

Table 2. Specification of gas injection device.

Parameter	Value
Outlet diameter (mm)	7
Valve lift (mm)	1.5
Injection pressure (MPa)	1.0

The CFD model of the in-cylinder mixing process for the CNG DI engine is shown in Figure 1, which includes the intake port, exhaust port, cylinder area and injection device cavity. An injection device study has been analyzed in the authors' previous research [20], and here, the injection device was arranged at the center of the cylinder at a 45° angle to the cylinder axis. There was large pressure and velocity near the nozzle throat and injection device. The CFD model was locally refined to a minimum grid size of 0.3 mm near the nozzle. In addition, the computational cost of CFD needed to be kept within acceptable limits. The grid size gradually increased away from the nozzle throat, reaching a maximum size of about 3 mm. The grid partition is shown in Figure 2. For convenience, the CNG composition was assumed to be 100% methane. The simulation calculation started from the intake valve opening (IVO) time and ended at the ignition time. That is, it started from 30° before the top dead center (BTDC), corresponding to a 330° crank angle (CA), and ended at CA700. The entire computational domain was assumed to be initially stationary. The RNG k–ε turbulence model and the non-equilibrium wall function were used in this study. The turbulent Schmid number took the fixed default value of 0.7. In this paper, we focused on the theoretical exploration of the mechanisms that influence the performance of in-cylinder mixing in engine working conditions.

2.2. Transient CFD Model Verification

The minimum mesh size near the nozzle of the CFD model was 0.3 mm, and the maximum mesh size of other positions was about 3 mm. The core area of the jet of the nozzle was an important position for the jet development of the CFD model, and the mesh needed to be verified. In the author's previous research [21], the mesh in the core region of the nozzle jet was validated. Experiments were carried out on the cumulative transient flow rate of the gas fuel injection device. Compared with the experimental results, the maximum error in the transient flow rate of the simulations was about 6.1%, validating the flow rate properties of the CFD model. In addition, the supersonic jet morphology was verified, and the simulation results were in good agreement with the experimental jet imaging results, as shown in Figure 3.

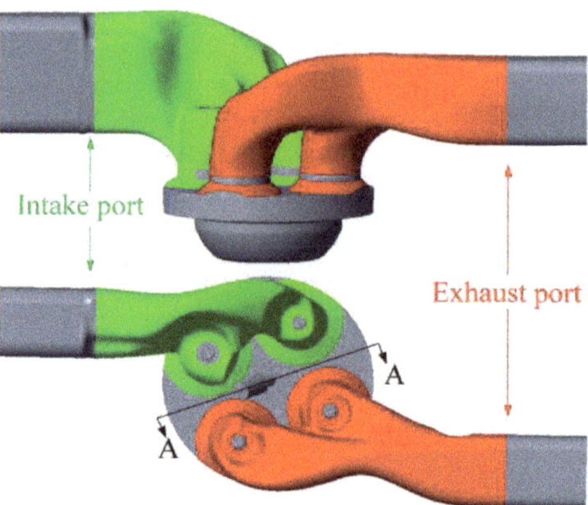

Figure 1. The CFD calculation domain for CNG DI engine.

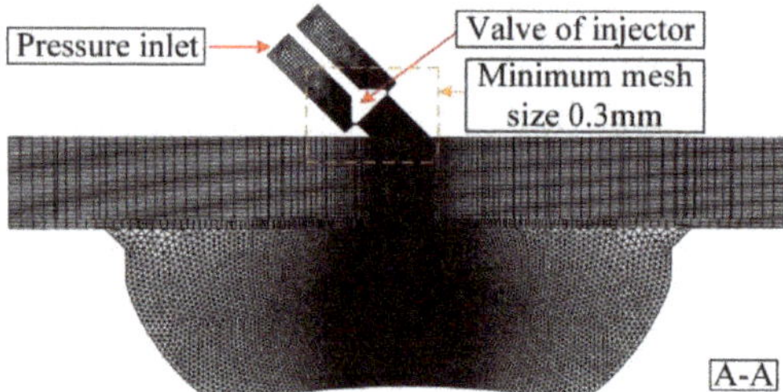

Figure 2. CFD calculation domain and mesh on the A-A section.

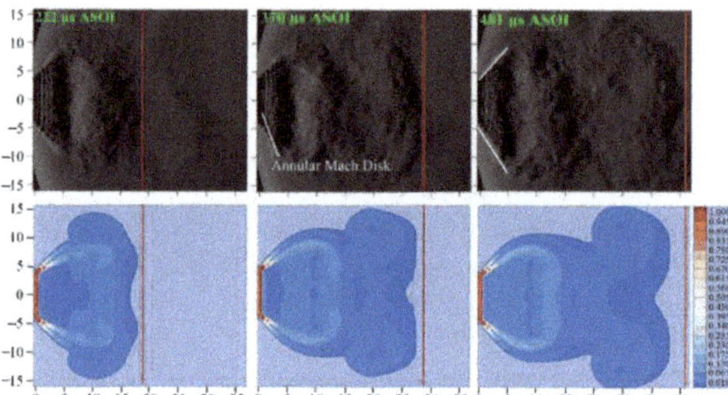

Figure 3. Experimental results (first row) and simulation results (second row) of jet morphology.

2.3. Research Method

To investigate the effects of different engine speeds and loads on the in-cylinder mixing performance, the research plan for this paper was determined, and the specific parameters are shown in Table 3. The main purpose of Cases 1–4 was to investigate the effect of different engine speeds (1200, 1500, 1800 and 2000 rpm) on the mixing performance under the same engine load (i.e., the engine had the same load of 100%). Cases 4–7 were used to explore the effect of different loads on the mixing performance under the same speed (rated speed of 2000 rpm). At different loads, the inlet pressure of the engine was adjusted accordingly to ensure a stoichiometric ratio of combustion in the cylinder. Different gas fuel demand was achieved by adjusting the start of injection (SOI), while the end of injection

(EOI) remained the same. Cases 8–11 built on Cases 4–7 and further explored the impact of load on the mixing performance at non-rated speeds.

Table 3. Parameters of each case.

Case	Load	Rotate Speed, rpm	Inlet Pressure, bar	SOI, ° CA	EOI, ° CA	DOI, ° CA
1	100%	1200	2.0	584	635	51
2	100%	1500	2.0	584	635	51
3	100%	1800	2.0	584	635	51
4	100%	2000	2.0	584	635	51
5	75%	2000	1.7	588	635	47
6	50%	2000	1.3	594	635	41
7	25%	2000	0.8	601	635	34
8	25%	1500	0.6	607	635	28
9	50%	1500	0.8	603	635	32
10	75%	1500	1.1	597	635	38
11	100%	1500	1.4	593	635	42

3. Evaluation Methods for Mixing Performance

In an ideal state, different working conditions of the engine correspond to different mixture distribution methods. To improve the rationality and accuracy of the research on mixture distribution, this paper proposed an objective classification method for the mixture distribution of gas engines. The mixture distribution of CNG engines was divided into homogeneous mixtures and non-homogeneous mixtures, and the non-homogeneous mixtures were further divided into reasonable stratification, unreasonable stratification and other mixture types.

$$\text{Mean} = \frac{\sum m_{f,i}}{i} \tag{1}$$

$$\text{SD} = \sqrt{\frac{\sum \left(m_{f,i} - \text{Mean} \right)^2}{i}} \tag{2}$$

To quantitatively analyze the influence mechanism of engine working conditions on the in-cylinder mixing performance, a phase space analysis method based on the MEAN and standard deviation (SD) of the gas fuel mass fraction (FMF) was designed, as shown in Figure 4. The objective classification method of the mixture was applied in the phase space. The MEAN and SD of the FMF were defined as evaluation indicators for measuring the in-cylinder mixing performance. The horizontal axis was taken as the MEAN of the in-cylinder FMF of the mixture, and the vertical axis was taken as the SD of the FMF of the mixture. MEAN and SD are defined in Equations (1) and (2). In the formula, $m_{f,i}$ represents the FMF in the i region.

A larger MEAN indicates that the distribution of gas fuel is not confined within a particular combustion chamber area. Conversely, a smaller MEAN indicates that some combustion chamber areas may not even have gas fuel. A higher SD indicates that the gas fuel distribution in the area containing gas fuel varies greatly (poor mixing), while a lower SD indicates that the gas fuel distribution in the area varies less (excellent mixing). The lower the SD and the larger the MEAN, the better the homogeneous mixing performance, while the higher the SD and the larger the MEAN, the better the stratified mixing performance.

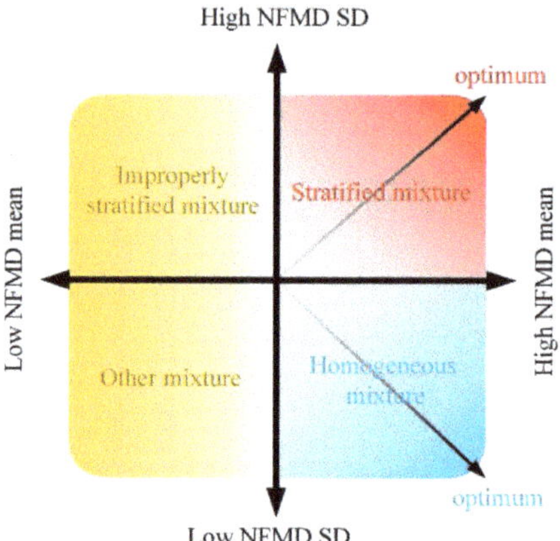

Figure 4. The phase space analysis method for the mixing performance.

4. Results and Discussion

The focus of this study was to explore the in-cylinder mixing process of gas fuel and the mixing performance at the end of the compression stroke. Considering that the average FMF is different under different speeds and loads, the FMF was normalized for the convenience of comparative research.

4.1. The Effect of Engine Speed

Firstly, regarding the effect of speed on the mixing process, we compared Cases 1–4 as shown in Figure 5, where the mixing process of gas fuel was divided into three stages based on the distribution of MEAN and SD in the phase space: injection, transition, and diffusion. The injection stage was from B (CA590) to C (CA630), where the valve of the gas fuel injection device was open and gas was continuously injected into the cylinder. The transition stage was from C to D (CA640), where the gas fuel injection valve was closed (CA635) in the middle moment of the transition phase. The diffusion stage was from D to F (CA700), which was the late stage of the mixing process between the gas fuel and air.

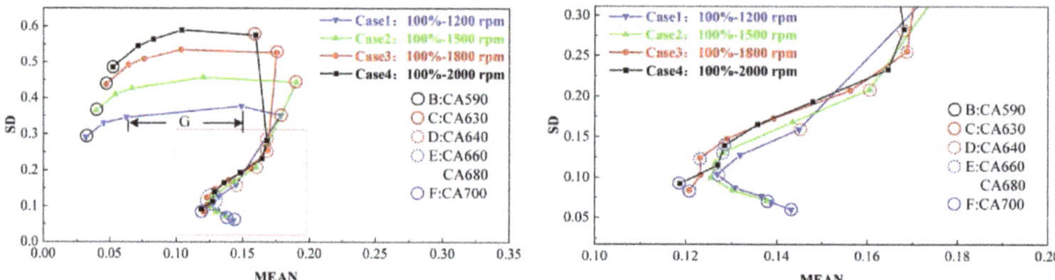

Figure 5. The mixing process of the mixture (**left**) and enlarged diagram (**right**) at different speeds under 100% load.

During the injection stage (B–C), the MEAN in the phase space rapidly increased with the duration of mixing, as shown in Figure 5. During the period of CA610–CA620 (defined as the G period), comparing Cases 1–4, the increase in MEAN was most significant in Case 1, and the increase in MEAN decreased in order in Cases 2–4. According to the DOI in Table 3 (SOI CA584~EOI CA635), the G period corresponded to the maximum lift period of the gas fuel injection device. The injection of gas fuel was the fastest and most stable, and the gas fuel quickly occupied the cylinder volume and mixed with the air in cylinder. Therefore, the MEAN increased rapidly with the duration of mixing. To analyze the reasons for the difference in the increase in MEAN among Cases 1–4 during the G period, an analysis was conducted based on the gas volume ratio. For Case 1, the gas volume ratio was approximately 16.7% at CA610 and increased to about 30.2% at CA620, which was a rise of approximately 13.5%. For Case 2, the increase was about 10.2% compared to CA610 at CA620. The increase for Case 3 was around 7.6% and around 6.7% for Case 4. During the G period, the gas volume ratio increase was highest for Case 1, followed by Case 2, Case 3 and Case 4. Thus, in the phase space, the increase in MEAN decreased in order in Cases 1–4. The in-cylinder FMF cloud map on the A-A section at CA620 was taken, as shown in Figure 6. It can be seen that, in Cases 1–4, the tumble flow was formed due to the wall blocking effect after the gas fuel jet impinged on the cylinder wall, and the gas fuel was directed towards the top of the piston. In Case 1, the gas fuel jet already crossed the in-cylinder centerline. While in Cases 2–4, the distance between the tail end of the gas fuel jet and the centerline increased in order, the space for gas fuel diffusion was significantly smaller than that in Case 1. Therefore, under the same load, a change in speed will lead to a difference in gas diffusion.

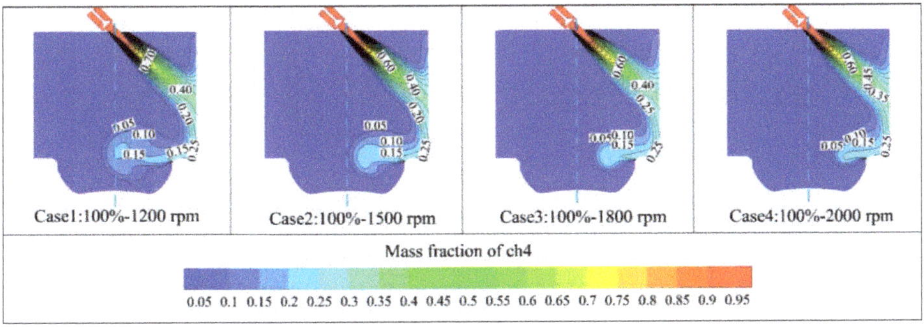

Figure 6. The in-cylinder methane mass fraction distribution at CA620.

During the transition stage (C-D), the MEAN in the phase space remained basically unchanged with the duration of mixing, as shown in Figure 5. The SD rapidly decreased with the duration of mixing, halving from its initial value, as shown in Figure 4. According to the DOI in Table 3 (Cases 1–4: EOI 635), the transition stage corresponded to the period when the valve of the gas fuel injection device was closed, and the duration of mixing was relatively short. Moreover, the FMF remained basically unchanged during the transition stage. Therefore, the MEAN of FMF remained basically unchanged in the phase space. To analyze the reason for the SD being halved from its initial value during the transition stage in Cases 1–4, the FMF cloud map on the A-A section at the D (CA640) moment was taken, as shown in Figure 7. It is obvious that the gas fuel jet ended at this moment. It can be found that the gas fuel impinging jets still existed because of an inertia effect. The gas fuel jet was directed to the other side of the cylinder. The gas fuel already crossed the centerline of the cylinder. The velocity vector map on the A-A section at the D (CA640) moment was taken, as shown in Figure 8. The gas fuel jet developed towards the left side of the cylinder and was affected by the tumble flow. The velocity in the left cylinder also developed. In the cylinder, the degree of mixing between gas fuel and air intensified, indicating that these mixtures were in a highly transient state. The above phenomenon led to halving of the SD in the phase space of Cases 1–4.

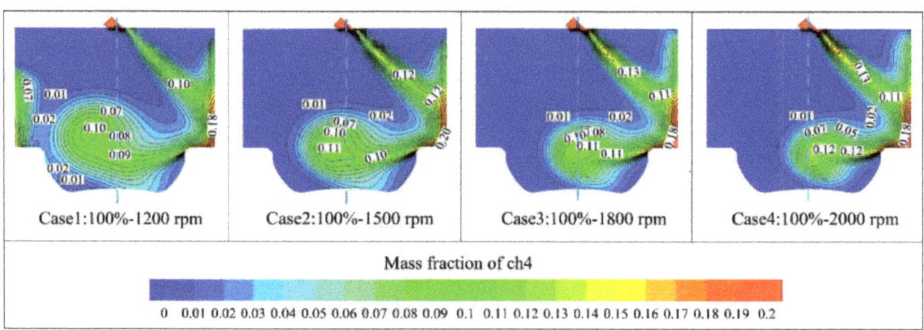

Figure 7. The in-cylinder methane mass fraction distribution at CA640.

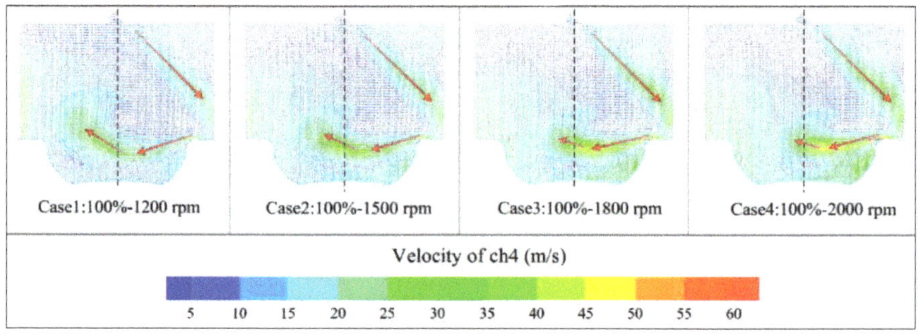

Figure 8. The in-cylinder velocity vector at CA640.

During the diffusion stage (D-F), the SD in the phase space continuously decreased with the duration of mixing, as shown in Figure 5. After the E (Cases 1 and 2: CA660, Cases 3 and 4: CA680) moment, the MEAN showed different trends, with Cases 1 and 2 increasing while Cases 3 and 4 continued to decrease. The diffusion stage corresponded to the later stage of the gas fuel mixing process. The gas fuel continuously occupied the cylinder volume and continuously mixed with the air. Therefore, the SD continuously decreased with the duration of mixing. To analyze the reasons for the different trends in MEAN after the E (defined as the characteristic inflection point) moment, the FMF cloud map on the A-A section at time E was taken, as shown in Figure 9. It can be found that the gas fuel was generally concentrated on the other side of the gas fuel jet impingement, and there was a gas-rich phenomenon on the left side of the cylinder. In the H region, the gas fuel impinged on the left cylinder wall, forming a phenomenon of pushing back towards the in-cylinder centerline. The mixing process in the cylinder experienced a brief buffering period, which may be the reason for the appearance of the characteristic inflection point. The characteristic inflection points of Cases 1 and 2 appeared earlier than those of Cases 3 and 4, which caused the MEAN of Cases 1 and 2 to increase after the E moment, while the MEAN of Cases 3 and 4 decreased.

4.2. The Effect of Engine Load

Secondly, regarding the influence of load on the mixing process, a comparison was made between Cases 4 and 7, as shown in Figure 10. The mixing trend of gas fuel with the duration of mixing in the phase space was very similar, showing a reverse "C" shape. The mixing process of gas fuel was also divided into three stages: injection, transition and diffusion.

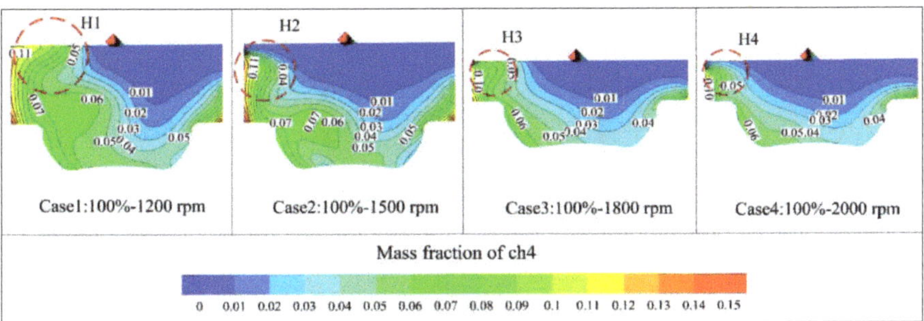

Figure 9. The in-cylinder methane mass fraction distribution at E moment.

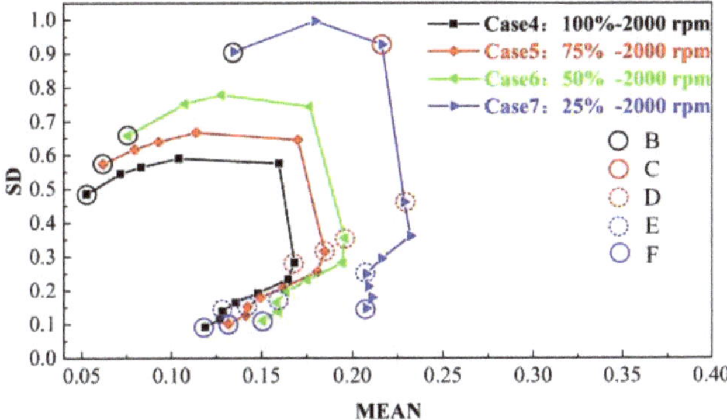

Figure 10. The mixing process of the mixture at different loads under 2000 rpm.

At the same time, the SD of Cases 4–7 gradually increased in the phase space, as shown in Figure 10. After the D (CA640) moment, the changes in the phase space began to decrease significantly, and the changes in MEAN and SD slowed down. After the D moment, the valve of the gas fuel injection device closed, and the mixing of gas fuel and air was no longer affected by the valve injection. Therefore, the changes in MEAN and SD in the phase space slowed down, and the mixture of fuel and air in the cylinder entered a stable development stage. To analyze the influence of load on the mixing process, the in-cylinder velocity–vortex cloud map at CA640 was taken, as shown in Figure 11. The vortex with a Q-criterion value of 2×10^5 was taken. Larger vortices make the structure and properties of the vortices more intuitively observable. It can be seen that the inertial jet was pushed towards the bowl piston top due to the lack of upstream injection pressure during the upward movement of the piston. Meanwhile, there were vortices in the bowl piston top, and the boundaries of the vortices crossed the center of the cylinder. The velocity difference near the boundary of the vortex was large in Cases 4–7. Case 4 had the highest load, and the highest point of the jet velocity was farthest from the vortex boundary, located at the center of the vortex. As the load decreased in Cases 5–7, the highest point of jet velocity gradually approached the vortex boundary. According to the energy dissipation rule of vortices, the jet velocity was proportional to the vorticity. The higher the velocity of the vortex boundary jet, the more difficult the energy dissipation of the vortex is, resulting in greater variation in the distribution of the gas fuel. Therefore, this phenomenon caused the SD in the same moment of Cases 4–7 to gradually increase in the phase space.

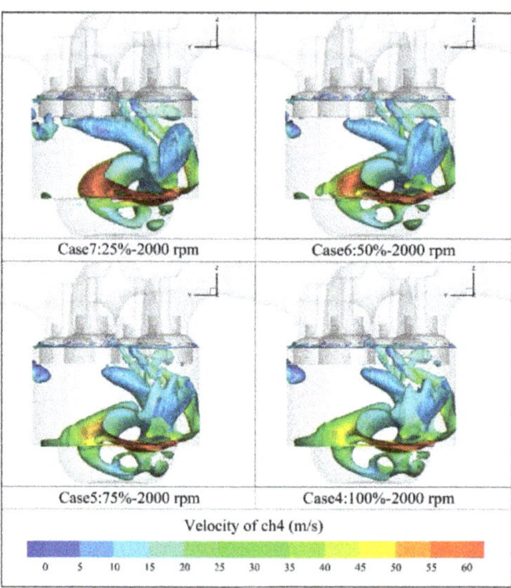

Figure 11. The in-cylinder velocity–vortex distribution at CA640.

4.3. The Effect of Engine Multiple Working Conditions

Finally, the influence of different working conditions on the mixing performance in the cylinder at the end of the compression stroke (CA700) was analyzed, and the corresponding laws were discussed. There were two different results in the phase space, as shown in Figure 12. In Cases 1–4, under the condition of 100% load, the lower the speed, the larger the MEAN of the FMF, and the smaller the SD. In Cases 4–7, at the rated speed of 2000 rpm, the smaller the load, the larger the MEAN of the FMF, and the larger the SD. For Cases 8–11, at a non-rated speed of 1500 rpm, the influence of load on the mixing performance was the same as that at the rated speed. It can be preliminarily concluded that, when the engine load was the same, the lower the speed, the better the in-cylinder homogeneous mixing performance, when the engine speed was the same, the lower the load, the better the in-cylinder stratification trend.

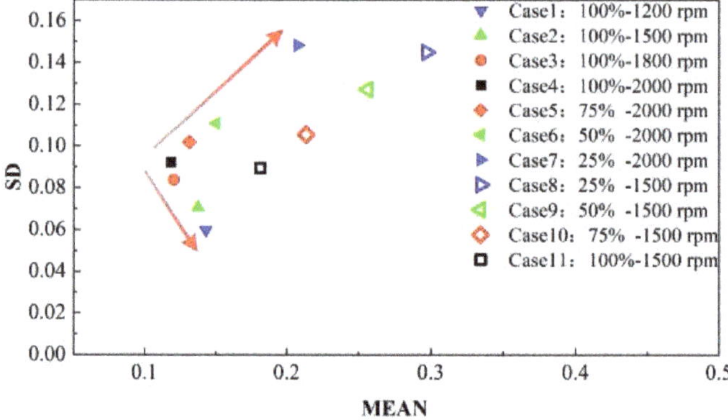

Figure 12. Mixing performance in the cylinder at CA700.

The influence of engine speed on the in-cylinder mixing performance at the end of the compression stroke was analyzed. The FMF distribution at the CA700 moment is shown in

Figure 13, where a thin mixture (FMF < 2%) was hidden for the purpose of a clear contrast. Obviously, in Cases 1–4, the gas fuel was concentrated near the opposite cylinder wall of the direction of injection. In Case 1, the gas fuel was distributed almost throughout the cylinder, followed by Case 2, in which the gas fuel was concentrated near the bowl piston top and the cylinder wall in Cases 3 and 4. Case 1 had the best distribution of gas fuel, while the distribution of gas fuel gradually became worse in Cases 2–4. These results are similar to the research of Zhang et al. [22]. The gas fuel mainly concentrated at the piston top or cylinder wall, which was unfavorable for the complete combustion of the gas mixture. It can be concluded that, when the load is the same, the lower the engine speed, the better the in-cylinder homogeneous mixing performance. The FMF distribution at the CA700 moment confirms the accuracy of the influence law of engine speed on the mixing performance in the phase space.

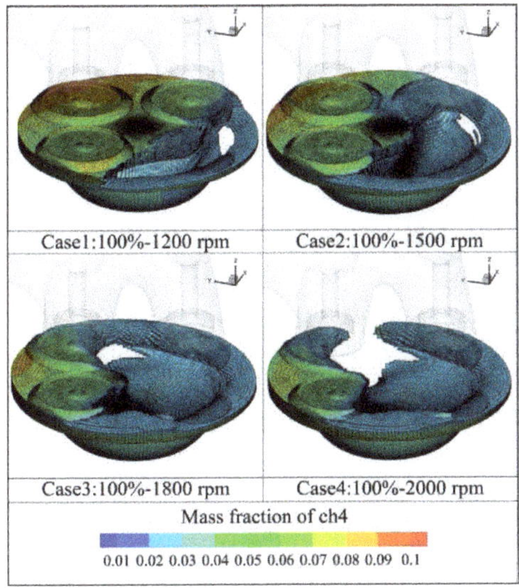

Figure 13. The in-cylinder FMF distributions of Cases 1–4 at CA700.

The influence of engine load on the in-cylinder mixing performance at the end of the compression stroke was analyzed. The probability distribution frequency (PDF) of the FMF at the CA700 moment is shown in Figure 14. It was calculated that the best mixture concentration region (BMCR) of the gas fuel in Cases 4–7 was consistent, and it was between 2.5% and 3%. Correspondingly, 2–2.5% was taken as the thinner interval, 3–3.5% was taken as the thicker interval, 0–2% was taken as the thinnest interval, and the FMF exceeding 3.5% was taken as the thickest interval. In order to achieve better stratified mixing performance, the proportion of the BMCR should be lower than that of other cases. Based on this, the proportion of the thicker interval should be reduced as much as possible, and the proportion of the thinner interval should be increased as much as possible. From Figure 13, the BMCR decreased as the load decreased, from 12.72 to 12.29. The thicker interval decreased from 9.64 to 8.61, while the thinner interval increased from 13.13 to 14.83. The PDF of FMF in Cases 4–7 was consistent with the characteristics of the separated mixture. The optimal concentration range of the separated mixture was low compared to the homogeneous mixture [9]. It can be concluded that at the same speed, the lower the load, the better the in-cylinder stratified mixing performance. The PDF of the FMF at the CA700 moment confirms the accuracy of the influence law of engine load on the mixing performance in the phase space.

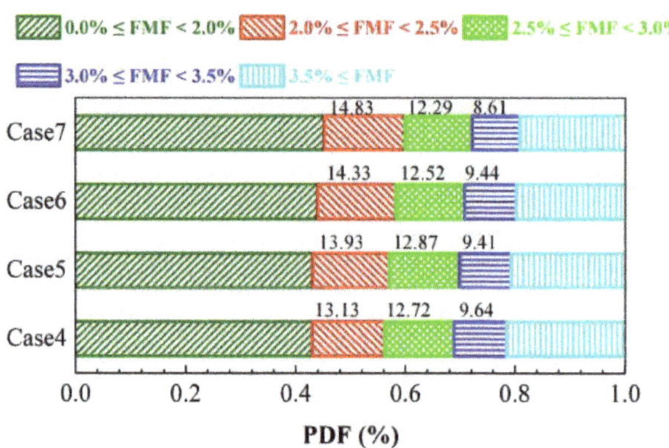

Figure 14. The PDF of gas fuel mass fraction under Cases 1–4 at CA700.

5. Conclusions

In this research, a CFD model for the DI mixing process of CNG engines was established. A quantitative evaluation mechanism for the in-cylinder mixing performance of CNG engines was proposed. The influence law of different engine speeds and loads on the in-cylinder mixing process and its mixing performance was researched. The main conclusions are as follows:

(1) Under the influence of engine speed, the characteristics of the gas fuel mixing process can be divided into three stages in the phase space. The gas fuel mixture rapidly occupies the cylinder volume in injection stage. During the transition stage, the gas fuel mixture is in a highly transient state. The diffusion stage is characterized by the continuous homogenization of the mixture.

(2) In the phase space, the diffusion stage of the mixing stage shows a characteristic inflection point after the E moment. The gas fuel impinges on the left cylinder wall, forming a phenomenon of pushing back towards the in-cylinder centerline. This may be the reason for the emergence of characteristic inflection points.

(3) The in-cylinder mixing process is influenced by the load factors too, and the mixing trend of the gas fuel in the phase space is the same. As the load decreases, the velocity of the gas fuel jet near the boundary of the vortex increases, making it more difficult for the energy of the vortex to dissipate. This leads to a greater variation in the gas fuel distribution in the cylinder.

(4) The in-cylinder mixing performance at the end of compression stroke under different working conditions can be reflected in the phase space. As the engine load decreases, the MEAN increases, while the SD also increases, and the gas fuel mixture approaches the stratified mixture. As the engine speed decreases, the MEAN increases, while the SD decreases, and the gas fuel mixture approaches the homogeneous mixture.

Author Contributions: Methodology and software, H.W. and T.W.; verification, L.Z. and Y.Z.; resources, L.Z. and J.C.; writing original draft preparation, H.W., L.L. and Y.S. All authors have read and agreed to the published version of the manuscript.

Funding: This work was supported by the National Natural Science Foundation of China (Grant No. 52105260 and No. 11802108); Changzhou Sci & Tech Program (Grant No. CE20225049); Natural Science Research Project of Higher Education Institutions in Jiangsu Province [Grant No. 21KJB460008, 22KJA580002]; Qinglan Engineering Project of Jiangsu Universities. The APC was funded by the National Natural Science Foundation of China, Changzhou Sci & Tech Program and Natural Science Research Project of Higher Education Institutions in Jiangsu Province (Grant No. 21KJB460008).

Data Availability Statement: Not applicable.

Conflicts of Interest: The authors declare no conflict of interest.

Nomenclature

CNG	compressed natural gas
DI	direct injection
PFI	port fuel injection
CFD	computational fluid dynamics
DOI	duration of injection
SOI	start of injection
EOI	end of injection
IVO	intake valve opening time
IVC	intake valve closing time
BTDC	before top dead center
ATDC	after top dead center
CA	crank angle
FMF	gas fuel mass fraction
PDF	probability distribution frequency
BMCR	best mixture concentration region

References

1. IEA. Gas. Available online: https://www.iea.org/fuels-and-technologies/gas (accessed on 14 March 2023).
2. Gideon, F.; Tarun, S.; Fionn, R. Exploring a case transition to low carbon fuel: Scenarios for natural gas vehicles in Irish road freight. *SSRN Electron. J.* 2021. Available online: https://papers.ssrn.com/sol3/papers.cfm?abstract_id=3855537 (accessed on 22 May 2023).
3. Abdullah, N.N.; Anwar, G. An Empirical Analysis of Natural Gas as an Alternative Fuel for Internal Transportation. *Int. J. Engl. Lit. Soc. Sci.* **2021**, *6*, 479–485. [CrossRef]
4. Tadeusz, D.; Przemysław, M.; Karolina, Z. Costs and benefits of using buses fuelled by natural gas in public transport. *J. Clean. Prod.* **2019**, *225*, 1134–1146.
5. Mahendar, S.K.; Erlandsson, A.; Adlercreutz, L. *Challenges for Spark Ignition Engines in Heavy Duty Application: A Review*; SAE Technical Paper 2018-01-0907; SAE International: Warrendale, PA, USA, 2018.
6. Abdullah, S.; Mahmood, W.; Aljamali, S.; Shamsudeen, S.A.A.A. Compressed Natural Gas Direct Injection: Comparison Between Homogeneous and Stratified Combustion. In *Advances in Natural Gas Emerging Technologies*; InTech Open: London, UK, 2017; pp. 153–169.
7. Kakaee, A.-H.; Nasiri-Toosi, A.; Partovi, B.; Paykani, A. Effects of piston bowl geometry on combustion and emissions characteristics of a natural gas/diesel RCCI engine. *Appl. Therm. Eng.* **2016**, *102*, 1462–1472. [CrossRef]
8. Chiodi, M.; Berner, H.-J.; Bargende, M. *Investigation on Different Injection Strategies in a Direct-Injected Turbocharged CNG-Engine*; SAE Technical Papers 2006-01-3000; SAE International: Warrendale, PA, USA, 2006.
9. Spiegel, L.; Spicher, U. Mixture Formation and Combustion in a Spark Ignition Engine with Direct Fuel Injection. In *International Congress and Exposition*; SAE International: Warrendale, PA, USA, 1992; pp. 24–28.
10. Yadollahi, B.; Boroomand, M. A Numerical Investigation of Combustion and Mixture Formation in a Compressed Natural Gas DISI Engine with Centrally Mounted Single-Hole Injector. *J. Fluids Eng.* **2013**, *135*, 091101. [CrossRef]
11. Sankesh, D.; Lappas, P. *Natural-Gas Direct-Injection for Spark-Ignition Engines—A Review on Late-Injection Studies*; SAE Technical Paper 2017-26-0067; SAE International: Warrendale, PA, USA, 2017.
12. Sankesh, D.; Edsell, J.; Mazlan, S.; Lappas, P. Comparative Study Between Early and Late Injection in a Natural-Gas Fueled Spark-Ignited Direct-Injection Engine. *Energy Procedia* **2017**, *110*, 275–280. [CrossRef]
13. Wang, T.; Zhang, X.; Zhang, J.; Hou, X. Numerical analysis of the influence of the fuel injection timing and ignition position in a direct-injection natural gas engine. *Energy Convers. Manag.* **2017**, *149*, 748–759. [CrossRef]
14. Yousefi, A.; Guo, H.; Birouk, M. Effect of diesel injection timing on the combustion of natural gas/diesel dual-fuel engine at low-high load and low-high speed conditions. *Fuel* **2018**, *235*, 838–846. [CrossRef]
15. Keskinen, K.; Kaario, O.; Nuutinen, M.; Vuorinen, V.; Künsch, Z.; Liavåg, L.O.; Larmi, M. Mixture formation in a direct injection gas engine: Numerical study on nozzle type, injection pressure and injection timing effects. *Energy* **2016**, *94*, 542–556. [CrossRef]
16. Addepalli, S.K.; Mallikarjuna, J.M. Quantitative Parametrization of Mixture Distribution in GDI Engines: A CFD Analysis. *Arch. Comput. Methods Eng.* **2018**, *26*, 639–662. [CrossRef]
17. Mohammad, A.; Bashar, Q. Evaluating the in-cylinder gas mixture homogeneity in natural gas HCCI free piston engine under different engine parameters using 3D-CFD analysis. *Energy Sources* **2018**, *40*, 1097–1113.
18. Baratta, M.; Misul, D.; Xu, J.; Fuerhapter, A.; Heindl, R.; Peletto, C.; Preuhs, J.; Salemi, P. Development of a High Performance Natural Gas Engine with Direct Gas Injection and Variable Valve Actuation. *SAE Int. J. Engines* **2017**, *10*, 2535–2551. [CrossRef]

19. Wang, T.; Chang, S. Effect of Injection Location and Multi-Hole Nozzle on Mixing Performance in a CNG-Fueled Engine with Port Gas Injection, Using CFD Analyses. *MATEC Web Conf. EDP Sci.* **2017**, *95*, 06003. [CrossRef]
20. Wang, T.; Zhang, L.; Li, L.; Wu, J.; Wang, H. Numerical Comparative Study on the In-Cylinder Mixing Performance of Port Fuel Injection and Direct Injection Gas-Fueled Engine. *Energies* **2022**, *15*, 5223. [CrossRef]
21. Wang, T.; Wang, H.; Zhang, L.; Zheng, Y.; Li, L.; Chen, J.; Gong, W. A Numerical Study on the Transient Injection Characteristics of Gas Fuel Injection Devices for Direct-Injection Engines. *Actuators* **2023**, *12*, 102. [CrossRef]
22. Zhang, X.; Wang, T.; Zhang, J. Numerical analysis of flow, mixture formation and combustion in a direct injection natural gas engine. *Fuel* **2019**, *259*, 116268. [CrossRef]

Article

Magnetic Poles Position Detection of Permanent Magnet Linear Synchronous Motor Using Four Linear Hall Effect Sensors

Bin Zhou and Cong Huang *

School of Mechanical and Electronic Engineering, Wuhan University of Technology, Wuhan 430070, China; zhb_8250@163.com
* Correspondence: 280905@whut.edu.cn

Abstract: Magnetic pole position detection is the core of the closed-loop control system of the permanent magnet linear synchronous motor (PMLSM), and its position estimation accuracy directly affects control performance and dynamic response speed. In order to solve the problem of the increased estimation error of magnetic pole position caused by magnetic field distortion at the end of PMLSM while also considering the cost of control hardware, the paper uses four linear Hall sensors as magnetic pole position detection components and adopts an optimized estimation algorithm to improve the dynamic performance of the motor. Firstly, a numerical simulation of the magnetic field of poles was conducted using Ansoft Maxwell software, and combined with theoretical analysis, the optimal installation position range of four linear Hall orthogonal placements relative to the motor was obtained. Meanwhile, based on the existing vector tracking position observer, an improved observer detection model is proposed. The Matlab/Simulink software was used to compare the Hall-based detection model with the Hall-based improved observer detection model, verifying the feasibility of the improved detection algorithm. Finally, the rationality of the spatial layout design of linear Hall and the feasibility of improving the estimation algorithm were verified through experiments.

Keywords: magnetic poles position detection; permanent magnet linear synchronous motor; linear Hall effect sensors; estimation algorithm

Citation: Zhou, B.; Huang, C. Magnetic Poles Position Detection of Permanent Magnet Linear Synchronous Motor Using Four Linear Hall Effect Sensors. *Actuators* 2023, 12, 269. https://doi.org/10.3390/act12070269

Academic Editors: Qinfen Lu, Xinyu Fan, Cao Tan and Jiayu Lu

Received: 1 June 2023
Revised: 20 June 2023
Accepted: 28 June 2023
Published: 30 June 2023

1. Introduction

The PMLSM has the characteristics of small structural volume, large thrust, and high efficiency. Compared to the linear feed motion provided by the ball screw in traditional machinery, the linear motor can be directly connected to the linear motion components rigidly, which can effectively improve friction, noise, and other problems in mechanical transmission. Therefore, the linear motor control system has better motion performance [1].

Magnetic pole position detection is one of the key links in linear motor control systems, and the accuracy of position estimation is crucial for the control accuracy of the motor. There are two main types of magnetic pole position detection: position sensor detection and sensorless detection. At present, the main detection methods for position sensors include grating, magnetic grating, and Hall sensors [2–4]. Although magnetic gratings and gratings have high detection accuracy, their high cost, large volume, and high requirements for the working environment severely limit the range of motor use. Low-resolution switch-type and lock-type Hall sensors, despite their low cost, have limited position estimation accuracy and are usually used in square wave-controlled permanent magnet brushless motors with low requirements for position detection accuracy [5,6].

Hall sensors can be divided into two types based on whether the output electrical signal has continuous amplitude changes over a continuous time range: switch-type Hall sensors and linear Hall sensors. A switch-type Hall sensor controls its output tube to turn off and on based on changes in the magnetic field. An ideal switch-type Hall sensor should respond quickly when triggered. However, due to the presence of non-ideal factors, the switching action of the output tube of the switch-type Hall sensor will have a significant delay, resulting in the output high and low levels not being able to flip in time. The switch-type Hall sensor is not suitable for working in situations that require high accuracy. The Hall voltage output by the linear Hall sensor is proportional to the magnetic flux density, which converts the weak magnetic signal into analog output. In the peripheral signal processing circuit of the linear Hall sensor, the corresponding compensation structure is designed to make the linear Hall sensor have high detection accuracy and can detect a high magnetic field range.

In recent years, a new type of sensor (Hall sensor) has rapidly developed in the field of information collection, which can be used as a position sensor instead of a photoelectric encoder in the control process of permanent magnet motors to detect rotor position information. Due to its advantages of small size, light weight, and low price, it is widely used in fields such as automatic monitoring, automatic control, and information detection. It plays an increasingly important role in daily life [7]. Compared with the switch Hall sensor, the linear Hall sensor can output a Hall voltage that is proportional to the magnetic flux density, so it is often used in displacement measurement, throttle detection, and other applications that need to respond quickly to changes in magnetic flux density and detect the displacement of objects and liquid surplus, according to the size of the magnetic flux density at different positions [8]. For example, in the control process of permanent magnet motors, linear Hall position sensors are used to accurately detect rotor position in order to reduce costs and improve reliability. However, due to the strong offset voltage that accompanies the output of the Hall voltage caused by the induction of magnetic field changes, the use of linear Hall sensors to achieve precise displacement and position measurement can seriously affect the output accuracy of the linear Hall sensor, resulting in a deviation from the actual measurement [9]. Therefore, in order to improve the accuracy of the linear Hall sensor in detecting magnetic flux density changes, improve the ability of the linear Hall sensor to convert magnetism into electricity, and reduce the interference of offset voltage and low-frequency noise in the chip on the output accuracy, it is the most important to design the linear Hall sensor.

Linear Hall sensors can output continuous signals, and when there are more than two sensors, they can uniquely and accurately determine the position information of each magnetic pole [10]. The reference [11] used 13 linear Halls for the position detection of permanent magnet linear synchronous motors and solved the phase deviation and harmonic problem caused by the increase in the number of Halls through the fast Fourier transform and fixed-point iteration methods. The reference [12] proposes an improved rotor position estimation method in a permanent magnet synchronous motor (PMSM) with low-resolution Hall effect sensors. This method promises to decrease the estimated position errors, which are caused by the Hall position offset. A linear interpolation based on the least squares method is used to estimate the rotor position.

At present, the position sensorless detection technologies include the sliding mode control method [13], back electromotive force calculation method [14], state observer method [15], Kalman and extended Kalman filter method [16], etc. However, these algorithms have a high dependence on motor parameters, poor control robustness, and poor universality, and most of them are complex, making them less suitable for engineering applications.

However, there is still a trend of increasing in the same direction between the measurement accuracy and cost of position sensor detection. Different types of sensors have effects on the structural volume, application scenarios, and control performance of motor control systems. When using low-cost and low-precision position sensors, it is also necessary to consider the impact of installation errors on measurement accuracy. At the same time, it is necessary to select a reasonable position estimation algorithm and optimize the magnetic pole position detection algorithm based on the motor operating conditions to improve the accuracy of motor position detection and thereby improve motor control performance.

This paper comprehensively considers factors such as control cost, position detection accuracy, and engineering applications. Linear Hall sensors are used as magnetic pole position detection components, and the focus is on researching the magnetic pole position detection method of PMLSM based on linear Hall. The installation layout of linear Hall relative to linear motors is analyzed, and an improved algorithm for position estimation is proposed to reduce the volume of the overall drive system of the motor and reduce the cost of control hardware, balancing the control accuracy and stability of the motor.

2. Detection Method of Magnetic Poles Position

When the Hall sensors are used for measuring the magnetic pole position of rotating motors, they are usually installed at the teeth or slots at the axial end of the stator to detect changes in the permanent magnet magnetic field on the mover or auxiliary extended mover. By demodulating the magnetic field value, the motor mover position and speed information are obtained. Due to the unique structure of permanent magnet linear synchronous motors, there are more ways to install Hall sensors. The more convenient method is to install the Hall sensor above the permanent magnet of the mover, which has a certain constraint relationship with the motor stator to ensure the accuracy of later control, simultaneously avoiding the impact of the magnetic field generated by the stator on the Hall detection permanent magnet. The installation position diagram of the linear Hall sensor is shown in Figure 1.

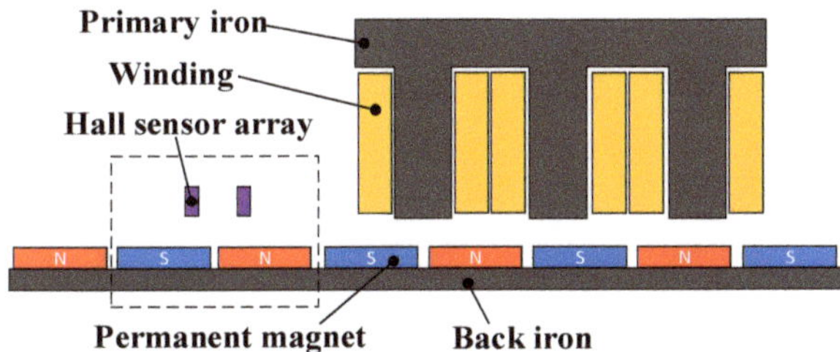

Figure 1. Schematic diagram of the installation position of linear Hall sensor.

The performance parameters of linear Hall sensors themselves have a significant impact on the accuracy of magnetic pole position detection signals. This paper selects the MLX90242 series linear Hall sensor produced by Melexis company. Figure 2 is the schematic diagram of the exterior of the Hall sensor. The sensor has three interfaces, namely the power supply input terminal, grounding terminal, and output voltage terminal. The prescription-shaped block in the center of the sensor is the effective monitoring area (Hall sensing chip). Figure 3 shows the schematic diagram of the detection output waveform of the corresponding magnetic pole of the linear Hall sensor during normal operation. When the surface of the Hall sensor faces the S pole of the permanent magnet, the output voltage shows an upward trend, while the output voltage shows a downward trend at the N pole. The position of the Hall sensing chip relative to the overall component has clear dimensional parameters, which should be taken into account during actual installation to achieve the most accurate installation accuracy possible. This Hall sensor has a small overall size, high sensitivity, small temperature drift, wide allowable temperature range, and a wide detectable magnetic field range. As shown in Figure 4, it contains an error correction circuit inside, which can roughly eliminate the analog compensation error caused by Hall effect devices. The partial magnetic specifications of the sensor are shown in Table 1.

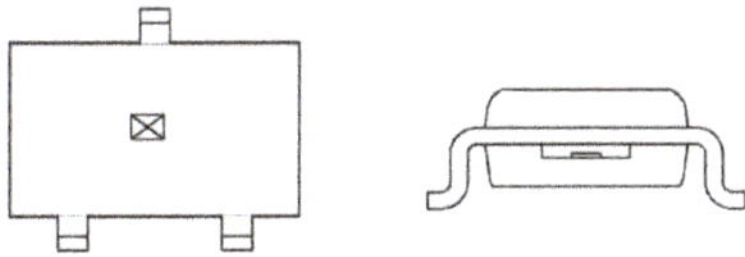

Figure 2. Outline diagram of MLX90242 series sensor.

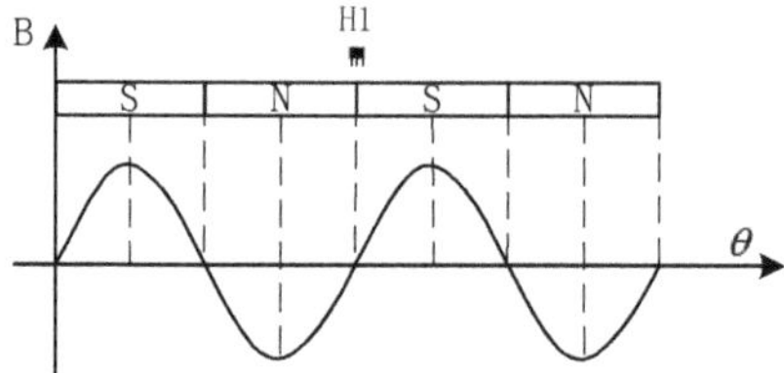

Figure 3. Detection waveform of MLX90242 series sensor.

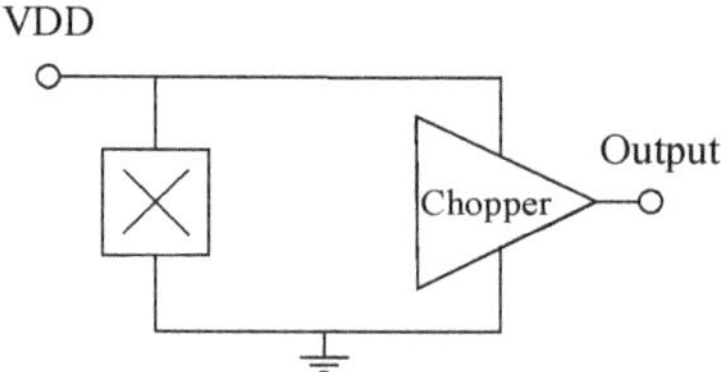

Figure 4. Internal electrical schematic diagram.

Table 1. Magnetic parameters of linear Hall sensors.

	Voltage Temperature Drift (mV)	Sensitivity (mV/mT)
Minimum		33.2
Typical	±25	39
Maximum		44.9

We placed two linear Hall sensors orthogonal, and their detection waveforms are shown in Figure 5. The two sine and cosine waveforms represent the output signals of two linear Hall sensors with a phase difference of 90°. The position angle can be obtained using the arctangent function, and the expression is as follows:

$$\begin{cases} u_\alpha = k\sin\theta \\ u_\beta = k\cos\theta \end{cases},$$ (1)

$$\begin{cases} \theta = arctan(u_\alpha/u_\beta) & u_\alpha \geq 0, u_\beta \geq 0 \\ \theta = arctan(u_\alpha/u_\beta) + \pi & u_\beta < 0 \\ \theta = arctan(u_\alpha/u_\beta) + 2\pi & u_\alpha < 0, u_\beta \geq 0, \end{cases}$$ (2)

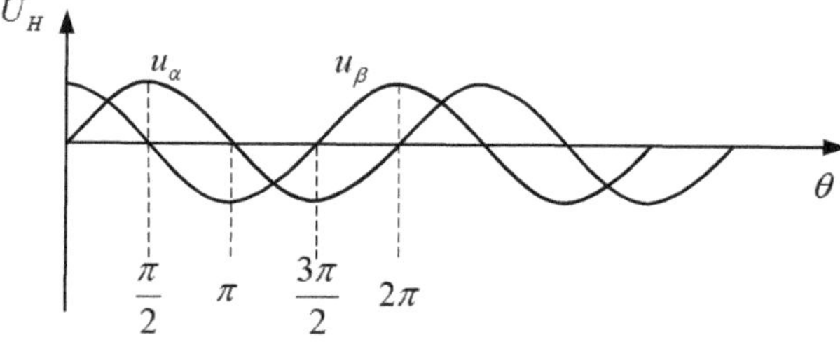

Figure 5. Linear Hall sensor output waveform.

By using four linear Hall sensors and ensuring a phase difference of 90° between adjacent Hall sensors, the signal can be amplified using two-signal differential processing, and the problem of zero offset caused by inaccurate sensor installation position can be effectively avoided.

The calculation formula for the output voltage of a linear Hall sensor is:

$$U_H = K_H I B \cos\alpha, \tag{3}$$

where K_H is the sensitivity coefficient, I is the input current of the Hall element, B is the magnetic flux density, and α is the electrical angle between the horizontal position of the sensor and the peak of the vertical magnetic field. When the selected sensor operates at 25 °C, $K_H I = 39$ (mV/mT).

2.1. Installation Position of Linear Hall Relative Motor in Horizontal Direction

The installation position of Hall is mainly related to the position of the magnetic electromotive force axis of the three-phase winding [17]. A PMLSM with 15 slots and 20 poles was selected as the research object. This type of motor is composed of five unit motors ($Z_0/P_0 = 3/2$, Z_0 is the number of slots in the unit motor and P_0 is the number of pole pairs in the unit motor), with a motor pole distance of 12 mm, slot width of 10 mm, and tooth width of 6 mm.

Considering the placement of the external linear Hall sensor of the motor based on the condition that the d axis coincides with the magnetic electromotive force FA axis, the installation layout strategy of the two orthogonal linear Hall sensors relative to the motor in the horizontal direction is shown in Figure 6.

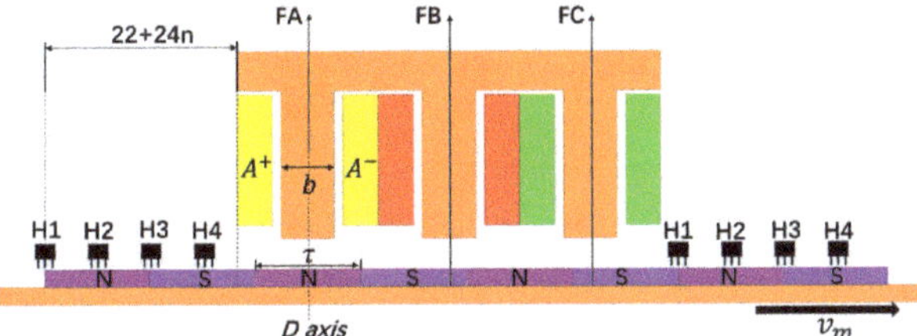

Figure 6. Installation diagram of linear Hall in the horizontal direction.

When the mover moves horizontally to the right, if the Hall sensor is installed on the left side of the motor stator, the first linear Hall sensor H1 should be placed at a horizontal distance of 22 + 24n (mm) from the end of the motor stator (where n is a positive integer) in order to ensure that the voltage signal output by the first linear Hall sensor is the same as the waveform of the opposite electromotive force of phase A and to achieve zero crossing control of the opposite electromotive force of phase A using the first linear Hall sensor. Similarly, when the Hall sensor is placed on the right side of the motor stator, the first linear Hall sensor H1 should be placed at a horizontal distance of 2 + 24n (mm) from the end of the motor stator.

2.2. Installation Position of Linear Hall Relative Motor in Vertical Direction

The installation position of the linear Hall in the vertical direction can be determined by Equation (3), which is mainly related to the magnetic field strength at the permanent magnet of the motor. The voltage output of the linear Hall is linearly proportional to the magnetic field strength of the permanent magnet at its installation height. This paper uses Ansoft Maxwell magnetic field simulation software to quantitatively analyze the magnetic field strength of the magnetic pole at vertical height. The permanent magnet adopts neodymium iron boron with brand N35. The size specification of a single permanent magnet is $20 \times 12 \times 2$ mm. It magnetizes according to the thickness direction of the magnet, and the magnetization directions of adjacent magnets are opposite. The back iron is composed of ferrite material with suitable magnetic conductivity, ensuring the complete closure of the entire magnetic circuit. The thickness of the back iron is 2 mm. The 2D model is shown in Figure 7.

Figure 7. Two-dimensional model diagram of PMLSM.

The simulation distribution of the magnetic field lines in Figure 8 was obtained through a software solution. The magnetic field lines in Region I of Figure 8 are distorted due to the proximity to the end of the motor stator, which is affected by the silicon steel sheet. Region II is located in the middle section of the overall stroke, and the distribution of magnetic field lines is relatively consistent, with rules to follow. Region III undergoes significant distortion of the magnetic field lines due to its proximity to the end of the mover.

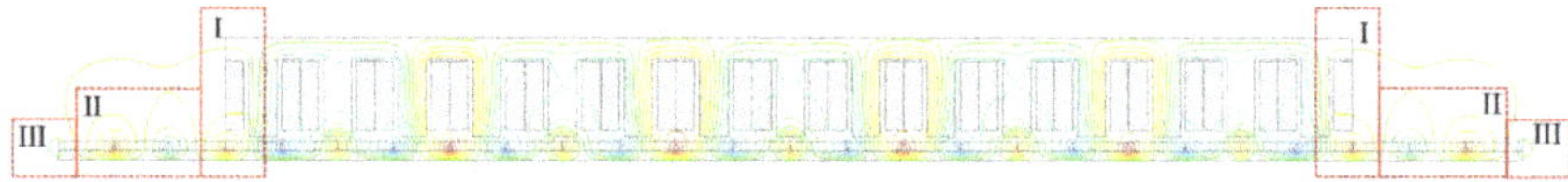

Figure 8. Two-dimensional magnetic field lines of PMLSM.

We took a pair of adjacent magnetic poles in Region II of Figure 8 and set a 23 mm length inspection path on the vertical line to observe the distribution of vertical magnetic flux density on this path. Figure 9 shows the magnetic flux density variation curves and local enlarged images of two permanent magnets with different polarities on their respective perpendicular lines. Considering the hardware circuit configuration, the maximum vertical magnetic flux density that the Hall sensor can detect is 21.15 mT. As shown in the partially enlarged image in Figure 9, the vertical heights of two permanent magnets with different magnetic properties from the surface of the permanent magnet when reaching the specified magnetic flux density are 12.19 mm and 11.97 mm, respectively. The difference between the two is 0.22 mm. Finally, the vertical height installation position of the linear Hall sensor is selected as 12 mm.

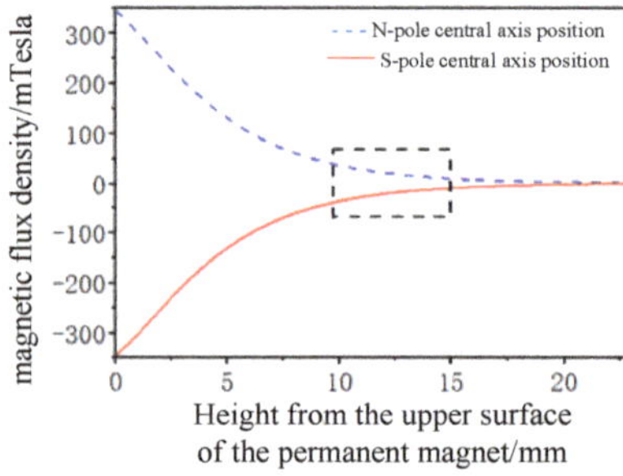
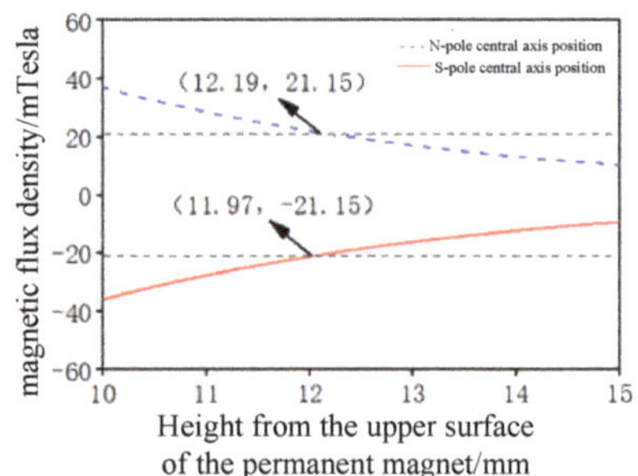

Figure 9. Distribution and local amplification of magnetic flux density at a vertical height of the magnetic pole.

3. Mathematical Model for Estimating Magnetic Pole Position

The position and speed information of the mover are obtained through Hall sensors, and the position detection error is obtained based on the back electromotive force of the motor. The error is promptly compensated to the negative feedback input of the closed-loop control, and real-time correction is carried out on the motor operating speed and magnetic pole position angle, thereby reducing the estimation error of speed and position. Figure 10 is the system structure diagram of the vector tracking position observer.

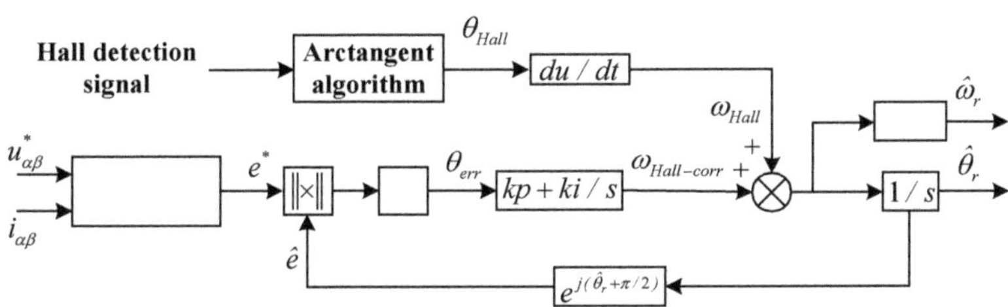

Figure 10. Observer system structure block diagram.

The voltage equation of the PMLSM can be represented by the following formula:

$$\begin{bmatrix} u^*_\alpha \\ u^*_\beta \end{bmatrix} - \begin{bmatrix} R_s + pL_s & 0 \\ 0 & R_s + pL_s \end{bmatrix} \begin{bmatrix} i_\alpha \\ i_\beta \end{bmatrix} = \hat{\omega}_r \psi_f \begin{bmatrix} -sin\hat{\theta}_r \\ cos\hat{\theta}_r \end{bmatrix}, \tag{4}$$

where u^*_α, u^*_β, i_α, i_β, $\hat{E}_\alpha$, and $\hat{E}_\beta$ are the observed values of stator voltage, current, and back electromotive force in the orthogonal $\alpha - \beta$ stationary coordinate system, respectively.

The angular velocity of the motor magnetic pole output by the observer is:

$$\hat{\omega}_r = \omega_{Hall} + \omega_{Hall\text{-}corr}, \tag{5}$$

where ω_{Hall} is the motor speed obtained by solving the linear Hall signal through arctangent operation, which serves as the feedforward input of the observer, and $\omega_{Hall\text{-}corr}$ is the speed error correction value obtained by using the back electromotive force.

In Figure 10, $\hat{e}$ and e^* are the back electromotive force values that are subjected to unit processing. The phase angles of the two are the same, but there may be some errors in the estimation results in actual situations. Therefore, the position estimation error can be obtained by detecting $\hat{e}$ and e^*:

$$\|e^* \times \hat{e}\| = \left\| \begin{bmatrix} -sin\theta^*_r \\ cos\theta^*_r \end{bmatrix} \times \begin{bmatrix} -sin\hat{\theta}_r \\ cos\hat{\theta}_r \end{bmatrix} \right\| = sin\left(\hat{\theta}_r - \theta^*_r\right), \tag{6}$$

Generally, the difference between the two phase angles is very small, which can be further simplified as the magnetic pole position error:

$$sin(\hat{\theta}_r - \theta^*_r) = \hat{\theta}_r - \theta^*_r, \tag{7}$$

Placing the above observer into the entire motor vector closed-loop control system results in a control block diagram, as shown in Figure 11.

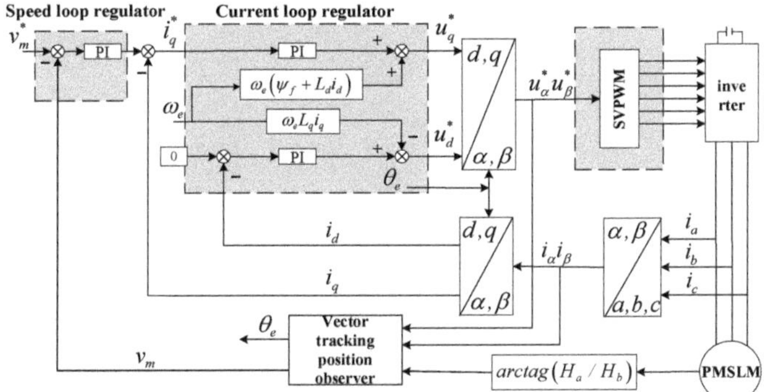

Figure 11. Control block diagram of position observer based on the linear Hall.

4. Simulation and Comparison of Magnetic Pole Position Estimation

We built a corresponding simulation model in Matlab/Simulink using the linear Hall-based PMLSM vector control system discussed above. A displacement length of 222 mm was selected as the operating stroke of the motor and reached the end of the motor at 222 mm to ensure full stroke operation.

Figure 12 shows the estimation error of the magnetic pole position and angle calculated using the linear Hall detection waveform. Figure 13 shows the estimation error diagram of the motor pole position and angle obtained from the simulation model constructed by the observer detection model. Comparing the two, it can be concluded that the actual and estimated positions of the magnetic poles have suitable consistency in the first half of motor operation. The overall angle estimation error of the observer model is smaller than that of the Hall detection model. During the last two magnetic pole cycles of the stroke, the angle estimation error only fluctuates within a small range, and the maximum angle estimation error does not exceed 0.125 rad, which is only 20% of the maximum angle estimation error of the Hall detection model in the same region.

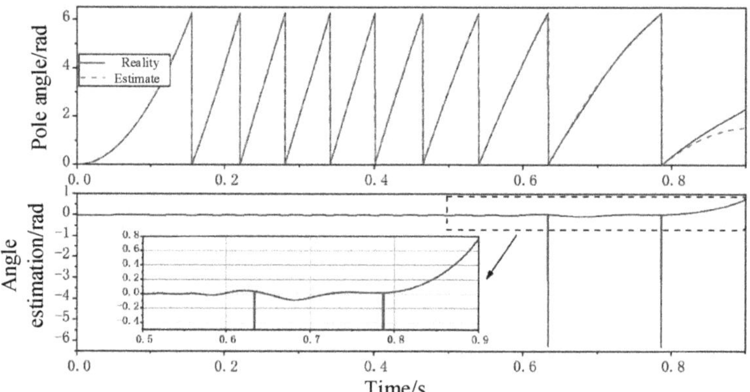

Figure 12. Hall detection model estimation angle and angle estimation error.

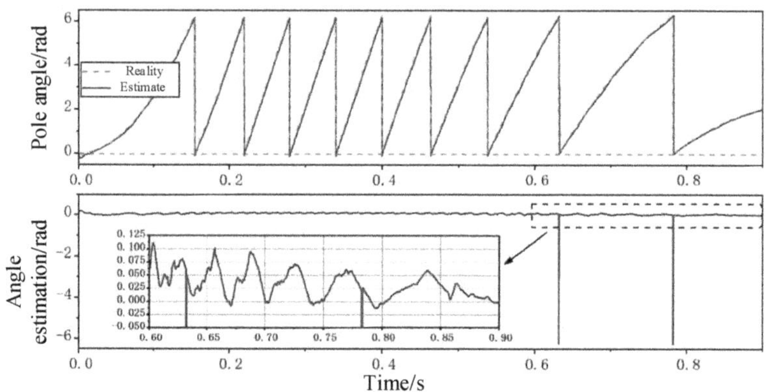

Figure 13. Observer model estimation angle and angle estimation error.

After simulation analysis and comparison of the two detection models, it was found that the observer pole position detection model based on the linear Hall has stronger adaptability to variable speeds and smaller position estimation error.

5. Experimental Verification of Magnetic Pole Position Estimation

This experimental platform uses Microchip's dsPIC33EP series main chip and inverter circuit to achieve motor control using a 15-slot, 20-pole, single-sided flat PMLSM with the same model as the simulation. We designed a three-degrees-of-freedom Hall adjustment device to meet the movement of a linear Hall in three installation directions and assist in using laser displacement sensors to detect the relative movement distance of the Hall installation position. The experimental setup is shown in Figure 14.

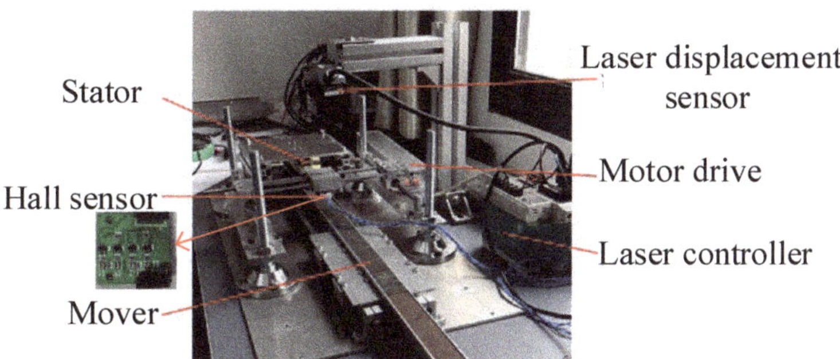

Figure 14. Experimental device platform.

Figure 15 shows the waveform signal measured at a vertical installation height of 12.8 mm in the motor Region II of the Hall circuit board. It can also be seen from the Lissajous figure that the orthogonality of the waveform is suitable. The experimental verification shows that the detection height value is in suitable agreement with the simulation height value. Figures 16–19 show the variation curves of motor speed and displacement in the Hall detection model and Hall-based position observer model observed in the MPLAB IDE simulation software under trapezoidal variable speed mode.

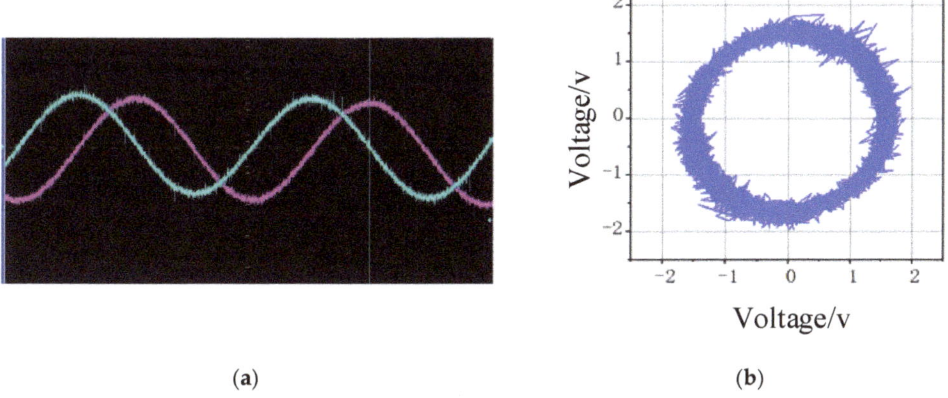

(**a**)   (**b**)

Figure 15. The best installation position of HALL circuit board oscilloscope waveform and Lissajous graph. (**a**) Board oscilloscope waveform; (**b**) Lissajous graph.

From Figures 16 and 17, it can be observed that within the first 200 mm of the motor stroke, the maximum error in speed estimation remains within 0.04 m/s. Due to magnetic field distortion at the end of the stroke, the actual speed of the motor is always too high, and the maximum error in speed estimation is 0.043 m/s. At the end of the stroke, the actual displacement of the motor is slightly greater than the estimated displacement, and the actual displacement of the motor exceeds the theoretical displacement by 3.18 mm. At the end of the stroke, the maximum value of the overall displacement error can reach 5.04 mm.

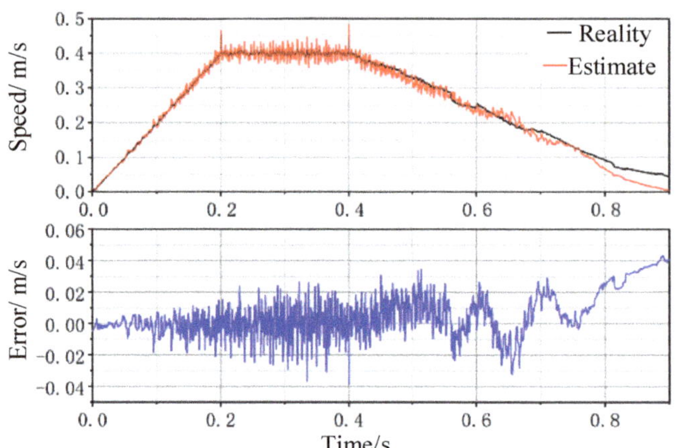

Figure 16. Motor speed and error variation curve of the Hall detection model.

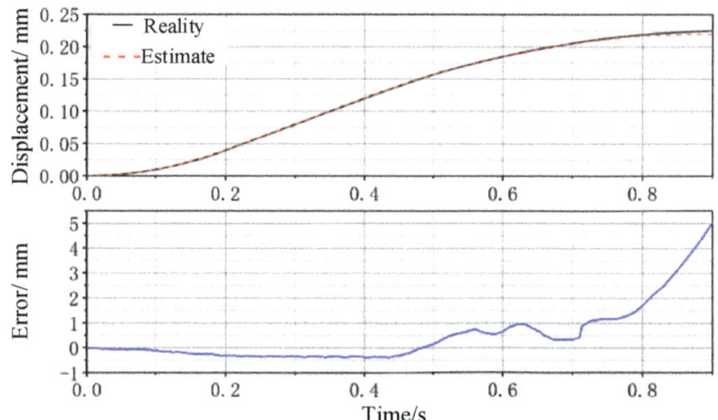

Figure 17. Motor displacement and error variation curve of the Hall detection model.

From Figure 18, it can be observed that the estimated speed of the observer and the actual speed of the motor always follow well throughout the entire motor operation process. Compared to the Hall detection model mentioned above, the speed fluctuation is smaller and smoother. Even at the end of the stroke, the speed estimation error does not exceed 0.02 m/s, and the maximum overall speed estimation error is 0.019 m/s, which is only 42% of the maximum speed estimation error of the Hall detection model mentioned above.

From the motor displacement diagram shown in Figure 19, it can also be seen that the observer estimated displacement follows the actual motor displacement well as a whole. At the end of the stroke, the actual displacement of the motor is slightly greater than the estimated displacement, and the actual displacement of the motor exceeds the theoretical displacement by 1.90 mm. The displacement error reaches a maximum value of 2.33 mm at the end of the stroke, which reduces the maximum displacement estimation error by more than half compared to the Hall detection model mentioned above.

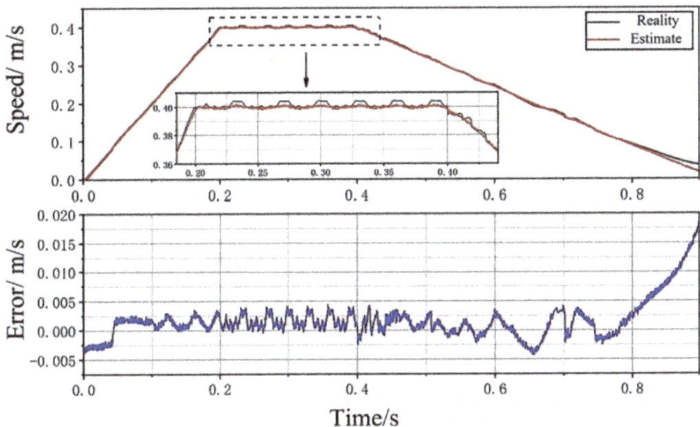

Figure 18. Motor speed and error variation curve based on linear Hall observer detection model.

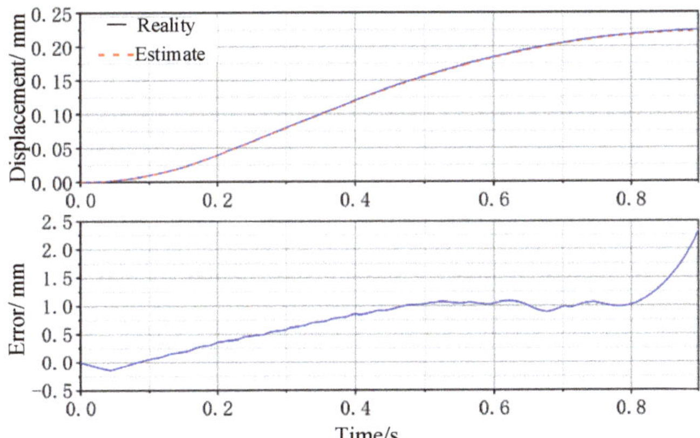

Figure 19. Motor displacement and error variation curve based on linear Hall observer detection model.

Comparing the experimental conclusion with the simulation results, it can also be seen that the results of the two conform to the consistency of the variation pattern. Through experimental comparison of two detection models, it can be found that the observer detection model that compensates and corrects the linear Hall detection signal has superior performance in both motor speed tracking and displacement detection.

6. Conclusions

This paper comprehensively considers factors such as control cost, position detection accuracy, and engineering application. Firstly, the linear Hall sensor and its internal magnetic characteristics are determined. The theoretical installation layout design of a relative 15-slot, 20-pole PMLSM in sine wave vector control is studied. The specific installation position parameters of the linear Hall are determined by building the corresponding motor model in Ansoft Maxwell.

To solve the problem of increased detection position estimation error caused by magnetic field distortion at the end pole of the motor mover during full stroke operation of PMLSM, a vector tracking position observer method based on the linear Hall is proposed to improve the motor pole position estimation ability.

Two types of magnetic pole position detection models were built in Matlab/Simulink and verified through experiments with the same parameters. The estimation performance of the two models for motor magnetic pole position angle, motor speed, and displacement in trapezoidal variable speed mode was discussed. After comparison, it was found that the observer magnetic pole position detection model based on the linear Hall has stronger adaptability to variable speed and smaller position estimation error.

Author Contributions: B.Z.: conception of the study, propose theory and method, supervisor; C.H.: literature search, manuscript preparation, and writing. All authors have read and agreed to the published version of the manuscript.

Funding: This research received no external funding.

Data Availability Statement: Not applicable.

Conflicts of Interest: The authors declare no conflict of interest.

References

1. Zhang, Y.W.; Huang, X.Z.; Ding, X.Z.; Wang, A.; Xu, J. Speed Control Strategy of Low-Cost Modular Permanent Magnet Synchronous Linear Motor Based on Sliding Mode Vector Tracking Observer. *IEEE Trans. Ind. Appl.* **2023**, *59*, 866–872. [CrossRef]
2. Zhang, C.L.; Li, B.R.; Ye, P.Q.; Zhang, H. Analog-Hall-Sensor-Based Position Detection Method with Temperature Compensation for Permanent-Magnet Linear Motor. *IEEE Trans. Instrum. Meas.* **2021**, *70*, 9512111. [CrossRef]
3. Zhao, J.; Wang, W.W.; Zhou, Y.; Zhao, J. Robust high precision multi-frame motion detection for PMLSMs? mover based on local upsampling moving least square method. *Mech. Syst. Signal Process.* **2021**, *159*, 107803. [CrossRef]
4. Hao, D.Q.; Zhu, L.Q.; Pan, Z.K.; Guo, Y.K.; Chen, Q.S. Permanent Magnet Linear Synchronous Motor Servo System Based on DSP. *Adv. Precis. Instrum. Meas.* **2012**, *103*, 388–393. [CrossRef]
5. Yan, L.Y.; Ye, P.Q.; Zhang, C.L.; Zhang, H. An Improved Position Detection Method for Permanent Magnet Linear Motor Using Linear Hall Sensors. In Proceedings of the 2017 20th International Conference on Electrical Machines and Systems, Sydney, NSW, Australia, 11–14 August 2017.
6. Hussain, H.A.; Toliyat, H.A. Field Oriented Control of Tubular PM Linear Motor Using Linear Hall Effect Sensors. In Proceedings of the 2016 International Symposium on Power Electronics, Electrical Drives, Automation and Motion, Capri, Italy, 22–24 June 2016; pp. 1244–1248.
7. Dwivedi, A.; Ramakrishnan, A.; Reddy, A.; Patel, K.; Ozel, S.; Onal, C.D. Design, Modeling, and Validation of a Soft Magnetic 3-D Force Sensor. *IEEE Sens. J.* **2018**, *18*, 3852–3863. [CrossRef]
8. Yariçi, İ.; Öztürk, Y. A new approach to linear displacement measurements based on Hall effect sensors. *Turk. J. Electr. Eng. Comput. Sci.* **2023**, *31*, 238–248. [CrossRef]
9. Uzlu, B.; Wang, Z.; Lukas, S.; Otto, M.; Lemme, M.C.; Neumaier, D. Gate-tunable graphene-based Hall sensors on flexible substrates with increased sensitivity. *Sci. Rep.* **2019**, *9*, 18059. [CrossRef] [PubMed]
10. Xiao, L.; Yunyue, Y.; Zhuo, Z. Study of the linear Hall-effect sensors mounting position for PMLSM. In Proceedings of the ICIEA 2007: 2nd IEEE Conference on Industrial Electronics and Applications, Harbin, China, 23–25 May 2007; Volume 1–4, pp. 1175–1178.
11. Kim, J.; Choi, S.; Cho, K.; Nam, K. Position Estimation Using Linear Hall Sensors for Permanent Magnet Linear Motor Systems. *IEEE Trans. Ind. Electron.* **2016**, *63*, 7644–7652. [CrossRef]
12. Zhang, X.H.; Zhang, W. An Improved Rotor Position Estimation in PMSM with Low-Resolution Hall-Effect Sensors. In Proceedings of the 2014 17th International Conference on Electrical Machines and Systems, Hangzhou, China, 22–25 October 2015; pp. 2722–2727.
13. Qiao, Z.W.; Shi, T.N.; Wang, Y.D.; Yan, Y.; Xia, C.; He, X. New Sliding-Mode Observer for Position Sensorless Control of Permanent-Magnet Synchronous Motor. *IEEE Trans. Ind. Electron.* **2013**, *60*, 710–719. [CrossRef]
14. Jiang, D.; Yu, W.X.; Wang, J.N.; Zhao, Y.; Li, Y.; Lu, Y. A Speed Disturbance Control Method Based on Sliding Mode Control of Permanent Magnet Synchronous Linear Motor. *IEEE Access* **2019**, *7*, 82424–82433. [CrossRef]
15. Hui, Y.; Chi, R.H.; Huang, B.; Hou, Z. Extended State Observer-Based Data-Driven Iterative Learning Control for Permanent Magnet Linear Motor with Initial Shifts and Disturbances. *IEEE Trans. Syst. Man Cybern.-Syst.* **2021**, *51*, 1881–1891. [CrossRef]
16. Yin, Z.G.; Gu, Y.X.; Du, C.; Gao, F. Research on Back-Stepping Control of Permanent Magnet Linear Synchronous Motor Based on Extended State Observer. In Proceedings of the 2018 IEEE International Power Electronics and Application Conference and Exposition, Shenzhen, China, 4–7 November 2018; pp. 179–183.
17. Lee, D.H.; Lee, J.H.; Ahn, J.W. Mechanical vibration reduction control of two-mass permanent magnet synchronous motor using adaptive notch filter with fast Fourier transform analysis. *IET Electr. Power Appl.* **2012**, *6*, 455–461. [CrossRef]

actuators

Article

Research on Stability Control Technology of Hazardous Chemical Tank Vehicles Based on Electromagnetic Semi-Active Suspension

Jianguo Dai *, Youning Qin, Cheng Wang, Jianhui Zhu and Jingxuan Zhu

Faculty of Transportation Engineering, Huaiyin Institute of Technology, Huai'an 223003, China; 1161504301@hyit.edu.cn (Y.Q.); wangcheng@hyit.edu.cn (C.W.); 11160036@hyit.edu.cn (J.Z.); 1184015224@hyit.edu.cn (J.Z.)
* Correspondence: djg619265809@hyit.edu.cn

Abstract: Liquid sloshing in the tank can seriously affect the stability of hazardous chemical tanker trucks during operation. To this end, this paper proposes a solution based on an electromagnetic semi-active suspension system to prevent chemical spills and ensure safe driving of hazardous chemical tank vehicles. A comprehensive investigation was conducted across four domains: theoretical research, simulation model establishment, co-simulation platform construction, and simulation data analysis. Three fuzzy controllers were used to suppress the vibration of the tank vehicles, and a simulation study of the stability control of the tank vehicles under electromagnetic semi-active suspension was carried out. The results show that the electromagnetic semi-active suspension can significantly reduce the vertical, pitch, and roll vibrations of the tank vehicles by 17.60%, 25.78%, and 27.86%, respectively. The research results of this paper are of great significance for improving the safety and stability of hazardous chemical tanker trucks.

Keywords: tanker truck; electromagnetic semi-active suspension; tank oscillation; control strategy; co-simulation

Citation: Dai, J.; Qin, Y.; Wang, C.; Zhu, J.; Zhu, J. Research on Stability Control Technology of Hazardous Chemical Tank Vehicles Based on Electromagnetic Semi-Active Suspension. *Actuators* **2023**, *12*, 333. https://doi.org/10.3390/act12080333

Academic Editor: Takeshi Mizuno

Received: 4 June 2023
Revised: 10 August 2023
Accepted: 15 August 2023
Published: 17 August 2023

1. Introduction

Road transportation using tank vehicles, distinguished by its large loading capacity and low transportation cost, serves as the main mode of transporting hazardous liquid chemicals in China [1]. The tanker truck is a heavy-duty transportation vehicle with a specialized tank structure, with high bearing capacity, high center of gravity, and large volume. Under non-full load conditions and in complex operational scenarios, the tanker truck's internal liquid is susceptible to significant oscillations, interacting intensively with the tank body. This interaction can alter the vehicle's center of gravity, precipitating a drastic shift in the axle load, posing risks of tilting or even rolling over. These dynamics seriously undermine the operational safety and stability of the vehicle.

In order to reduce the occurrence of rollover accidents, many measures have been proposed to improve the lateral stability of tank vehicles. Yim et al. [2–5] controlled the active lateral stabilizer bar based on different control algorithms, using lateral load transfer rate as the control objective to reduce vehicle oscillation. Xu et al. [6,7] established an active steering-based anti-rollover control system for vehicles, which can effectively reduce the vehicle's roll angle according to experimental results. Hu et al. [8–11] used the lateral-sway-angle speed as the control variable, and determined the additional lateral sway moment using different control calculation methods. They employed differential braking to apply the lateral sway moment, thereby controlling the vehicle's stability. Through simulation experiments, they confirmed that such a method effectively suppresses the vehicle's oscillation influence. However, the methods mentioned above have not effectively solved the contradiction between vehicle comfort and handling stability. Lateral stabilizer bars cannot adjust the roll-angle stiffness in real time, which may cause excessive vehicle roll during high-speed turning. Using differential braking and active steering not only introduces safety hazards during high-speed driving, but also contributes to driver fatigue and insecurity, negatively impacting the driving experience.

To address these issues, this paper proposes the use of controllable suspension technology to improve the driving stability of liquid tankers. The proposed controllable suspension can directly control the body sway of the liquid tanker, adjust the body posture in real time, and diminish the tank sway, all without compromising the driver's experience. The proposition considers both the comfort and handling stability of the vehicle. Among controllable suspensions, the semi-active suspension has the advantages of low energy consumption, low cost, and similar control effect compared with the active suspension. Moreover, the electromagnetic semi-active suspension, which adopts the electromagnetic principle to change the damping characteristics of the damper, has a faster response speed and higher reliability. The electromagnetic semi-active suspension is composed of a sensor, an actuator, a controller, and a power supply. The sensors are designed to detect body posture and road information. The actuators, composed of electromagnetic actuators and dampers, along with the controller, are tasked with calculating the damping of the electromagnetic semi-active suspension. The controller sends control signals to the actuators, ensuring semi-active control of the suspension. When the electromagnetic linear actuator is inactive, the damper leverages hydraulic oil from the working cylinder and damping springs to achieve damping. Conversely, when the electromagnetic linear actuator is operational, the controller modifies the device's electromagnetic impedance by altering the circuit equivalent resistance. This flexible adjustment of the suspension damping creates a controlled damping force, enabling semi-active control of the vehicle. By adopting electromagnetic semi-active suspension technology, the liquid tanker can attain enhanced driving stability and an enriched driving experience for the driver, while ensuring the comfort and handling stability of the vehicle. Such an advancement contributes significantly to reducing the risk associated with dangerous chemical transport accidents, thereby improving overall road safety.

2. Working Characteristics of Electromagnetic Semi-Active Suspension

2.1. Structural Principle of Electromagnetic Semi-Active Suspension

The suspension system is a key component of the vehicle and an important device to ensure the smooth running and stable handling of the vehicle. Passive suspension refers to the suspension whose stiffness and damping coefficient do not change with the external state. Semi-active suspension is a controllable suspension system that can adjust damping parameters to improve vehicle ride comfort and stability. Compared with semi-active suspension, passive suspension has the advantages of a simple structure and low cost. However, there is no energy supply device in the passive suspension system, and its stiffness and damping cannot be artificially controlled and adjusted during the driving process, so it is difficult for the passive suspension to take into account the requirements of vehicle driving, comfort, and handling stability, and it is increasingly unable to meet the high performance and high-energy efficiency needs of the rapid development of vehicle technology. Therefore, electromagnetic semi-active suspension technology has gradually become a research hotspot.

As shown in Figure 1a, the electromagnetic semi-active suspension device is mainly composed of an outer magnetic yoke, a permanent magnet, a moving coil, and an inner core. The shock absorber piston rod is used as the inner yoke of the electromagnetic semi-active suspension, wherein the coil skeleton is fixed to the piston rod. When the suspension vibration and shock absorber piston reciprocate, the shock absorber will follow the synchronous movement of the coil, and cut off the magnetic induction line to generate an induction current. The generated induction current can be used to supply the semi-active control of the device, but also can be stored for other electrical equipment's energy supply.

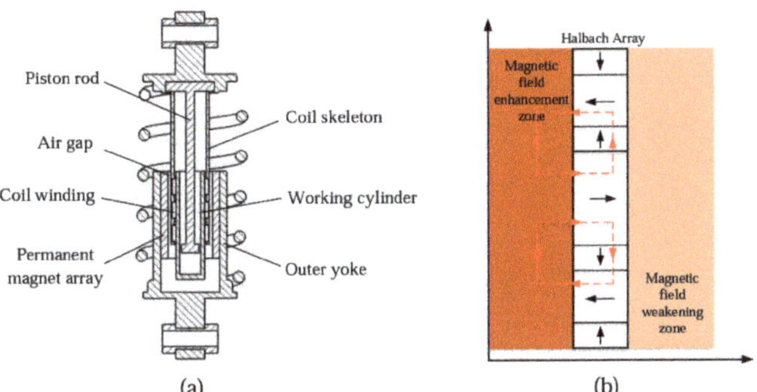

Figure 1. Structure (**a**) and schematic diagram (**b**) of electromagnetic semi-active suspension.

The electromagnetic semi-active suspension used in this paper is based on a cylindrical damper structure, with the addition of an electromagnetic linear actuator. The electromagnetic linear actuator is embedded in the suspension damper, as shown in Figure 1a [12]. The permanent magnet of the electromagnetic actuator adopts the Halbach array structure [13,14], which can effectively improve the electromagnetic characteristics of the device, as shown in Figure 1b. The left side of the Halbeck array is the area where the field is enhanced, which is also the area where the coil is active, which can generate a greater induced current, and on the right side is the area where the field is weakened.

The electromagnetic semi-active suspension retains the traditional suspension's piston hydraulic cylinder, which can passively absorb shock through the hydraulic cylinder. In addition, an added electromagnetic linear actuator can also provide semi-active control of the vehicle body. During semi-active control, the vehicle control module executes a preset suspension control strategy based on the vehicle's posture as detected by sensors. By controlling a supercapacitor to provide a corresponding current to the electromagnetic linear actuator, the electromagnetic damping force of the actuator is adjusted to improve the vehicle's posture and enhance driving stability.

Therefore, the electromagnetic damping force can be expressed by the thrust coefficient and the back electromotive force coefficient of the electromagnetic actuator. In electromagnetic semi-active suspension, the thrust coefficient of the electromagnetic actuator is denoted as $k_i = B_i L_i$; the back electromotive force coefficient of the electromagnetic actuator is denoted as $k_\delta = B_\delta L_\delta$. Therefore, the electromagnetic damping force can be expressed as follows:

$$F_a = \frac{k_i k_\delta}{R + r} v_a \tag{1}$$

Let the damping of the electromagnetic actuator be denoted as $C_a = \frac{k_i k_\delta}{R+r}$. Then, the electromagnetic damping force of the electromagnetic semi-active suspension can be written as:

$$F_a = C_a \cdot v_a \tag{2}$$

2.2. Control Strategy for Electromagnetic Semi-Active Suspension

For a more refined analysis of the driving dynamics of a tanker truck, the vehicle is abstracted into a seven degree of freedom model, as depicted in Figure 2. This model thoroughly considers the vibration characteristics of the tanker truck body in vertical, pitch, and roll directions. It also takes into account the vertical vibration characteristics of the four electromagnetic semi-active suspensions bridging the vehicle body. To mitigate body vibrations in the vertical, pitch, and lateral tilt directions, this paper employs a fuzzy control method [15,16] for adjusting the damping of the electromagnetic semi-active suspension. Characterized by its adaptability and robustness, fuzzy control effectively addresses the challenges posed by the electromagnetic semi-active suspension in maintaining liquid tanker stability.

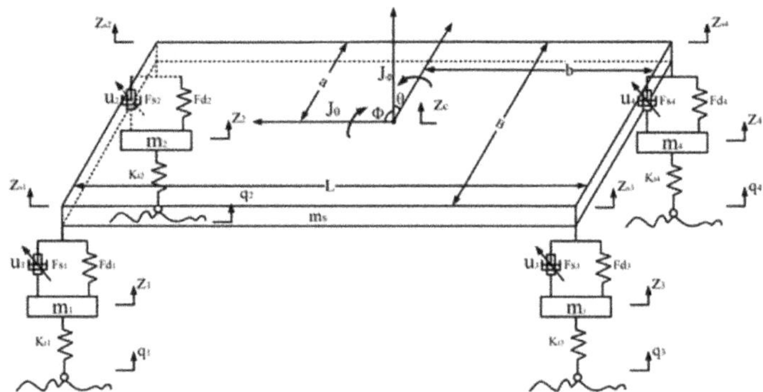

Figure 2. Seven degree of freedom vehicle model.

According to Newton's second law, the dynamic equation of the mass on the vehicle spring is as follows:

$$\begin{cases} m_s\ddot{z}_c + F_{s1} + F_{s2} + F_{s3} + F_{s4} + F_{t1} + F_{t2} + F_{t3} + F_{t4} - U_z = 0 \\ J_\phi\ddot{\phi} - (F_{s1} + F_{s2} + F_{t1} + F_{t2})b + (F_{s3} + F_{s4} + F_{t3} + F_{t4})(L - b) - M_\phi = 0 \\ J_\theta\ddot{\theta} - (F_{s1} + F_{s3} + F_{t1} + F_{t3})(B - a) + (F_{s2} + F_{s4} + F_{t3} + F_{t4})a - M_\theta = 0 \end{cases} \tag{3}$$

The dynamic equations for the four suspensions are as follows:

$$m_i\ddot{z}_i - F_{s1} - F_{t1} + K_{ti}(z_i - q_i) + u_i = 0 \tag{4}$$

where m_s and $(i = 1, 2, 3, 4)$ represent the sprung mass of the vehicle and the unsprung mass of the four wheels, respectively; z_c represents the vertical displacement of the vehicle body; ϕ and θ represent the pitch and roll angles, respectively; J_ϕ and J_θ represent the roll and pitch moments of inertia of the vehicle, respectively; $F_{si}(i = 1, 2, 3, 4)$, $F_{ti}(i = 1, 2, 3, 4)$ and $K_{ti}(i = 1, 2, 3, 4)$ represent the passive damping force, spring force, and tire stiffness coefficient of each suspension, respectively; $z_i(i = 1, 2, 3, 4)$ represents the displacement of each suspension mass; $q_i(i = 1, 2, 3, 4)$ represents the road roughness excitation; U_z represents the vibration control force in the vertical direction; M_θ and M_φ represent the roll and pitch vibration moments of the vehicle body, respectively; B and L represent the wheelbase and track width of the tanker truck, respectively; b represents the distance from the center of mass of the tanker truck to the front axle; a represents the distance from the center of mass of the tanker truck to the wheels; $u_i(i = 1, 2, 3, 4)$ represents the control force of each electromagnetic semi-active suspension.

Given the interference between the control forces of the four suspensions on the vehicle body during actual driving, and the intertwined control objectives during the control process, direct adjustments to the control forces of the four suspensions may not yield significant control effects. Therefore, this paper first designs three fuzzy controllers to suppress the vibration of the liquid tanker truck body in the vertical, pitch, and roll directions, respectively. Subsequently, the requisite control forces and moments of the fuzzy controllers are equivalently calculated and allocated as the damping forces for the four electromagnetic semi-active suspensions. Finally, these suspensions feed the corresponding damping forces into the vehicle model, enabling comprehensive control of the vehicle's motion. The principle of this control strategy is shown in Figure 3.

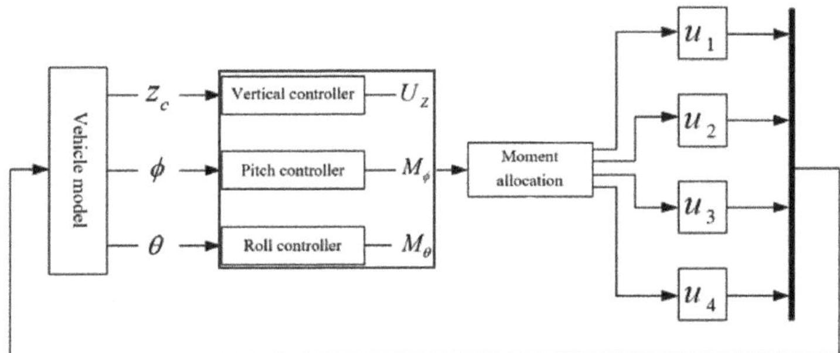

Figure 3. Schematic diagram of the control strategy principle.

The damping forces of the four suspensions can be calculated using the following moment-allocation formula:

$$\begin{cases} U_z = u_1 + u_2 + u_3 + u_4 \\ M_\phi = -(u_1 + u_2)b + (u_3 + u_4)(L - b) \\ M_\theta = (u_2 + u_4)a - (u_1 + u_3)(B - a) \end{cases} \tag{5}$$

3. Construction of the Vehicle Model for the Hazardous Chemical Liquid Tanker Truck

This paper studies the driving stability of the liquid tanker truck under different road conditions. For the precise simulation of complex road conditions, we utilized the TruckSim heavy vehicle simulation software and established a pendulum equivalent model of the tank body using Simulink to account for the effect of liquid sloshing. This integrative modeling approach enhances the accuracy of the simulation, yielding more realistic and dependable results, thereby better fulfilling the research requirements.

3.1. Analysis of Liquid Sloshing in the Tank Body

This paper aims to study the characteristics of liquid sloshing in the liquid tanker truck during driving, where the filling ratio is an important factor affecting the liquid sloshing. A tank with a filling ratio of 60% was selected for the study, as it demonstrates the dynamics of liquid sloshing and is a more common configuration in actual transportation. The tank body of the liquid tanker truck was modeled in Fluent and features an elliptical cross-section, with a major axis of 1 m, a minor axis of 0.8 m, and a length of 6 m. The liquid sloshing under the same longitudinal excitation with different filling ratios was simulated. The tank body with a filling ratio of 60% was selected to simulate the lateral excitation of the liquid tanker truck during turning, and the simulation time was 5 s. The longitudinal sloshing force and moment of the tank body over time were obtained, and the simulation results are shown in Figure 4.

This paper uses an equivalent pendulum model [17,18] to simulate the liquid sloshing inside the tank. The schematic diagram of the equivalent pendulum model is shown in Figure 5.

The dynamic equation is:

$$\ddot{\gamma} + \frac{c_l}{m_p}\dot{\gamma} + \frac{g}{l_p}\gamma = \frac{a_p}{l_p} \tag{6}$$

The lateral sloshing force of the liquid:

$$F_y = m_p l_p \ddot{\gamma} - m_l a_p \tag{7}$$

The lateral tilting moment of liquid sloshing on the center of the tank bottom is:

$$M_y = -h_p m_p l_p \ddot{\gamma} + m_l h_l a_p + m_p l_p g \gamma \tag{8}$$

In the liquid-equivalent pendulum model diagram shown in Figure 5, m_0 is the fixed mass of the liquid, in kg; m_p is the mass of the equivalent pendulum of the liquid, in kg; h_0 is the height from the center of mass of the liquid's fixed mass to the tank bottom, in m; h_p is the height from the center of mass of the pendulum to the tank bottom, in m; c_l is the equivalent damping of the liquid; γ is the swing angle of the equivalent pendulum; l_p is the length of the equivalent pendulum, in m; a_p is the lateral acceleration of the tank body.

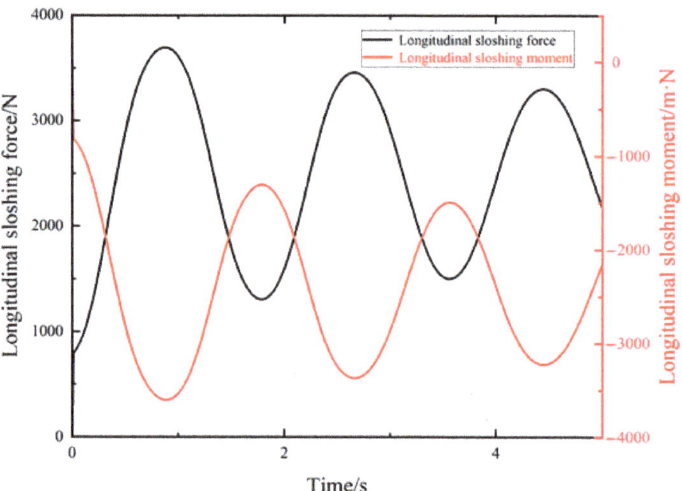

Figure 4. Longitudinal sloshing force and moment of the tank body.

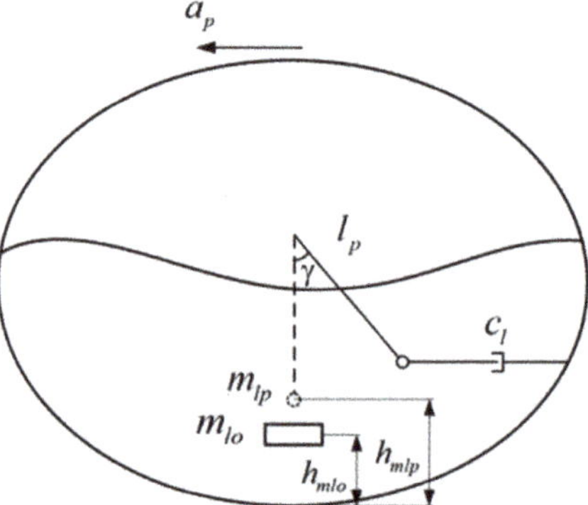

Figure 5. Schematic diagram of the equivalent pendulum model.

3.2. Construction of a Complete Vehicle Mode for the Liquid Tanker Truck Based on TruckSim

In order to visually obtain a dynamic simulation of the vehicle, this paper uses TruckSim to establish a complete vehicle model. Based on the parameters of a selected liquid tanker truck model, the models for the vehicle body, suspension system, tires, steering system, powertrain system, braking system, and aerodynamics are set in TruckSim. The main parameters of the complete vehicle model are shown in Table 1.

Table 1. Main parameters of the complete vehicle model.

Vehicle Parameters	Symbol	Configuration
Overall length/width/height/mm	$L/W/H$	5200/2550/3200
Height of center of mass/mm	h	1175
Distance from center of mass to front axle/mm	l_f	1450
Wheelbase/mm	A	4950
Overall vehicle mass/kg	m_s	4460
Inertia around y-axis/kg*m^2	I_y	35,467.7
Inertia around x-axis/kg*m^2	I_x	2185.4
Inertia around z-axis/kg*m^2	I_z	34,966.4

3.3. Establishment of Simulink Suspension Control Model

Based on the analysis of the forced sloshing of the liquid inside the tank in Section 2.1, MATLAB was employed to identify the parameters of the pitch sloshing force and lateral-tilting moment curves of the tank body, and to derive the parameters of the equivalent pendulum model. Following this, the tank-equivalent pendulum model was established in Simulink. The lateral sloshing force and lateral tilting moment of the liquid on the tank bottom, obtained from the Fluent numerical simulation, were compared with those derived from the tank-equivalent pendulum model in Simulink, as shown in Figure 6.

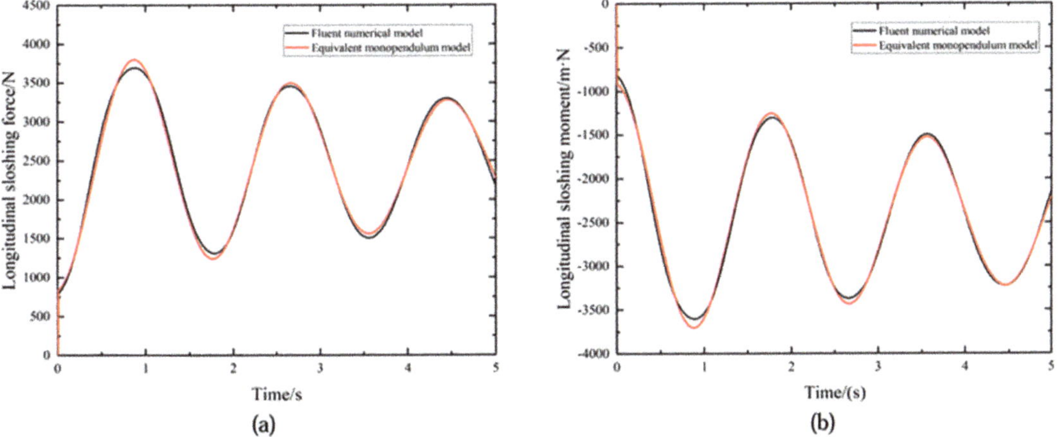

Figure 6. Comparison between the equivalent pendulum model and the Fluent numerical model for the longitudinal sloshing force (**a**) and the longitudinal sloshing moment (**b**).

From Figure 6, it can be seen that the simulation results of the liquid sloshing pendulum model built by Simulink software fit well with those of the Fluent numerical model, which verifies the reliability and accuracy of the established Simulink liquid-equivalent pendulum model, laying a basis for the establishment of the next step: a joint model of the liquid tanker truck.

Based on the designed suspension control strategy, the suspension control module was established in Simulink, as shown in Figure 7.

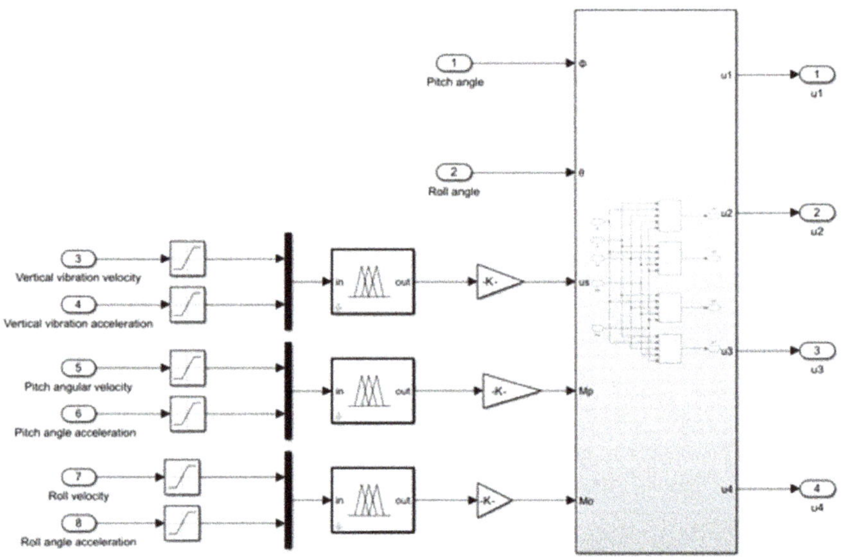

Figure 7. Suspension control module.

3.4. Construction of the TruckSim–Simulink Co-Simulation Platform

TruckSim can be connected to Simulink models through a data interface, and the established TruckSim complete vehicle model can be connected to the Simulink suspension control module and tank-equivalent pendulum model through the input and output variables. The variables output from TruckSim to Simulink include the vertical vibration speed of the vehicle body $\dot{z}_c$, the vertical vibration acceleration of the vehicle body $\ddot{z}_c$, the pitch angle of the vehicle body ϕ, the pitch angle velocity of the vehicle body $\dot{\phi}$, the pitch angle acceleration of the vehicle body $\ddot{\phi}$, the roll angle of the vehicle body θ, the roll angle velocity of the vehicle body $\dot{\theta}$, the roll angle acceleration of the vehicle body $\ddot{\theta}$, and the lateral acceleration a_y. The variables input from Simulink to TruckSim include the damping force of the four electromagnetic semi-active suspensions u_i, the pitch sloshing force F_y, and the sloshing moment M_y of the tank-equivalent pendulum model. The established co-simulation platform of the liquid tanker truck using TruckSim–Simulink is shown in Figure 8.

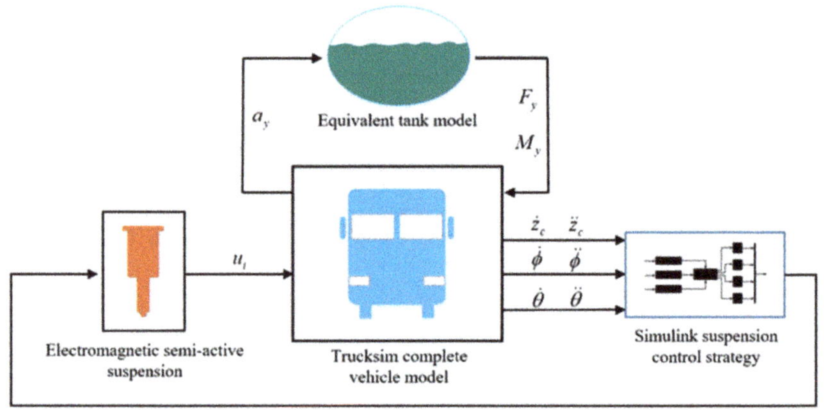

Figure 8. TruckSim–Simulink co-simulation model.

3.5. Validation of the Complete Vehicle Model

To verify whether the TruckSim–Simulink co-simulation model established in this paper can accurately portray the vehicle's dynamic characteristics, we opted for a step input test as a means of model validation. In compliance with the regulations of the GB/T 12534 Road Vehicle Test Method General Rules, we conducted a step input test for the steering wheel angle. The vehicle speed was set at 60 km/h, the steering wheel angle was adjusted to 180°, the road adhesion coefficient was set at 0.8, and the duration was fixed at 10 s. The resulting steering-wheel-angle step input curve is illustrated in Figure 9. For ease of analysis, a seven degree of freedom vehicle model was established in Simulink as a reference model for conducting the same steering-wheel-angle step input simulation test. The roll angle and yaw angle of the liquid tanker truck obtained from the co-simulation model were compared with those of the reference model, as shown in Figure 10.

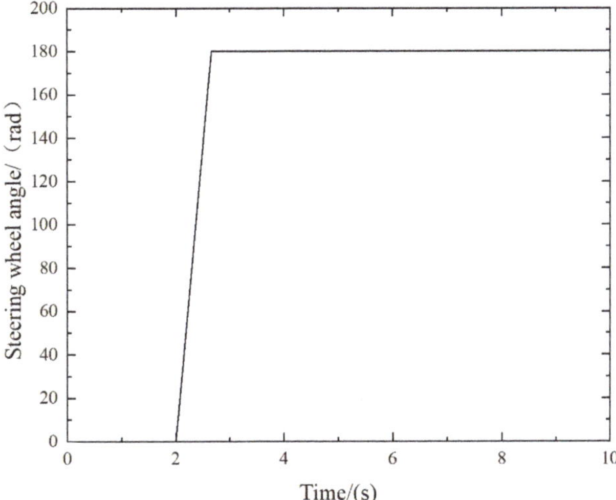

Figure 9. The steering wheel angle under step input.

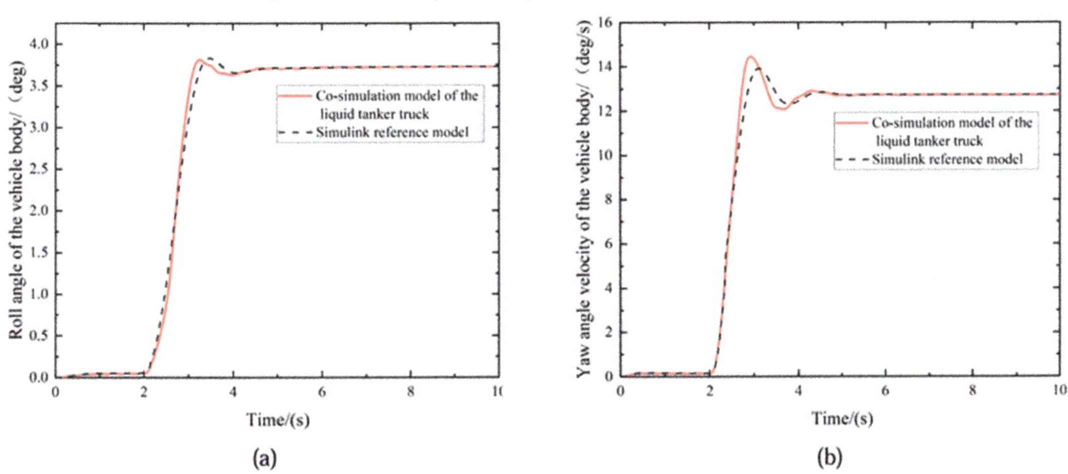

Figure 10. Roll angle (**a**) and yaw angle velocity (**b**) of the vehicle body.

As inferred from Figure 10, although there exists some discrepancy between the Simulink reference model and the co-simulation model, and their respective peak values differ, the overall trends of their parameter curves are essentially consistent. Moreover, the steady-state value error between the two models remains less than 5%. This indicates that the established co-simulation model of the liquid tanker truck can accurately simulate the basic motion characteristics of the liquid tanker truck.

4. Stability Control Simulation of the Hazardous Materials Tank Truck

To verify the feasibility of using electromagnetic semi-active suspension to control the stability of the liquid tanker truck, this paper implements the double lane change test condition. The input steering wheel angle is shown in Figure 11, where the vehicle speed is set to 60 km/h, and the road adhesion coefficient is 0.85. The simulation results of the vertical vibration acceleration, pitch angle acceleration, and roll angle acceleration of the liquid tanker truck body are comparatively analyzed between the passive suspension system and the electromagnetic semi-active suspension control system.

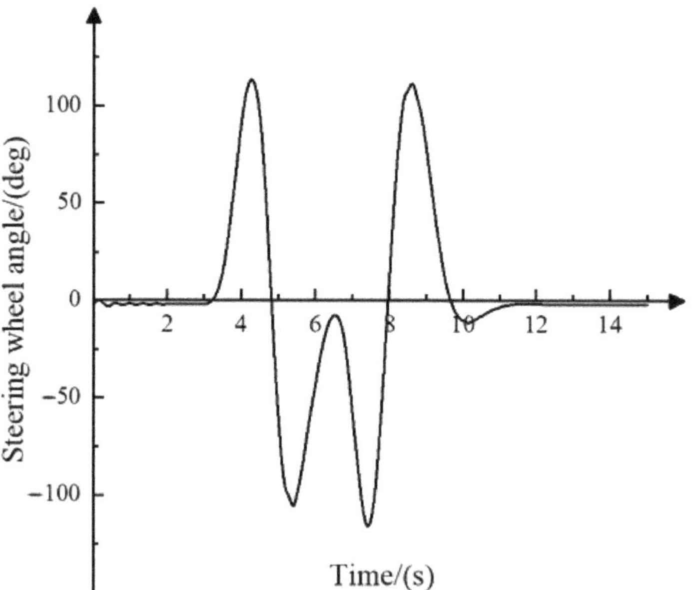

Figure 11. Steering wheel angle.

The curves of the vertical vibration acceleration and pitch angle acceleration of the liquid tanker truck body are shown in Figures 12 and 13. They show that the vertical vibration and pitch of the body change sharply from 0 to 4 s, indicating that the startup acceleration condition of the liquid tanker truck has a significant effect on the vertical vibration and pitch of the body. Figure 14 shows that the startup acceleration of the liquid tanker truck has little effect on the roll angle, but the roll angle of the body starts to change sharply when the liquid tanker truck changes lanes at 3 s. By comparing the root-mean-square values in Table 2, it can be concluded that the liquid tanker truck controlled by the electromagnetic semi-active suspension has significantly improved vibration performance in the vertical, pitch, and roll directions of the vehicle body. The roll angle exhibited a remarkable performance improvement of 27.86%, highlighting the significant impact of the electromagnetic energy-fed suspension in suppressing the roll angle of the liquid tanker truck.

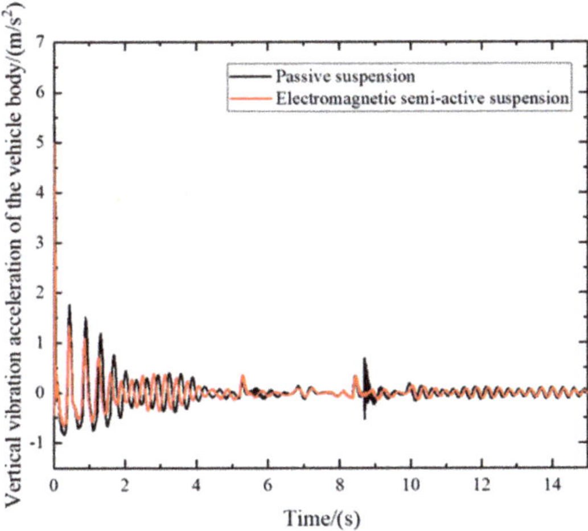

Figure 12. Vertical vibration acceleration of the vehicle body.

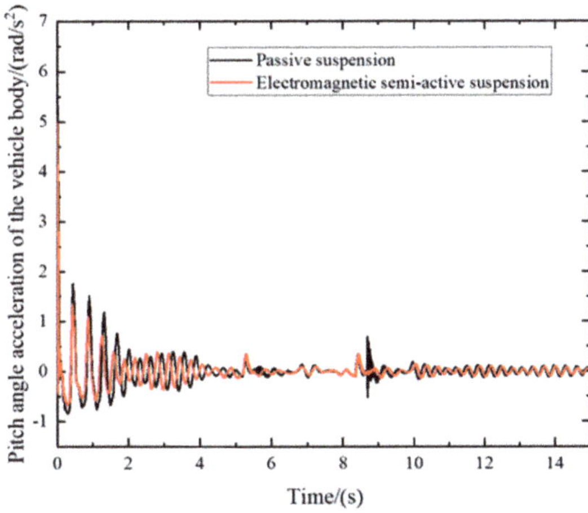

Figure 13. Pitch angle acceleration of the vehicle body.

Table 2. Comparison of root-mean-square values of control performance between the electromagnetic semi-active suspension and the passive suspension for the entire vehicle.

Suspension Type	Vertical Vibration Acceleration	Pitch Angle Acceleration	Roll Angle Acceleration
Passive suspension	0.12732	0.29526	0.27277
Electromagnetic semi-active suspension	0.10491	0.21914	0.19678
Performance improvement	17.60%	25.78%	27.86%

The root-mean-square equation of vertical vibration acceleration is:

$$\ddot{z}_{c\ rms} = \sqrt{\frac{\sum_{i=1}^{15} \ddot{Z}_{c\ i}{}^{2}}{i}} \tag{9}$$

The root-mean-square equation of pitch angle acceleration is:

$$\ddot{\varphi}_{rms} = \sqrt{\frac{\sum_{i=1}^{15} \ddot{\varphi}_i{}^2}{i}} \tag{10}$$

The root-mean-square equation of roll angle acceleration is:

$$\ddot{\theta}_{rms} = \sqrt{\frac{\sum_{i=1}^{15} \ddot{\theta}_i{}^2}{i}} \tag{11}$$

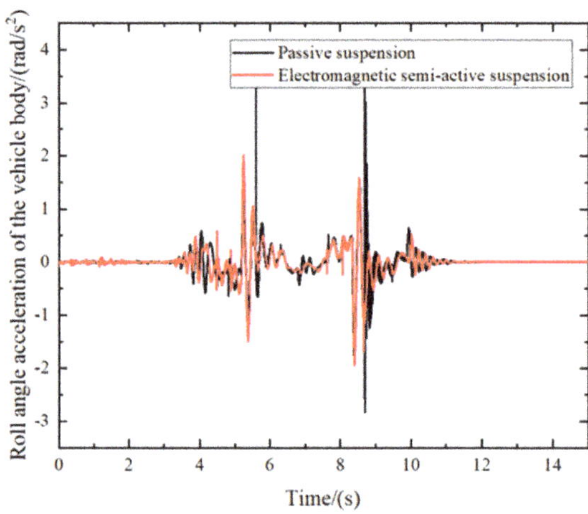

Figure 14. Roll angle acceleration of the vehicle body.

5. Conclusions

This paper aims to solve the problem of liquid tank vehicles being prone to rollover during transportation due to liquid sloshing in the tank. To this end, the proposed solution is to use electromagnetic semi-active suspension to reduce vehicle body sway. By analyzing the characteristics of liquid sloshing in the tank, a TruckSim vehicle model and a single-pendulum equivalent model for simulating liquid sloshing are established. An electromagnetic semi-active suspension control strategy is developed, and variables are set to connect the two software models. A TruckSim–Simulink co-simulation model is constructed with the goal of improving the stability of the liquid tanker truck. Moreover, simulation studies of the electromagnetic semi-active suspension control technology are conducted. The following conclusions are mainly obtained through the above research:

1. This paper proposes an innovative solution to improve the stability of liquid tank vehicles by using electromagnetic semi-active suspension. The structure and working principle of the electromagnetic semi-active suspension are analyzed, and the feasibility of the proposed solution is investigated.

2. This paper integrates and innovates existing simulation technologies and proposes a TruckSim–Simulink co-simulation model. A vehicle model is established in Truck-Sim, and a fuzzy control module for the electromagnetic semi-active suspension and an equivalent model of the tank are established in Simulink. The two models are connected through input and output variables to achieve co-simulation.

3. Steering-wheel-angle step input simulations are performed on both the TruckSim–Simulink co-simulation model and the reference vehicle model established in Simulink. Based on the simulation results, it is found that the error in the vehicle dynamics

characteristics expressed by the two models is less than 5%, proving the accuracy of the co-simulation model.

4. By comparing the simulation results of the electromagnetic semi-active suspension and the passive suspension for controlling the stability of the liquid tanker truck, it is found that the electromagnetic semi-active suspension can effectively control the vibration of the liquid tanker truck in the vertical, pitch, and roll directions, and the performance improvement in roll angle reached 27.86%, significantly higher than other control objectives, indicating that the electromagnetic semi-active suspension is more effective in controlling the roll angle of the liquid tanker truck.

This paper provides new ideas and methods for the safety and stability control of hazardous material liquid tank vehicles, which is of great theoretical and practical significance for ensuring road traffic safety and preventing hazardous material leaks. Future research can further explore the control algorithm and system design of the electromagnetic semi-active suspension to improve its control performance and reliability.

Author Contributions: Conceptualization, J.D. and Y.Q.; methodology, J.D. and Y.Q.; software, formal analysis, and investigation, J.D., Y.Q., C.W. and J.Z. (Jianhui Zhu); resources, J.D., Y.Q. and C.W.; data curation, J.D. and J.Z. (Jingxuan Zhu); writing—original draft preparation, J.D., Y.Q., J.Z. (Jianhui Zhu) and J.Z. (Jingxuan Zhu); writing—review and editing, J.D. and Y.Q. All authors have read and agreed to the published version of the manuscript.

Funding: This work was supported by the National Natural Science Foundation of China (No. 51605183, 51975239), and the Natural Science Foundation of Jiangsu Province for Universities (Grant No. 21KJB480005).

Institutional Review Board Statement: Not applicable.

Informed Consent Statement: Not applicable.

Data Availability Statement: Not applicable.

Conflicts of Interest: The authors declare no conflict of interest.

References

1. Zhu, Z.; Li, S. Analysis of the operation status of atmospheric pressure tank trucks for road transportation of liquid dangerous goods. *Chem. Eng. Des. Commun.* **2020**, *46*, 201+217.
2. Seongjin, Y.; Kwangki, J.; Kyongsu, Y. An investigation into vehicle rollover prevention by coordinated control of active anti-roll bar and electronic stability program. *Int. J. Control Autom. Syst.* **2012**, *10*, 275–287.
3. Sampson, D.J.M. *Active Roll Control of Articulated Heavy Vehicles*; Cambridge University: Cambridge, UK, 2010.
4. Cronjé, P.; Els, P. Improving off road vehicle handling using an active anti-roll bar. *J. Terramech.* **2010**, *47*, 179–189. [CrossRef]
5. Li, Z. *Dynamics Analysis and Rollover Prevention Control of Liquid Sloshing in Tanker Trucks*; Chongqing Jiaotong University: Chongqing, China, 2016.
6. Xu, Y. Research on rollover prevention control of vehicles based on active steering technology. *Automot. Eng.* **2005**, *27*, 518–521.
7. Yu, Z.; Li, J.; Cheng, X.; Dai, F.; Li, S. Optimal control simulation of stability for liquid tanker truck. *Oil Gas Storage Transp.* **2019**, *38*, 885–891+918.
8. Hu, X.; Zhao, Z. Stability control of semi-trailer tanker truck based on phase plane partition. *J. Highw. Transp. Res. Dev.* **2015**, *32*, 151–158.
9. Solmaz, S.; Akar, M.; Shorten, R. Adaptive rollover prevention for automotive vehicles with differential braking. *IFAC Proc. Vol.* **2008**, *41*, 4695–4700. [CrossRef]
10. Li, J.; Yu, Z.; Cheng, X.; Dai, F.; Li, S. Simulation of stability control for liquid tanker truck under vehicle-liquid coupling response. *Oil Gas Storage Transp.* **2020**, *39*, 188–194+221.
11. Zhao, W.; Feng, R.; Zong, C. Rollover prevention control strategy for liquid tanker trucks based on equivalent sloshing model. *J. Jilin Univ. (Eng. Technol. Ed.)* **2018**, *48*, 30–35.
12. Jin, H.; Dai, J.; Xia, J.; Wang, C.; Jiang, C.; Xue, C.; Yin, L.; Zhang, S.; Peng, S.; Shi, J.; et al. A New Type of Electromagnetic Linear Energy-Feeding Semi-Active Suspension. CN214281185U, 2021.
13. Li, Z.; Wu, Q.; Liu, B.; Gong, Z. Optimal Design of Magneto-Force-Thermal Parameters for Electromagnetic Actuators with Halbach Array. *Actuators* **2021**, *9*, 231. [CrossRef]
14. Liu, X.; Xiao, L.; Cui, H.; Huang, S. Torque analysis and optimization of Halbach axial permanent magnet coupling. *Micromotors* **2021**, *54*, 9. [CrossRef]

15. Li, S.; Zhang, P.; Yang, J. T-S fuzzy control research of active suspension system driven by hub motor for electric vehicle. *J. Vib. Shock* **2022**, *41*, 9.
16. Dong, S.; Meng, J.; Song, C. Fuzzy control strategy research of valve-controlled semi-active damper. *J. Lanzhou Jiaotong Univ.* **2022**, *41*, 6.
17. Di, Y.; Chu, J. Comparative research on two equivalent mechanical models of liquid sloshing in tanks. *Int. J. Eng. Syst. Model. Simul.* **2018**, *10*, 159–168.
18. Huang, Z.; Wu, W.; Zhou, F.; Gao, C.; Li, C. Dynamics modeling and rigid-liquid coupling characteristics research of tanker truck. *Mod. Manuf. Eng.* **2020**, *8*, 20–26.

Article

Design and Analysis of Brake-by-Wire Unit Based on Direct Drive Pump–Valve Cooperative

Peng Yu [1,2], Zhaoyue Sun [2,*], Haoli Xu [2], Yunyun Ren [3] and Cao Tan [1,2,*]

1   State Key Laboratory of Automotive Simulation and Control, Jilin University, Changchun 130022, China; yupeng199910@163.com
2   School of Transportation and Vehicle Engineering, Shandong University of Technology, Zibo 255000, China; utxuhl@yeah.net
3   Shuntai Automobile Co., Ltd., Zibo 255000, China; renyy123123@163.com
*   Correspondence: chsunzy@outlook.com (Z.S.); njusttancao@yeah.net (C.T.);
    Tel.: +86-13053308225 (Z.S.); +86-0533-2786837 (C.T.)

Abstract: Aiming at the requirements of distributed braking and advanced automatic driving, a brake-by-wire unit based on a direct drive pump–valve cooperative is proposed. To realize the wheel cylinder pressure regulation, the hydraulic pump is directly driven by the electromagnetic linear actuator coordinates with the active valve. It has the advantages of rapid response and no deterioration of wheel side space and unsprung mass. Firstly, by analyzing the working characteristics and braking performance requirements of the braking unit, the key parameters of the system are matched. Then, in order to ensure the accuracy of the simulation model, the co-simulation model of the brake unit is established based on the Simulink-AMESim co-simulation platform. Then, the influence law of key parameters on the control performance is analyzed. Finally, the experimental platform of the brake unit is established. The accuracy of the co-simulation model and the feasibility of the brake-by-wire unit based on direct drive pump–valve cooperative are verified through the pressure control experiment and ABS simulation, which shows that the braking unit has good dynamic response and steady-state tracking effect.

Keywords: electric vehicle; brake-by-wire; electro-hydraulic co-simulation; parameter matching; characteristic analysis

Citation: Yu, P.; Sun, Z.; Xu, H.; Ren, Y.; Tan, C. Design and Analysis of Brake-by-Wire Unit Based on Direct Drive Pump–Valve Cooperative. *Actuators* **2023**, *12*, 360. https://doi.org/10.3390/act12090360

Academic Editor: Ioan Ursu

Received: 26 July 2023
Revised: 8 September 2023
Accepted: 12 September 2023
Published: 14 September 2023

1. Introduction

With the development of the vehicle chassis toward the direction of X-by-wire chassis and slide chassis, the demand for distributed braking systems is becoming more and more intense. At the same time, high-level automatic driving puts forward higher requirements for the response speed and control accuracy of the drive-by-wire system [1–4].

The Electronic Hydraulic Brake (EHB) System has been widely favored due to its advantages, such as rapid response, high power density, and brake structure compatibility [5–9]. In 2013, Bosch launched the I-Booster, a motor servo booster that is independent of vacuum booster, which is a typical motor servo EHB system. TRW introduced a high-pressure accumulator-based EHB system called the SCB, which has a brake master cylinder consisting of front and rear chambers and parallel pistons. Since the pistons can be moved fore and aft and the front part is connected to the front wheel, the pressure of the system can be adjusted through the rear part of the pistons [10]. In 2010, Hitachi launched the EHB system named e-ACT, which uses an electric motor to drive an actuator that pushes a master cylinder piston, and the rotational force of the motor is converted into linear motion by a ball screw [11]. In 2021, Continental launched a new generation of EHB system MKC2 based on MKC1, which uses a multi-logic framework with independent partitions, successfully reducing the number of brake system components, and its modular and scalable design provides ideas for future dynamic control of brake systems.

At present, the EHB system is still the mainstream of the development of brake-by-wire systems. In order to take into account the needs of the market and efficiency, it is of great significance to explore a new brake-by-wire structure to improve the performance of brake-by-wire and simplify the system structure [12–16]. Sun designed and developed an integrated master cylinder and decoupling electro-hydraulic composite braking system and proposed the hydraulic braking force distribution strategy [17]. Wu designed an integrated EHB system. By controlling the solenoid valve and designing the working mode switching control strategy, the working mode switching was realized to achieve high redundancy and independent control of the four-wheel cylinders of the braking system [18]. EHB with different characteristics has been proposed to promote the development of brake-by-wire technology, but it is difficult to directly apply to distributed braking systems [19–23]. It is necessary to design a distributed brake-by-wire unit to meet the needs of intelligent driving and meet the requirements of new energy vehicles for precise control of braking force and lightweight. Fu designed a new type of electromechanical brake with an automatic wear adjustment function and adopted a worm gear mechanism for power transmission, which has the advantage of low delay and small size [24]. Yu designed a distributed electro-hydraulic brake-by-wire system, which uses a motion conversion mechanism to change the motion of the rotating motor into linear motion and then drive the hydraulic wheel cylinder [25]. Chang designed an EHB unit that uses a linear motor to drive the hydraulic piston, amplifying the driving force of the motor through unequal-diameter hydraulic pistons and wheel cylinder pistons [26]. The existing EHBs are often improved based on proven hydraulic braking systems, which still retain complex hydraulic pipelines [27–30]. Distributed brake-by-wire systems often require the entire brake unit to be integrated into the caliper, deteriorating wheel side space and unsprung mass. Therefore, this paper designed a brake-by-wire unit based on direct drive pump-valve cooperative. The electromagnetic linear actuator (EMLA) is directly driven by the hydraulic pump and the active valve coordination to achieve wheel cylinder pressure regulation.

The main contributions are summarized as follows: (1) A brake-by-wire unit based on direct drive pump–valve cooperative is proposed, which has the advantages of rapid response and no deterioration of wheel side space and unsprung mass; (2) Based on the Simulink-AMESim co-simulation platform, a co-simulation model of the brake unit is established and verified; (3) The characteristics of the key parameters are analyzed, and the influence law of each parameter on the control performance is investigated. The rest of this work is organized as follows. Parameter matching design of the key parameters is carried out in Section 3. The dynamic model of the brake-by-wire unit based on direct drive pump–valve cooperative is established in Section 4. The feasibility is verified by pressure experiments and ABS simulations in Section 5. Finally, a conclusion is made in Section 6.

2. System Scheme and Principle

In response to the requirements of distributed braking and high-level automatic driving, this paper designed a brake-by-wire unit based on a direct drive pump–valve cooperative. The structural diagram is shown in Figure 1, which mainly includes an EMLA, hydraulic pump, active valve group, brake wheel cylinder based on the existing structure, and replenishment oil tank. The EMLA, hydraulic pump, active valve group, and replenishment oil tank are integrated and installed on the vehicle frame. The oil pipe is connected to the brake wheel cylinder based on the existing structure, which effectively simplifies the brake pipe and realizes distributed braking without deteriorating the wheel side space and unsprung mass. The EMLA directly drives the reciprocating motion of the piston of the hydraulic pump, combined with the switch control of the active valve group, to realize the rapid adjustment of the driving wheel cylinder pressure. The active valve group is composed of two solenoid valves and the oil refill valve is a normally closed valve, which controls the continuity of the oil circuit between the replenishment oil tank and the direct drive pump, and the pressure maintaining valve is a normally open valve that controls the oil circuit from the pump to the brake wheel cylinder. The diameter of the piston in the direct drive pump is smaller than the diameter of the wheel cylinder piston, and the driving force of the actuator is amplified by the principle of unequal diameter hydraulic amplification, thus obtaining sufficient braking force using a smaller actuator. As the power source of the brake unit, the EMLA completes the conversion of electrical energy to mechanical energy. The detailed working principle of using high-power density moving coil electromagnetic linear actuators is shown in references [31,32].

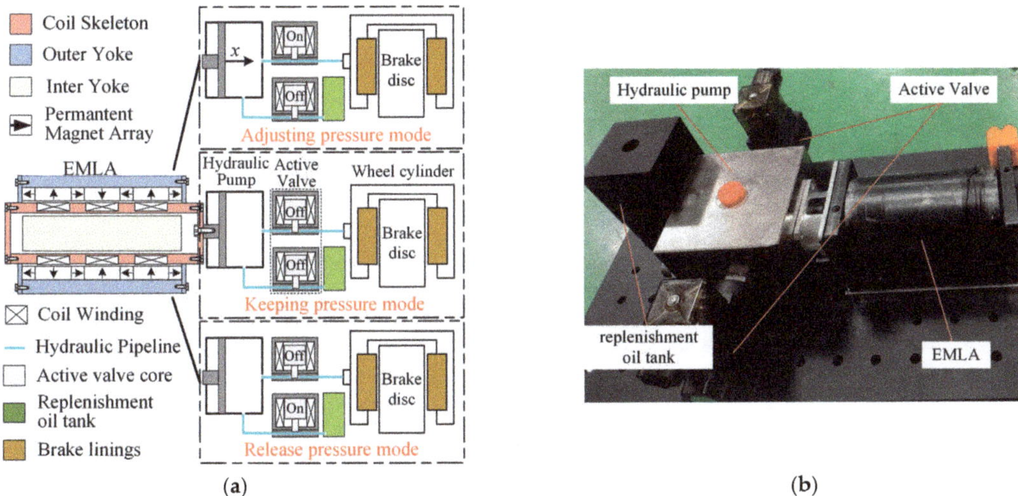

Figure 1. The structure of brake-by-wire unit based on direct drive pump–valve cooperative: (**a**) The schematic diagram; (**b**) The prototype.

The structure of brake-by-wire unit based on direct drive pump–valve cooperative is shown in Figure 1. The typical working process is divided into adjusting pressure mode, keeping pressure mode, and releasing pressure mode. In the releasing pressure mode, the pressure maintaining valve is opened, and the oil refill valve is closed. Under the action of brake wheel cylinder pressure and brake wheel cylinder return spring, the brake wheel cylinder hydraulic flow back hydraulic pump, actuator, and hydraulic pump piston return to the initial position. If the linear actuator does not return to the initial position due to hydraulic oil leakage or other reasons, the pressure maintaining valve is closed, the oil refill valve is opened, and the actuator drives the hydraulic pump piston back to the initial position. In the adjusting pressure mode, the pressure maintaining valve is opened, and the oil refill valve is closed. The EMLA drives the hydraulic piston, quickly discharging hydraulic oil from the hydraulic pump head into the brake wheel cylinder, pushing the brake wheel cylinder piston to eliminate brake clearance, and then quickly adjusting the brake wheel cylinder pressure by controlling the output force of the linear actuator. In the keeping pressure mode, the pressure maintaining valve and oil refill valve are closed, the actuator does not work, and brake wheel cylinder pressure is maintained. The brake unit relieves the working burden of the EMLA and reduces the working energy consumption under the keeping pressure mode.

The advantages of the brake-by-wire unit based on direct drive pump–valve cooperative include the following: (1) the realization of distributed braking while simplifying the hydraulic line and not deteriorating the unsprung mass; (2) the use of the line actuator to directly drive the piston, instead of choosing the form of motor and motion conversion mechanism, to improve the system response speed; (3) the direct drive pump does not work in the keeping pressure mode, effectively reducing the working energy consumption.

3. Parameter Matching Design

The key components of a brake-by-wire unit based on direct drive pump–valve cooperative are the actuator, direct drive pump, and active valve, which have a great influence on the braking performance. The following takes an X-by-wire chassis as an example to match key parameters, and X-by-wire chassis parameters are shown in Table 1.

Table 1. X-by-wire chassis parameters.

Item	Value	Unit
Mass at full load	300	kg
Wheel radius	228	mm
Brake type	Disc brake	
Cross-sectional radius of the piston	19	mm
Effective radius of friction plate	110	mm
Friction coefficient of friction plate	0.38	

To ensure that the electromagnetic force of the EMLA can be effectively amplified, the area of the plunger in the pump should be small to ensure sufficient amplification. However, if the plunger area is too small, it will result in too much pushrod travel and increase its axial size. Considering the magnification and axial size, the plunger area S_1 is set to 28 mm^2 so that the theoretical magnification ε is

$$\varepsilon = \frac{S_2}{S_1} \tag{1}$$

where S_2 is the cross-sectional area of the piston. The brake unit adopts a disc brake, and the braking torque comes from the caliper clamping force; the caliper clamping force F_c is

$$2F_c r_{c1} f = f_{b1} r_b \tag{2}$$

where r_{c1} is the effective radius of brake lining; f is the friction coefficient of brake lining; f_{b1} is the maximum braking force of a single wheel, and r_b is the wheel radius. The relationship between the caliper clamping force and the maximum pressure of the brake wheel cylinder is

$$F_c < P_{\max} S_2 \tag{3}$$

where $P_{\max}$ is the maximum pressure of brake wheel cylinder, calculated as 4.5 MPa. The maximum thrust of the actuator $F_{\max}$ is

$$F_{\max} = P_{\max} S_1 + F_p \tag{4}$$

where F_p is the pre-tightening force of the wheel cylinder piston. During the working process, electrical and electromechanical time constants are usually used to determine the dynamic performance of EMLA. The specific expression is

$$\begin{cases} t_e = \frac{L}{R} \\ t_M = \frac{MR}{K_m^2} \end{cases} \tag{5}$$

where K_m is the electromagnetic force coefficient; M is the dynamic mass of the actuator; R and L are the resistance and inductance of coils; t_e and t_m are the electrical time constant and the electromechanical time constant, respectively. The specific actuator parameters are shown in Table 2.

Table 2. Parameters of brake-by-wire unit based on direct drive pump–valve cooperative.

Components	Item	Value	Unit
Hydraulic pump	Plunger area	28	mm^2
	Length of pump chamber	16	mm
EMLA	Coil resistance	1.40	Ω
	Coil inductance	0.91	mH
	Back EMF constant	24.61	Vs/m
	Electromagnetic force coefficient	24.61	N/A
Active valve	Valve core diameter	4	mm
	Valve hole diameter	8	mm
	Opening time	2	ms

When the brake-by-wire unit based on direct drive pump–valve cooperative works, the two active valves connected with the pump chamber work alternately to cooperate with the adjusting pressure, keeping pressure, and releasing pressure mode in time. Therefore, the structural parameters of the active valve have an effect on the dynamic performance of the brake-by-wire unit based on the direct drive pump–valve cooperative.

Under the control signal, the valve needs to continuously achieve the opening and closing motion state in a short time, and the valve core stroke is short. Therefore, a ball valve type valve core with a simple structure, less wear, and a good sealing effect is adopted. Using an electromagnet as the core component of the active valve can shorten the opening and closing time of the valve core and increase its opening and closing frequency. A detailed analysis of the active valve can be found in reference [33]. The parameters of the selected active valve are shown in Table 2.

4. System Modeling

4.1. Modeling of the EMLA

The EMLA used in the brake-by-wire unit based on direct drive pump–valve cooperative is composed of the outer and inner yoke, the coil skeleton, the coil winding, and the permanent magnet array. The EMLA is directly connected to the hydraulic plunger, subject to friction, hydraulic resistance, etc. The coupling model of its mechanical, magnetic, and electrical subsystems is as follows:

$$\begin{cases} M\frac{d^2x}{dt^2} = F_m - F_f - P_1S_1 - F_{dis} \\ F_m = NB_el_eI = K_mI \\ u = IR + \frac{dI}{dt}L + K_ev \end{cases} \tag{6}$$

where N is the coil turns; B_e is the magnetic field strength; l_e is the single turn coil length; I is the coil current; x is the displacement of the actuator; F_f is the friction force affected by the actuator movement; F_{dis} is the uncertainty error and interference; u is the supply voltage; K_e is the back EMF constant, and v is the moving speed of the actuator. In order to improve the accuracy of modeling, a simple and effective expression of friction force is established, which is regarded as a static nonlinear function of velocity. The specific expression is as follows [34]:

$$F_f = B_1\dot{x} + A_f\arctan(\beta\dot{x}) \tag{7}$$

where B_1 is the viscosity coefficient; A_f is the Coulomb friction coefficient, and the traditional symbolic function is represented by the smoothing function arctan. β is a constant, and β is set large enough so that the function retains the properties of a symbolic function and makes the expression of friction more realistic.

4.2. Hydraulic System Modeling

The direct drive pump needs to work under high-pressure conditions, so the compressibility of the liquid must be considered. To simplify the mathematical model of the direct drive pump, it is assumed that there is no pressure loss along the flow path or local pressure loss when the liquid is flowing; the direct drive pump chamber and hydraulic piston do not deform; the pressure in the pump chamber is equal, and the entire brake unit is well sealed with no hydraulic oil leakage [35]. The pressure change model of the direct drive pump is

$$\dot{P}_1 = \beta_e \frac{S_1 \dot{x} - Q}{S_1(l - x)} \tag{8}$$

where β_e is the effective bulk elastic modulus of hydraulic oil; Q is the flow rate of direct drive pump, and l is the direct drive pump chamber length.

The active valve plays the role of flow distribution in the brake-by-wire unit based on direct drive pump–valve cooperative. Assuming that the active valve is fully open, the pressure difference between the two ends of the active valve spool is small, and the flow rate of the valve hole is simplified as follows:

$$Q = k_0 c_0 A_0 \sqrt{\frac{2}{\rho} \Delta P} \tag{9}$$

where k_0 is the flow linearization coefficient; c_0 is the flow coefficient of the valve port; A_0 is the valve port area; ΔP is the hydraulic pressure difference in the valve port, and ρ is the hydraulic oil density.

The brake wheel cylinder realizes the conversion of hydraulic pressure to braking force. Due to the small displacement of the piston x_h and the fact that the piston displacement remains basically unchanged when the pressure of the wheel cylinder increases, $\dot{x}_h$ can be ignored. The pressure change model of the wheel cylinder is

$$\dot{P}_2 = \beta_e \frac{Q}{S_2(l_h + x_h)} \tag{10}$$

where l_h is the length of the wheel cylinder chamber.

4.3. Quarter-Car Dynamics Model Establishment

This paper needs to study the slip rate control under emergency braking so a 1/4 Vehicle dynamics model is established. The vehicle longitudinal dynamics model is

$$F_x = -m_c \dot{v}_c = \mu_c m_c g \tag{11}$$

where F_x is the ground braking force; m_c is a quarter of the vehicle's weight; v_c is the tire's longitudinal speed, and μ_c is the longitudinal friction coefficient of the tire. The wheel dynamics model is

$$J\dot{\omega} = F_x r_b - T_b \tag{12}$$

where T_b is the braking torque, and J is the wheel's moment of inertia. For the tire longitudinal friction coefficient μ, the Burckhardt tire model is used to calculate

$$\mu(s) = c_1[1 - \exp(-c_2 s)] - c_3 s \tag{13}$$

where s is slip rate, and c_1, c_2, and c_3, respectively, represent the peak parameter, shape parameter, and difference parameter of the road adhesion coefficient curve corresponding to the typical road surface.

4.4. Co-Simulation Model Construction

The brake-by-wire unit based on direct drive pump–valve cooperative is a nonlinear time-varying system with mechanical, electrical, magnetic, and hydraulic coupling. Considering the necessary requirements of system testing and actual working conditions, AMESim is used to model the direct drive pump, active valve, wheel cylinder, hydraulic pipes, and other components. At the same time, the EMLA and the controller model are established in Matlab/Simulink. The co-simulation model is shown in Figure 2. Based on the interface technology between AMESim and Simulink, a co-simulation interface is established in AMESim, and corresponding S-functions are set in Simulink, fully utilizing the advantages of AMESim in nonlinear dynamic hydraulic system modeling and powerful functions of Matlab in complex controller mathematical model building and data calculation processing, to obtain fast and real-time results.

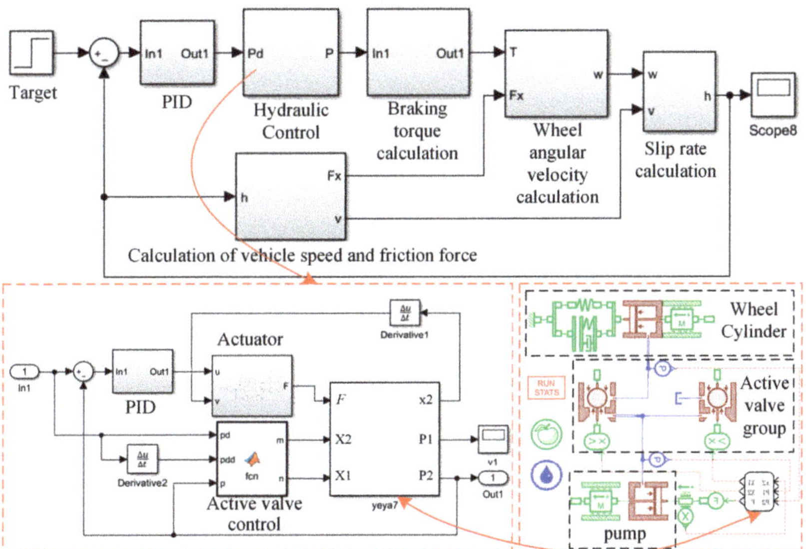

Figure 2. Electro-hydraulic co-simulation model.

5. Results and Analysis

5.1. The Experimental Platform

The experimental platform of the brake-by-wire unit based on direct drive pump–valve cooperative, as shown in Figure 3, is established. The personal computer transmits the control signal to the controller. Then, the controller controls the power drive module to control the driving voltage of EMLA and the switch of the active valve, which controls the pressure of the brake cylinder. At the same time, the pressure sensor and the position sensor collect the signal and transmit it to the personal computer through the controller, which forms the signal feedback. The controller adopts RTU-BOX, a rapid control prototype system, and its digital controller adopts TMS320C28346, a 32-bit floating point digital signal processor with a main frequency of 300 MHz. The position sensor with 0.01 mm resolution provides the position feedback of the electromagnetic linear actuator, and the pressure sensor with 0.01 MPa resolution provides the hydraulic pressure feedback of the wheel cylinder.

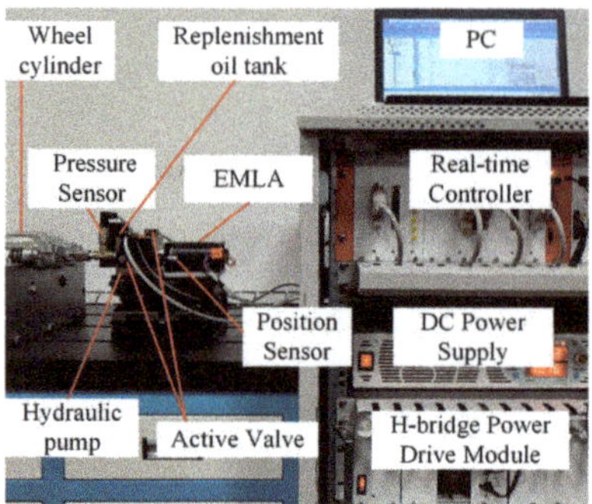

Figure 3. The experimental platform.

5.2. Analysis of Hydraulic Pressure Response Performance

We simulated the emergency braking condition of the target vehicle and set the step response target as 4 MPa. The test results are shown in Figure 4. Under the PID control algorithm, the brake unit can quickly achieve stability without overshooting, and the response time of simulation and experiment is 0.023 s and 0.026 s, respectively. The simulated curve and response time are close to experimental results, which verifies the validity of the co-simulation model.

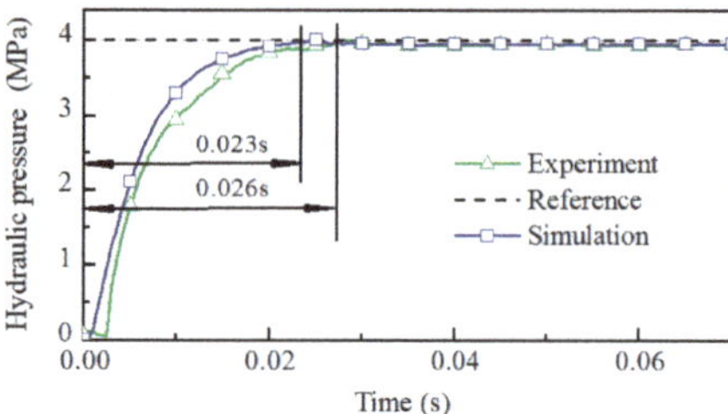

Figure 4. Step response curve.

The experimental results of the EMLA and active valves under the PID controller are shown in Figure 5. When the system reaches the target pressure, and the working mode is switched to the pressure-maintaining mode, the pressure-maintaining valve closes, causing small fluctuations in the hydraulic pressure. At the same time, the actuator control signal is 0, and the current value rapidly decreases. Under the pressure in the pump chamber, the actuator moves toward the reset direction for a period of displacement.

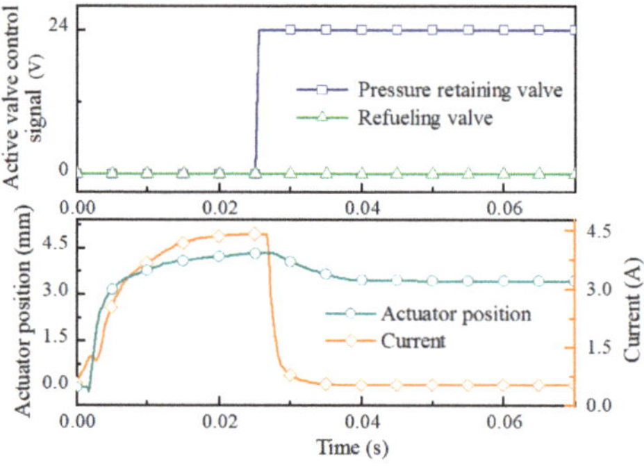

Figure 5. Status of the actuator and the active valves.

5.3. Analysis of Hydraulic Pressure Tracking Performance

To verify the hydraulic pressure tracking performance of the brake-by-wire unit based on direct drive pump–valve cooperative, the target sinusoidal pressure signal frequency is set to 2.5 Hz and amplitude to 2.5 MPa, and the hydraulic pressure tracking curve and the system state in the experiment are shown in Figure 6. PID control can track a sinusoidal signal well. In the process of sine wave target tracking, the average error of simulation and experiment PID control is 0.197 MPa and 0.218 MPa, and the amplitude attenuation of simulation and experiment PID control is 0.106 MPa and 0.121 MPa, respectively. The results show that the braking unit can effectively control the hydraulic pressure under a sine wave target. However, due to the limitation of DC power supply voltage and sensor measurement accuracy, as well as the difficulty of system oil filling in actual conditions, the actual pressure curve differs from the simulation.

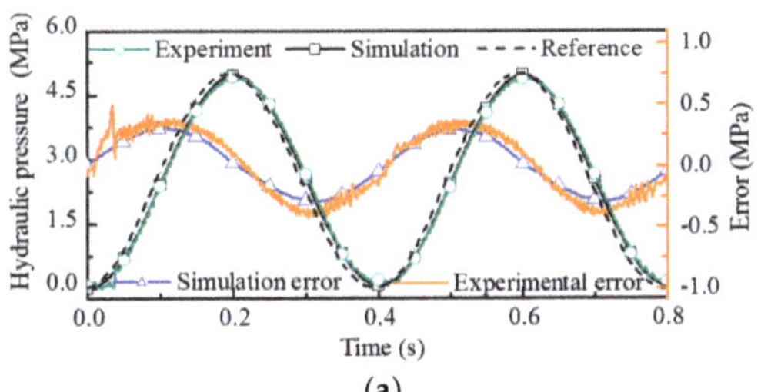
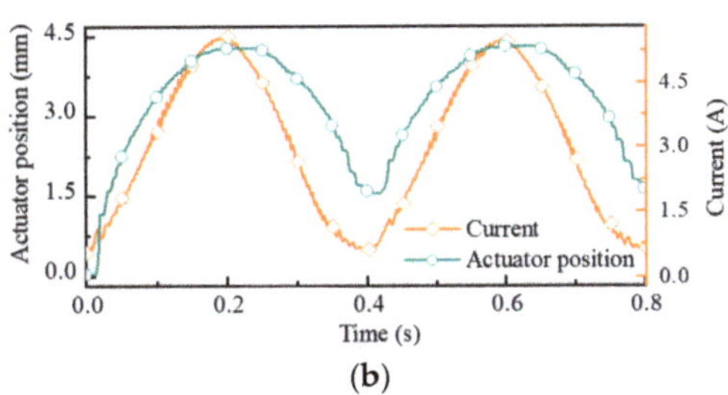

Figure 6. Experimental and simulation results under the sinusoidal target: (**a**) Hydraulic pressure tracking curve for simulation and experiment; (**b**) The system state in the experiment.

To further simulate the pressure tracking performance, we set the target signal of derivative mutation; the target hydraulic pressure frequency of the triangular wave braking was set to 2.5 Hz, and the amplitude to 2.5 MPa. The hydraulic pressure tracking curve and the system state in the experiment are shown in Figure 7. At 0.2 s and 0.4 s, the derivative of the target signal suddenly changes, but the PID controller can still track the triangular wave signal well. Due to the hydraulic oil in the system not being fully filled, the actuator position at 0.4 s is higher than the initial state. In the process of triangular wave target tracking, the average error of simulation and experiment PID control is 0.141 and 0.203, respectively. The results show that the proposed brake unit has good accuracy when maintaining a certain rate of increasing or decreasing hydraulic pressure, which verifies its good tracking performance.

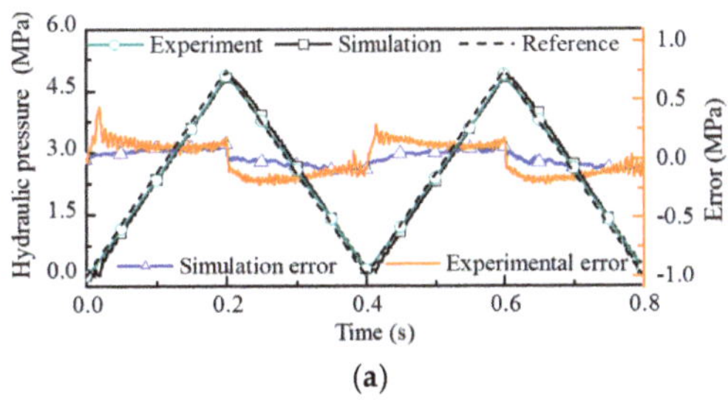
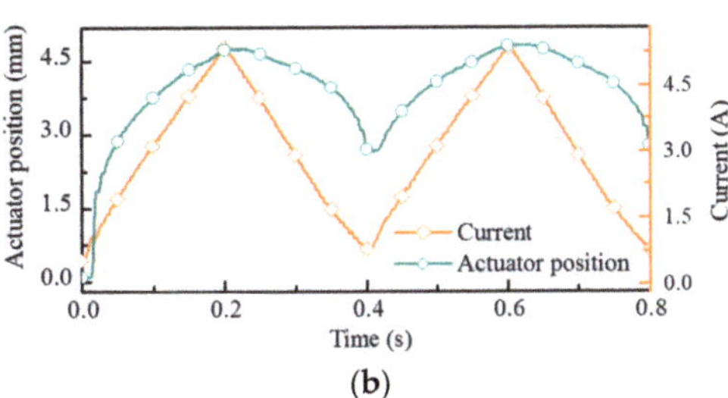

Figure 7. Experimental and simulation results under the triangular wave target: (**a**) Hydraulic pressure tracking curve for simulation and experiment; (**b**) The system state in the experiment.

5.4. Analysis of the Impact of Parameters on System Performance

The structure parameters that can be optimized were analyzed to obtain the influence trend of parameter changes on the system performance. The average tracking error and amplitude attenuation of the sinusoidal signal were used to express the influence of structural parameters on system performance changes. For this purpose, the change degree curves of independent variables are drawn in this paper, as shown in Figures 8–10. Through the curve changes, the impact of parameter changes on system performance is intuitively shown.

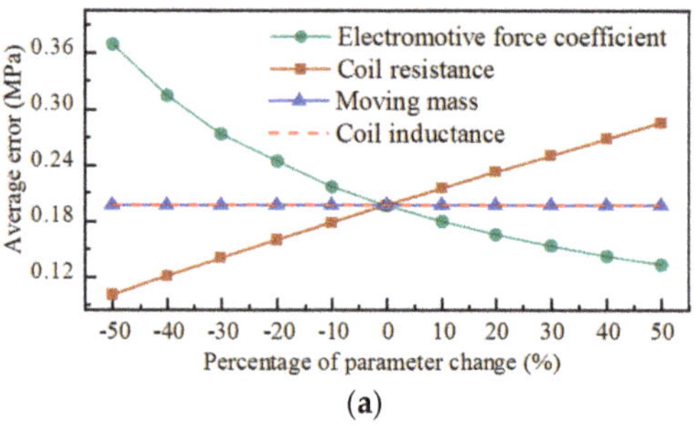

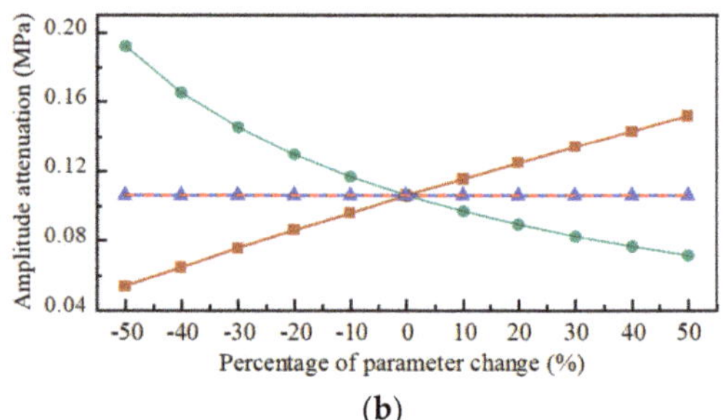

Figure 8. The influence of actuator parameters on system performance: (**a**) The influence on average error; (**b**) The influence on amplitude attenuation.

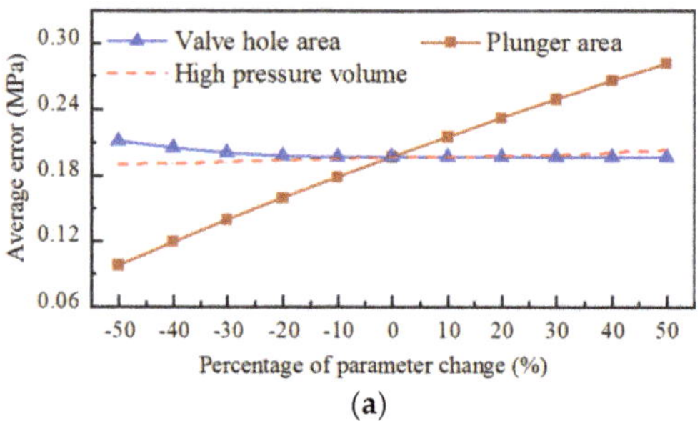

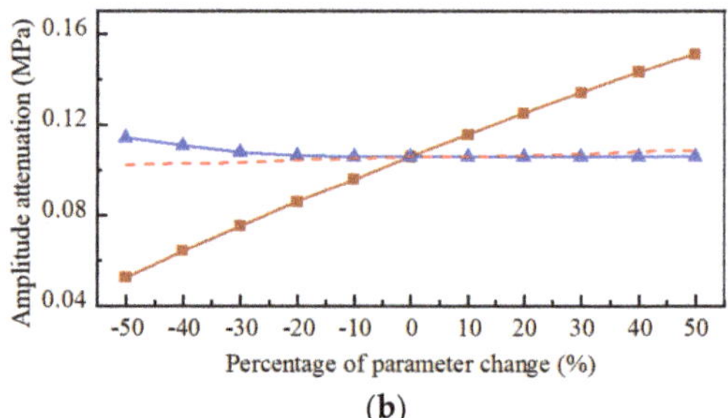

Figure 9. The influence of hydraulic system parameters on system performance: (**a**) The influence on average error; (**b**) The influence on amplitude attenuation.

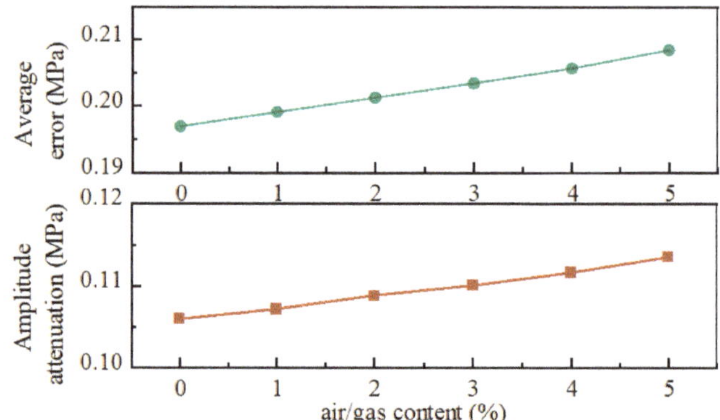

Figure 10. The influence of air content on average error and amplitude attenuation.

5.4.1. Actuator Parameter

As can be seen from Figure 8, for moving mass, including the mass of the actuator and the mass of the plunger inside the pump, although reducing the moving mass can reduce the load on the actuator, compared to the load on the actuator caused by hydraulic pressure, the smaller dynamic mass has a smaller impact on the system response speed and tracking accuracy. Therefore, there is no need to lightweight the dynamic mass of the actuator. For the electromagnetic force coefficient, an increase in the electromagnetic force coefficient will increase the electromagnetic force under the same current, and the target tracking and response can be carried out with a smaller current for the same hydraulic load, which will improve the precision of pressure control. Therefore, the electromagnetic force coefficient should be increased under the premise of a comprehensive consideration of installation space and cost. For the resistance, as the controller signal is used as the input voltage of the system, the smaller the resistance under the same voltage, the larger the current, and the larger the electromagnetic force under the same actuator state. Under the same hydraulic load, a smaller control signal is needed to track the target, which will improve the pressure control accuracy. For inductance, because the inductance term in Equation (6) has a smaller value than the other two terms, the change in inductance makes the change in current smaller under the same voltage, so the pressure control accuracy is almost unchanged.

5.4.2. Hydraulic System Parameter

As can be seen from Figure 9, for the plunger area, when the plunger area is reduced, the same electromagnetic force can generate greater hydraulic pressure. However, when the area is too small, the actuator displacement increases under the same hydraulic pressure target. So, it is necessary to reduce the plunger area while considering the actuator stroke. For the high-pressure volume, the increase in the high-pressure volume will result in a slower system pressure response and a decrease in pressure control accuracy. Additionally, it will also increase the actuator movement displacement, which limits the actuator stroke. Therefore, the hydraulic pipeline should be reduced as much as possible to reduce the high-pressure volume. For the valve port area, an increase in the valve port area will increase the instantaneous flow rate, but if it is too large, it will cause severe flow oscillation, resulting in excessive hydraulic pressure error in the keeping pressure mode, so the valve port area can be increased without increasing the error.

As can be seen from Figure 10, the elastic modulus of the oil is a comprehensive performance parameter that is related to factors such as gas content in the oil, system pressure, and oil temperature. When the oil is mixed with 1% air, the bulk elastic modulus will drop to about 5% of the pure oil. The pressure control precision of the system can be increased by increasing the bulk elastic modulus of the oil. Therefore, hydraulic oil with a large bulk elastic modulus should be selected, and measures should be taken to prevent air from being mixed into the oil.

5.5. ABS Performance Analysis

It is known from the above section that the proposed brake-by-wire unit based on direct drive pump–valve cooperative in this paper responds quickly and can accurately control the pressure in the cylinder, making it easier to realize the anti-lock braking function. This section further conducts a simulation study on the anti-lock braking performance of the brake-by-wire unit based on direct drive pump–valve cooperative and sets the initial speed at 100 km/h. Dry asphalt (c_1 = 1.280, c_2 = 23.99, c_3 = 0.52) is selected for the road surface. According to the formula, when the road surface adhesion coefficient μ derivative is zero, the function is at the maximum point, and the slip rate here is the optimal slip rate

$$s_{opt} = \frac{1}{c_2} \ln \frac{c_1 c_2}{c_3} \tag{14}$$

The optimal slip rate is 0.17, and the wheel anti-lock braking system starts when the simulation starts. The simulation results are shown in Figure 11. As the brake unit can accurately control the wheel cylinder pressure, it can accurately control the wheel slip rate. The response time of the slip rate is 0.422 s; the speed needs 2.404 s from the initial speed of 100 km/h to the stop, and the braking distance is 34 m, which meets the braking safety requirements.

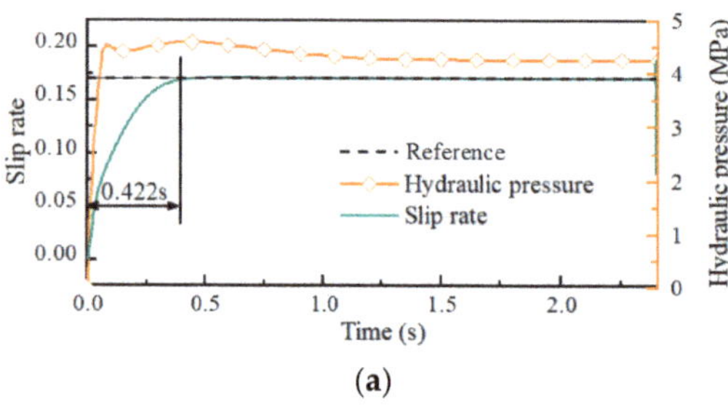
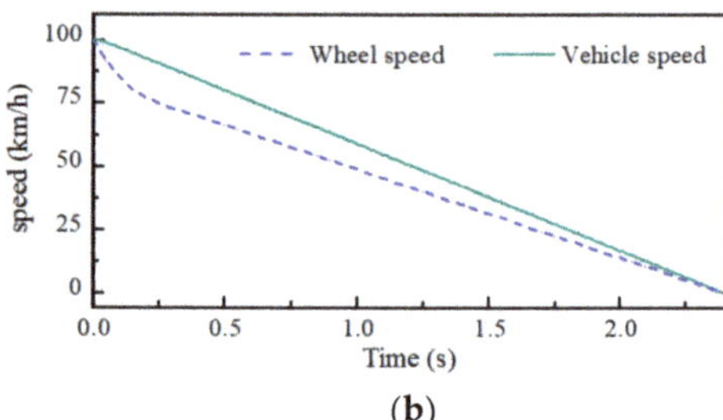

Figure 11. Simulation results of slip rate control: (**a**) Slip rate and hydraulic pressure curve; (**b**) Vehicle speed and wheel speed curve.

6. Conclusions

This paper proposes a brake-by-wire unit based on a direct drive pump–valve cooperative, which has the advantages of rapid response and no deterioration of wheel side space and unsprung mass. The co-simulation model of the brake unit is established based on the Simulink-AMESim co-simulation. The pressure experiment shows that it has a good pressure control performance, and the experimental curve is close to the simulation curve, which proves the accuracy of the simulation model. The feasibility of the simulation is further verified by ABS simulation, which provides a new implementation scheme for automotive distributed braking system. The characteristics of the key parameters are analyzed, and the influence of each parameter on the control performance is researched by changing the percentage of parameter change. This lays the foundation for further research. The structure optimization of EMLA and the design of the hydraulic control algorithm are the keys to improving the performance of brake units and also the main research direction in the future.

Author Contributions: Conceptualization: P.Y. and C.T.; methodology: P.Y., Z.S. and C.T.; software: P.Y. and H.X.; formal analysis: P.Y. and C.T.; supervision: C.T.; writing—original draft preparation: P.Y.; writing—review and editing: P.Y., Z.S., Y.R., H.X. and C.T. All authors have read and agreed to the published version of the manuscript.

Funding: This work was supported by Shandong Natural Science Foundation Project (Grant No. ZR2023ME178), Open Fund of State Key Laboratory of Automobile Simulation and Control of Jilin University (Grant No. 20210224), Shandong Province Science and Technology Small and Medium Enterprises Innovation Ability Enhancement Project (Grant No. 2023TSGC0404), Innovation team project of "Qing-Chuang science and technology plan" of colleges and universities in Shandong Province (Grant No. 2019KJB027 and Grant No.2022KJ232) and Young Technology Talent Supporting Project of Shandong Province (Grant No. SDAST2021QT20).

Data Availability Statement: Not applicable.

Conflicts of Interest: The authors declare no conflict of interest.

References

1. Ji, Y.; Zhang, J.; He, C.; Ma, R.; Hu, X. Constraint performance pressure tracking control with asymmetric continuous friction compensation for booster based brake-by-wire system. *Mech. Syst. Signal Process.* **2022**, *174*, 109083. [CrossRef]
2. Zhao, Z.; Zhou, L.; Zhang, J.; Zhu, Q.; Hedrick, J.K. Distributed and self-adaptive vehicle speed estimation in the composite braking case for four-wheel drive hybrid electric car. *Veh. Syst. Dyn.* **2017**, *55*, 750–773. [CrossRef]
3. Mingde, G.; Hailong, W. Full power hydraulic brake system based on double pipelines for heavy vehicles. *Chin. J. Mech. Eng.* **2011**, *24*, 790.
4. Todeschini, F.; Corno, M.; Panzani, G.; Fiorenti, S.; Savaresi, S.M. Adaptive cascade control of a brake-by-wire actuator for sport motorcycles. *IEEE/ASME Trans. Mechatron.* **2014**, *20*, 1310–1319. [CrossRef]

5. Tang, Q.; Yang, Y.; Luo, C.; Yang, Z.; Fu, C. A novel electro-hydraulic compound braking system coordinated control strategy for a four-wheel-drive pure electric vehicle driven by dual motors. *Energy* **2022**, *241*, 122750. [CrossRef]
6. Xiong, L.; Han, W.; Yu, Z. Adaptive sliding mode pressure control for an electro-hydraulic brake system via desired-state and integral-antiwindup compensation. *Mechatronics* **2020**, *68*, 102359. [CrossRef]
7. Han, W.; Xiong, L.; Yu, Z. Braking pressure control in electro-hydraulic brake system based on pressure estimation with nonlinearities and uncertainties. *Mech. Syst. Signal Process.* **2019**, *131*, 703–727. [CrossRef]
8. Todeschini, F.; Formentin, S.; Panzani, G.; Corno, M.; Savaresi, S.M.; Zaccarian, L. Nonlinear pressure control for BBW systems via dead-zone and antiwindup compensation. *IEEE Trans. Control Syst. Technol.* **2015**, *24*, 1419–1431. [CrossRef]
9. Pei, X.; Pan, H.; Chen, Z.; Guo, X.; Yang, B. Coordinated control strategy of electro-hydraulic braking for energy regeneration. *Control Eng. Pract.* **2020**, *96*, 104324. [CrossRef]
10. Savitski, D.; Ivanov, V.; Schleinin, D.; Augsburg, K.; Pütz, T.; Lee, C.F. Advanced control functions of decoupled electro-hydraulic brake system. In Proceedings of the 2016 IEEE 14th International Workshop on Advanced Motion Control, Auckland, New Zealand, 22–24 April 2016.
11. Oshima, T.; Fujiki, N.; Nakao, S.; Kimura, T.; Ohtani, Y.; Ueno, K. Development of an electrically driven intelligent brake system. *SAE Int. J. Passeng. Cars-Mech. Syst.* **2011**, *4*, 399–405. [CrossRef]
12. Li, L.; Zhang, T.; Sun, B.; Wu, K.; Sun, Z.; Zhang, Z.; Lin, L.; Xu, H. Research on electro-hydraulic ratios for a novel mechanical-electro-hydraulic power coupling electric vehicle. *Energy* **2023**, *270*, 126970. [CrossRef]
13. Pugi, L.; Pagliai, M.; Nocentini, A.; Lutzemberger, G.; Pretto, A. Design of a hydraulic servo-actuation fed by a regenerative braking system. *Appl. Energy* **2017**, *187*, 96–115. [CrossRef]
14. Gong, X.; Ge, W.; Yan, J.; Zhang, Y.; Gongye, X. Review on the development, control method and application prospect of brake-by-wire actuator. *Actuators* **2020**, *9*, 15. [CrossRef]
15. Tan, C.; Ren, H.; Li, B.; Lu, J.; Li, D.; Tao, W. Design and analysis of a novel cascade control algorithm for braking-by-wire system based on electromagnetic direct-drive valves. *J. Frankl. Inst.* **2022**, *359*, 8497–8521. [CrossRef]
16. Picasso, B.; Caporale, D.; Colaneri, P. Braking control in railway vehicles: A distributed preview approach. *IEEE Trans. Autom. Control* **2017**, *63*, 189–195. [CrossRef]
17. Liu, Y.; Sun, Z.; Wang, M. Brake force cooperative control and test for integrated electro-hydraulic brake system. *J. Jilin Univ.* **2016**, *46*, 718–724.
18. Liu, Z.; Long, Y.; Li, Y.; Tan, X.; Wu, G. Study on the Design and Parameter Matching of Integrated Electro-Hydraulic Braking System. *Automot. Eng.* **2022**, *44*, 1416–1424.
19. Joa, E.; Yi, K.; Sohn, K.; Bae, H. Four-wheel independent brake control to limit tire slip under unknown road conditions. *Control Eng. Pract.* **2018**, *76*, 79–95. [CrossRef]
20. Lee, C.F.; Manzie, C. Active brake judder attenuation using an electromechanical brake-by-wire system. *IEEE/ASME Trans. Mechatron.* **2016**, *21*, 2964–2976. [CrossRef]
21. Zhang, L.; Yu, L.; Wang, Z.; Zuo, L.; Song, J. All-wheel braking force allocation during a braking-in-turn maneuver for vehicles with the brake-by-wire system considering braking efficiency and stability. *IEEE Trans. Veh. Technol.* **2015**, *65*, 4752–4767. [CrossRef]
22. Yu, D.; Wang, W.; Zhang, H.; Xu, D. Research on anti-lock braking control strategy of distributed-driven electric vehicle. *IEEE Access* **2020**, *8*, 162467–162478. [CrossRef]
23. Henderson, J.P.; Plummer, A.; Johnston, N. An electro-hydrostatic actuator for hybrid active-passive vibration isolation. *Int. J. Hydromechatron.* **2018**, *1*, 47–71. [CrossRef]
24. Fu, Y.F.; Hu, X.H.; Wang, W.R.; Ge, Z. Simulation and experimental study of a new electromechanical brake with automatic wear adjustment function. *Int. J. Automot. Technol.* **2020**, *21*, 227–238. [CrossRef]
25. Wang, Z.; Yu, L.; You, C.; Wang, Y.; Song, J. Fail-safe control allocation for a distributed brake-by-wire system considering the driver's behaviour. *Proc. Inst. Mech. Eng. Part D J. Automob. Eng.* **2014**, *228*, 1547–1567. [CrossRef]
26. Gong, X.; Qian, L.; Ge, W.; Yan, J. Research on electronic brake force distribution and anti-lock brake of vehicle based on direct drive electro hydraulic actuator. *Int. J. Automot. Eng.* **2020**, *11*, 22–29. [CrossRef]
27. Xiao, F.; Gong, X.; Lu, Z.; Qian, L.; Zhang, Y.; Wang, L. Design and control of new brake-by-wire actuator for vehicle based on linear motor and lever mechanism. *IEEE Access* **2021**, *9*, 95832–95842. [CrossRef]
28. Zheng, H.Y.; He, R.; Zong, C.F. Research of electric vehicle regenerative braking control strategy based on EHB system. *Adv. Mater. Res.* **2013**, *724*, 1436–1439. [CrossRef]
29. Han, W.; Xiong, L.; Yu, Z. Analysis and optimization of minimum hydraulic brake-by-wire system for wheeled vehicles based on queueing theory. *IEEE Trans. Veh. Technol.* **2021**, *70*, 12491–12505. [CrossRef]
30. Ji, Y.; Zhang, J.; Zhang, J.; He, C.; Hou, X.; Han, J. Constraint performance slip ratio control for vehicles with distributed electrohydraulic brake-by-wire system. *Proc. Inst. Mech. Eng. Part D J. Automob. Eng.* **2023**. [CrossRef]
31. Lu, Y.; Tan, C.; Ge, W.; Zhao, Y.; Wang, G. Adaptive disturbance observer-based improved super-twisting sliding mode control for electromagnetic direct-drive pump. *Smart Mater. Struct.* **2022**, *32*, 017001. [CrossRef]
32. Wu, Y.; Liu, M.; Zhang, G.; Hou, Z.; Lin, Q.; Yuan, H. Study on high frequency response characteristics of a moving-coil-type linear actuator using the coils combinations. *Int. J. Hydromechatron.* **2022**, *5*, 226–242. [CrossRef]

33. Li, B.; Liu, Y.; Tan, C.; Qin, Q.; Lu, Y. Review on electro-hydrostatic actuator: System configurations, design methods and control technologies. *Int. J. Mechatron. Manuf. Syst.* **2020**, *13*, 323–346. [CrossRef]
34. Chen, Z.; Yao, B.; Wang, Q. Accurate motion control of linear motors with adaptive robust compensation of nonlinear electromagnetic field effect. *IEEE/ASME Trans. Mechatron.* **2012**, *18*, 1122–1129. [CrossRef]
35. Shi, Q.; He, L. A Model Predictive Control Approach for Electro-Hydraulic Braking by Wire. *IEEE Trans. Ind. Inform.* **2022**, *19*, 1380–1388. [CrossRef]

MDPI
St. Alban-Anlage 66
4052 Basel
Switzerland
www.mdpi.com

Actuators Editorial Office
E-mail: actuators@mdpi.com
www.mdpi.com/journal/actuators